Particle Physics: A Primer

Particle Physics: A Primer

Triptesh De

LEVANT
Levant Books
India

Particle Physics: A Primer
by **Triptesh De**

First Published 2011
Reprint 2017

Published by
Levant Books
27C, Creek Row, Kolkata 700 014, India

Exclusively Distributed by
Sarat Book Distributors
18B, Shyama Charan De Street, Kolkata 700 073, India
Email: orders@saratbookhouse.com

Printed and bound at Sarat Impressions Pvt. Ltd., Andul, Howrah.

ISBN: 978-93-80663-21-0

to my students

Preface

The book is largely based on the course of lectures the author gave to the Post-M.Sc. (Physics) students at the Saha Institute of Nuclear Physics, Kolkata and M.Sc. (Physics) students at Calcutta University. It gives an overview of the development in our understanding of the physics of the fundamental particles and their interactions from the early days when symmetries and invariance principles guided the study. Theoretical considerations played a dominant role in the more recent past when studies of gauge theories as possible candidates for the dynamics of particle interactions took over. A number of important milestones in the experimental observations and their theoretical interpretations led finally to the emergence of the standard model of particle physics which became well-established after the precision tests performed at various collider experiments. The topics covered in the book include Lie groups and Lie algebra, the quark model, V - A theory of weak interactions, deep-inelastic lepton-hadron scattering, electron-positron annihilation into hadrons, quark-flavors, chromodynamics, electroweak unification and the Glashow-Salam-Weinberg model. A number of topics discussed in other text books (for example, Dispersion relations, Regge pole theory and current algebra) have been left out. However, a brief discussion of some of the beyond the standard model theories, currently being studied, have been included. It is hoped that the book will be useful to M.Sc. and Ph.D. students as well as advanced undergraduate students.

Contents

Chapter 1

Preliminaries

Broadly speaking, Particle Physics involves the study of fundamental particles and their interactions. The subject is the outgrowth of the physicist's quest for the ultimate structure (building blocks) of matter and his ambition for having a unified understanding of the basic forces and phenomena observed in nature in terms of as few basic concepts and entities as possible. Particle Physics as a separate subject evolved out of studies in Nuclear Physics and Cosmic Ray Physics. It is often said that Particle Physics mostly deals with short distance and high energy phenomena. One implies the other, because, in order to probe short distances one needs to use high energy collisions. The subject is thus also known as High Energy Physics. The theoretical framework for studying particle physics necessarily involves both Quantum Mechanics and the Special Theory of Relativity. The specific combination used is known as the Relativistic Quantum Field Theory (RQFT). It is thus convenient to work in natural units ($\hbar = c = 1$). Moreover, symmetry and invariance principles have played a very important role in the development of our understanding of particle physics. The mathematical framework needed to study symmetries is Group Theory.

1.1 Particles and Interactions

So far, four basic interactions have been identified in nature. They are (in decreasing order of strength)—strong, electromagnetic, weak and gravitational. Strong interaction is the strongest and gravitational interaction is the weakest of the four. Electromagnetic and gravitational interactions have infinite range and their effects are well recognized in our everyday macroscopic world. On the other hand, strong and weak interactions have extremely short ranges and hence belong entirely to the microscopic world of atoms and nuclei. However, their effects are important to the macroscopic world as well. Strong interaction is responsible for the stability of matter and the weak interaction is responsible for the source of energy in the sun that makes life possible on earth.

The particles that have so far been observed in cosmic rays and accelerator experiments are classified into three broad categories—hadrons, leptons and mediators (gauge bosons). Hadrons are further divided into two groups—baryons and mesons. Although the initial classification was in terms of masses, it turns out that mass is not the only distinguishing property of a particle. Hadrons are strongly interacting particles. Leptons do not participate in strong interactions. All hadrons and leptons have weak interactions. However, only charged particles can interact through electromagnetic interactions. Mediators mediate particular interactions. For example, the photon mediates electromagnetic interactions.

1.2 Quantum Numbers and Conservation Laws

Particles are classified by certain attributes called quantum numbers. Quantum numbers are eigenvalues of dynamical variables that are constants of motion. Similarly, interactions are characterized by conservation laws. Some of the conservation laws are absolute, i.e. they are respected by all known interactions. There are also conservation laws which are only approximate, i.e. they are obeyed by strong interactions and in most cases by electromagnetic interactions but are violated in weak interactions. Examples of particle quantum numbers are mass, spin, intrinsic parity, electric charge, baryon number, lepton number, i-spin, strangeness, charm and colour. Baryons are half-integral-spin particles (fermions) and have non-zero baryon number ($B \neq 0$), whereas mesons are integral-spin particles (bosons) and have zero baryon number ($B = 0$). Leptons are spin-$\frac{1}{2}$ fermions and have non-zero lepton number ($L \neq 0$). Energy, linear momentum, angular momentum and electric charge are conserved in all known interactions. I-spin is not conserved in electromagnetic interaction. I-spin, strangeness and charm quantum numbers are not conserved in weak interactions. Weak interactions do not respect parity (space inversion) invariance and charge conjugation invariance. However, their product is respected in most weak processes.

1.3 Antiparticles

Associated with the concept of particles is the concept of antiparticles. Relativistic kinematics imply that a free particle can have both positive and negative energies ($E = \pm\sqrt{\vec{p}^{\,2} + m^2}$). In a relativistic quantum field theoretic description the negative energy states of a particle can be interpreted as the positive energy states of some other particle with same mass, spin and lifetime as the original particle but with "charges" opposite in sign to those of the particle. Thus the antiparticle of the electron, called the positron, has electric charge $+e$ (the electron has electric charge $-e$). Neutral particles (all "charges" zero) like the photon and the neutral pion π^0 are self-conjugate, that is the particles and antiparticles are identical. The idea of antiparticles was first introduced by Dirac in order to find a resolution to the problem of negative energy states in his relativistic theory of the electron. The concept naturally appears in the RQFT as a consequence of charge conjugation invariance.

1.4 Symmetries & Invariance Principles

There is a one-to-one correspondence between quantum numbers and conservation laws on the one hand, and symmetry and invariance properties of a physical system on the other. There are three aspects of a symmetry. Whenever, due to aesthetic or some other reason, a physical quantity is unmeasurable or two physical entities are indistinguishable, there exists a set of transformations which leave the system unchanged and as a consequence there are one or more observables called the generators of the symmetry transformations, which are constants of motion. For example, due to homogeneity of space, absolute distance is not measurable and hence the properties of the physical system must not depend on the origin of coordinates which means that the system is invariant under translation of space. As a consequence, the components of the linear momentum, which are the generators of spatial translations, are constants of motion.

The symmetries of physical systems that have so far been recognized are of four types:

(a) Space-time symmetries: These include symmetries under time translation, spatial translations, spatial rotations and Lorentz velocity or boost transformations.

(b) Discrete symmetries: There are three discrete symmetries: charge conjugation (C) invariance or particle-anti-particle symmetry, space inversion or parity invariance (P) and time reversal (T) invariance.

(c) Exchange symmetries: These symmetries refer to symmetries under exchange or permutation of particles and apply to a system of identical particles. As a consequence of exchange symmetries, the state function of a system of particles is either even or odd under exchange or permutation of particles. The former is the case for a system of integral-spin particles called bosons and the latter is the case for a system of half-integral-spin particles called fermions. Thus because of exchange symmetry, particles are either bosons or fermions. Baryons and Leptons are fermions, whereas meson and mediators are bosons.

(d) Internal symmetries: These symmetries correspond to invariance under transformations in some abstract or internal space (different from ordinary space-time). The most familiar example of an internal symmetry is the symmetry that is responsible for conservation of electric charge. An internal symmetry is named according to the nature of the underlying symmetry group. Examples of internal symmetries are $U(1)_Q$, $SU(2)_I$, $SU(3)_f$, $SU(3)_c$, $SU(2)_L \times U(1)_Y$, $SU(2)_L \times SU(2)_R$. Internal Symmetries are also known as unitary symmetries, because they correspond to invariance under unitary transformation in an abstract vector space as opposed to Minkowski space-time manifold.

1.5 The Standard Model

Experimental studies of particle physics phenomena and simultaneous attempts to understand them theoretically over the years have culminated in the emergence of what is known as the **Standard Model (SM)** of particle physics. The SM is based on three well-established postulates. These are: (1) All hadrons are composite, being bound states of elementary entities called 'quarks' (q) and their anti particles called 'antiquarks' ($\overline{q}$). In particular, baryons are bound states of three quarks ($B = qqq$), and mesons are bound states of one quark and one antiquark ($M = q\bar{q}$). (2) The dynamics of all basic interactions (strong, electromagnetic and weak interactions) are determined by the principle of local gauge invariance. As a consequence, the mediators are all spin-1 quanta, called the gauge bosons. (3) In order to have a consistent (renormalized) theory of the weak interaction, it has to be unified with the electromagnetic interaction. The unified theory of weak and electromagnetic interactions is known as the **Electroweak (EW) Theory**.

Quarks and leptons are the fundamental particles of the SM. Both are spin-$\frac{1}{2}$ fermions. As far as present knowledge goes, there are three pairs of leptons and also three pairs of quarks.They together form three fermion generations, each generation consisting of a pair of quarks and a pair of leptons. However, quarks differ from leptons in having non-zero values of a quantum number called "color" which may therefore be taken as the "charge" of strong interaction. The gauge theory of strong interaction of quarks mediated by massless spin-1 quanta, called "gluons", with $SU(3)_c$ as the underlying gauge group, is known as Quantum Chromo Dynamics (QCD). QCD is very similar to QED in many ways. However, the gluons are colored objects (i.e., they carry strong interaction charge). In QED there is only one photon and it is electrically neutral. This fact is responsible for most of the differences between QED and QCD. The electroweak part of the standard model, also known as the Glashow-Salam-Weinberg (GSW) model, involves the minimal gauge group $SU(2)_L \times U(1)_Y$. Thus there are four gauge bosons in GSW model W^+, W^-, Z^0 and the photon. Local gauge invariance implies that the gauge bosons must be massless. However, the vector bosons which mediate weak interaction must have high masses in order to account for the extreme short range of weak interactions. They acquire masses in the SM by spontaneous breaking of electroweak gauge symmetry $SU(2)_L \times U(1)_Y$ down to $U(1)_Q$. The mechanism of spontaneous breaking of gauge symmetry results in the appearance of a spin-zero particle called Higgs-boson in the SM. Similarly, although the exact gauge symmetry requires the gluons to be massless, strong interactions between hadrons are prevented from being of infinite range by a phenomena called quark confinement, which perhaps results from (although yet unproved) the nonabelian nature of the QCD gauge group which also accounts for the "asymptotic freedom" property of strong interactions of quarks.

More recently, particle physics has been successfully applied to cosmology which deals with the origin, evolution and the structure formation in the universe. Thus the unification of the world of the very big with the world of the very small is being achieved and a new subject, Astroparticle physics has emerged.

Table 1.1

Interaction	Strength	Range	Example
Strong	$\alpha_s = \frac{g_s^2}{4\pi} \sim 1 - 10$	10^{-13} cm	Nuclear Force
Electromagnetic	$\alpha = \frac{e^2}{4\pi} = \frac{1}{137}$	∞	H atom
Weak	$G_F = \frac{10^{-5}}{M_P^2}$	10^{-15} cm	Nuclear β decay
Gravitational	$G_N = \frac{10^{-39}}{M_P^2}$	∞	planetary motion

Table 1.2

Observed Particles

- Hadrons
  - Baryons $(N, \Lambda, \Sigma, \Xi, \Omega)$
  - Meson $(\pi, K, \eta, \eta', \rho, \omega, \phi)$
- Leptons $(\nu_e, e^-), (\nu_\mu, \mu^-), (\nu_\tau, \tau^-)$
- Mediators $(\gamma, W^\pm, Z^0)$

Table 1.3

Fundamental Particles (SM)

- Quarks $\begin{pmatrix} u \\ d \end{pmatrix}, \begin{pmatrix} c \\ s \end{pmatrix}, \begin{pmatrix} t \\ b \end{pmatrix}$
- Leptons $\begin{pmatrix} \nu_e \\ e^- \end{pmatrix}, \begin{pmatrix} \nu_\mu \\ \mu^- \end{pmatrix}, \begin{pmatrix} \nu_\tau \\ \tau^- \end{pmatrix}$
- Gauge bosons $\begin{pmatrix} \text{gluons}, \gamma \\ W^\pm, Z^0 \end{pmatrix}$

Table 1.4

Invariance	Associated Obeservable	Whether conserved/obeyed		
		Strong	Electromagnetic	Weak
1. Space-time translation	Energy-Momentum	Yes	Yes	Yes
2. Spatial rotation	Angular Momentum	Yes	Yes	Yes
3. Lorentz transformation (Boost.)	—	Yes	Yes	Yes
4. Space inversion (P)	Intrinsic Party	Yes	Yes	No
5. Time reversal (T)	—	Yes	Yes	?
6. Charge conjugation(C)	C-Parity	Yes	Yes	No
7. CPT	—	Yes	Yes	Yes
8. $U(1)_Q$	Electric charge	Yes	Yes	Yes
9. $SU(2)_I$	I-spin	Yes	No	No
10. $U(1)_Y$	Strangeness	Yes	Yes	No
11. G-conjugation	G-parity	Yes	No	No

Table 1.5

Interaction	Cross Section	Life time (τ)	Width (Γ)
1. Strong	10^{-25}–10^{-27} cm^2	10^{-22}–10^{-24} sec	10–150 MeV
2. Electromagnetic	10^{-30}–10^{-32} cm^2	10^{-16}–10^{-18} sec	10 eV–1 KeV
3. Weak	10^{-36}–10^{-38} cm^2	$\geq 10^{-13}$ sec ($10^{-8} - 10^{-10}$ sec)	—

Problems

1. Show that a scalar meson cannot decay into three pseudoscalar mesons.

2. Show that the reaction $\pi^- + d \to n + n + \pi^0$ cannot occur for both pions at rest.

3. State which of the following processes are allowed and which are forbidden, giving reasons in terms of conservation laws and stating the dominant interaction in the case of allowed processes:

   i) $\pi^0 \to e^- + e^+$ ii) $e^- + p \to \nu_e + n$

   iii) $\pi^- + p \to \Sigma^+ + K^-$ iv) $\mu^- \to e^- + \gamma$

   v) $\gamma + n \to \pi^- + p$ vi) $\Lambda \to \pi^0 + n$

vii) $K^- + p \to \Omega^- + K^+ + K^0$ viii) $p + \bar{p} \to \pi^+ + \pi^- + \pi^0$

ix) $p + n \to p + \Sigma^+ + K^-$ x) $\nu_\mu + n \to e^- + p$

xi) $K^+ \to \pi^+ + \gamma$ · xii) $K^- + p \to \Lambda + \pi^0$

xiii) $p \to \pi^0 + e^+$ xiv) $\nu_\mu + e^- \to \mu^- + \nu_e$

xv) $e^- + p \to e^- + n + \pi^+$ xvi) $\Omega^- \to \Xi^0 + K^-$

xvii) $e^- + e^+ \to \mu^- + \mu^-$ xviii) $\Omega^- \to \Lambda + K^-$

xix) $\bar{\nu}_e + p \to e^+ + n$ xx) $\Sigma^+ \to n + e^+ + v_e$

4. The cross section for the process $e^- + e^+ \to \mu^- + \mu^+$ at very high energies is given by the expression (is natural units)

$$\alpha = \frac{4\pi\alpha^2}{3s},$$

where $\alpha = \frac{e^2}{4\pi} = \frac{1}{137}$ is the electromagnetic fine structure constant and $W = \sqrt{s}$ = total CM energy in the process. Calculate σ in nb ($1nb = 10^{-33}$ cm^2) for $W = 1$ GeV.

5. The lifetime τ of the muon in its rest frame is given by the expression (in natural units)

$$\frac{1}{\tau} = \frac{G_F^2 m_\mu^5}{192\pi^3},$$

where $G_F = \frac{10^{-5}}{M_p^2}$ is the weak decay coupling constant, M_p is the proton mass and m_μ is the muon mass. Calculate τ in sec. (m_p = 938.2 MeV, $m_\mu = 105.6$ MeV).

6. Consider CP, the combined operation of charge conjugation (C) and space inversion or parity transformation (P). Show that

   (a) $CP|\pi^+\pi^-\rangle = +|\pi^+\pi^-\rangle$

   (b) $CP|\pi^0\pi^0\rangle = +|\pi^0\pi^0\rangle$

   (c) $CP|\pi^0\pi^0\pi^0\rangle = -|\pi^0\pi^0\pi^0\rangle$ if the 3π system has total angular momentum $J = 0$

   (d) $CP|\pi^+\pi^-\pi^0\rangle = -|\pi^+\pi^-\pi^0\rangle$ if $\ell_1 = \ell_2 = 0$, where ℓ_1 and ℓ_2 are the orbital angular momenta of the 2 independent relative motions of a 3-body system.

Chapter 2

Symmetries, Conservation Laws and Quantum Numbers

The study of underlying symmetries have played a very important role in the evolution of our knowledge and understanding of particle physics. Symmetries are useful for classification of particles in terms of quantum numbers and characterization of interactions in terms of conservation laws. Symmetries lead to mass degeneracy among particles, and conversely, mass degeneracy implies the existence of an underlying symmetry. Many useful results concerning cross sections and decay rates can be derived by symmetry considerations alone without going into details of the dynamics. Symmetries limit the possible choices of the forms of interactions and sometimes determine the dynamics exactly.

2.1 Symmetries in Classical Mechanics

In classical mechanics, in the Lagrangian formulation, the basic quantity is the Lagrangian $L(q, \dot{q})$ and one has the Lagrangian equations of motion,

$$\frac{d}{dt}\left(\frac{\partial L}{\partial \dot{q}_i}\right) - \frac{\partial L}{\partial q_i} = 0 \qquad i = 1, 2, \ldots n \tag{2.1}$$

If the Lagrangian is independent of a generalized coordinate q_j so that $\partial L/\partial q_j = 0$ then from (2.1) $dp_j/dt = 0$, where $p_j = \partial L/\partial \dot{q}_j$ and therefore p_j is a constant of motion.

2.2 Symmetries in Quantum Mechanics

In non-relativistic quantum mechanics, the basic entities are the Hamiltonian $H = H(q, p)$, with $p_j = \frac{\hbar}{i}\frac{\partial}{\partial q_j}$ and the state vector ψ which satisfies the Schrodinger equation

$$i\hbar\frac{d}{dt}\psi(q, t) = H\psi(q, t) \tag{2.2}$$

Under a transformation $U, \psi \rightarrow \psi' = U\psi$. For equation (2.2) to remain unchanged, $H \rightarrow H' = UHU^{-1}$. The system is said to be invariant under the transformation if $H = H' = UHU^{-1}$ or $[U, H] = 0$.

In quantum mechanics, one is mostly interested in those transformations which keep the norm of the state vector unchanged,

$[\langle\psi|\psi\rangle = \langle\psi'|\psi'\rangle = \langle U\psi|U\psi\rangle]$

This implies $U^\dagger U = UU^\dagger = 1$, that is, U is unitary. One encounters two kinds of symmetries: discrete and continuous.

Discrete Symmetries: For such symmetries $U^2 = 1$. This means $U = U^{-1} = U^\dagger$. U is both unitary, and hermitean. Since U is hermitean, its eigenvalues are real and hence U itself is an observable and its eigenvalues can be used for classification of states. An example of such discrete symmetries is the parity or space inversion.

Continuous Symmetries: For such symmetries, U is a function of r parameters which can take on continuous values over certain ranges.

$$U = U(\theta_1, \ldots, \theta_r); \qquad \alpha_i \leq \theta_i \leq \beta_i, \quad i = 1, n \tag{2.3}$$

Moreover, any finite continuous transformation can be built up as products of infinitesimal transformations. An infinitesimal transformation differs from the identity transformation by an infinitesimal amount. An infinitesimal transformation is thus of the form.

$$U = U(\delta\theta_1, \ldots \delta\theta_r) = 1 + i \sum_{a=1}^{r} \delta\,\theta_a F_a \tag{2.4}$$

U is unitary means F_a's are hermitean. $F_a = F_a^\dagger$

$$[U, H] = 0 \Rightarrow [F_a, H] = 0 \tag{2.5}$$

Then the Heisenberg equation of motion gives

$$\frac{d}{dt}F_a = \frac{1}{i}[F_a, H] = 0 \Rightarrow F_a\text{'s are constants of motion.} \tag{2.6}$$

(2.5) means that F_a and H can have simultaneous eigenstates and thus the eigen values of F_a's can be used to remove the degeneracy of energy eigenstates. F_a's do not commute in general.

$$[F_a, F_b] = i \sum_{c} f_{abc} F_c \tag{2.7}$$

and one has to find the maximal commuting set.

Translation:

$$\vec{x} \xrightarrow{T} x' = T\vec{x} = \vec{x} - \vec{a} \tag{2.8}$$

For infinitesimal transformation $\vec{a} = \vec{\epsilon}$. Under translation,

$$\psi(\vec{x}) \xrightarrow{T} \psi'(x) = T\psi(x) \tag{2.9}$$

$\psi'(x') = \psi(x)$ gives

$$\begin{aligned}\psi'(x) &= \psi(T^{-1}x) = \psi(x+\epsilon) = \psi(x) + \sum_{j=1}^{3} \epsilon_j \frac{\partial}{\partial x_j}\psi(x) \\ &= \left[1 + \frac{i}{\hbar}\sum_j \epsilon_j P_j\right]\psi(x) \end{aligned} \tag{2.10}$$

where $P_j = \dfrac{\hbar}{i} \cdot \dfrac{\partial}{\partial x_j}$

From (2.9) and (2.10) we get

$$T(\vec{\epsilon}) = 1 + \frac{i}{\hbar}\sum \epsilon_j P_j = 1 + \frac{i}{\hbar}\vec{\epsilon}\cdot\vec{P} \tag{2.11}$$

for finite transformation $T(a) = exp\left(\frac{1}{\hbar}\vec{a}\cdot\vec{P}\right)$. P_j's satisfy the communication relations

$$[P_j, P_k] = 0 \tag{2.12}$$

Rotation:

$$\vec{x} \xrightarrow{R} x' = R\vec{x}, \qquad x'_i = \sum_j R_{ij}x_j \tag{2.13}$$

$$\sum_i x_i^2 = \sum_i x'^2_i \Rightarrow R\tilde{R} = \tilde{R}R = 1 \tag{2.14}$$

for infinitesimal rotation $\vec{x'} = \vec{x} - \delta\vec{\theta} \times \vec{x}$. This gives

$$\begin{aligned}\psi'(\vec{x}) &= \psi(R^{-1}\vec{x}) = \psi(\vec{x} + \delta\vec{\theta} \times \vec{x}) \\ &= \psi'(\vec{x}) + \delta\vec{\theta} \times \vec{x} \cdot \vec{\nabla}\psi(\vec{x}) \\ &= [1 + \delta\theta \cdot \vec{x} \times \vec{\nabla}]\psi(\vec{x}) \\ &= \left[1 + \frac{i}{\hbar}\delta\vec{\theta}\cdot\vec{L}\right]\psi(\vec{x}), \quad L = \frac{\hbar}{i}\vec{x} \times \vec{\nabla} \\ &= R(\delta\theta)\psi(\vec{x}) \end{aligned} \tag{2.15}$$

where $R(\delta\vec{\theta}) = 1 + \dfrac{i}{\hbar}\delta\vec{\theta}\cdot\vec{L}$ (2.16)

For finite rotation, $R(\vec{\theta}) = exp(\frac{i}{\hbar}\vec{\theta}\cdot\vec{L})$. L_a's satisfy the commutation relations

$$[L_a, L_b] = i\hbar \sum_c \epsilon_{abc} L_c \tag{2.17}$$

Just as the components of the momentum operator (P_x, P_y, P_z) are the generators of spatial translation, the components of the angular momentum operator (L_x, L_y, L_z) are the generators of spatial rotation. In order to include spin, one defines a general angular momentum $\vec{J} = (J_1, J_2, J_3)$ which satisfies the same commutation relations (CR)'s as the L_a's, components of orbital angular momentum.

$$[J_a, J_b] = i\hbar \sum_c \epsilon_{abc} J_c \tag{2.18}$$

$$\text{Defining } J^2 = \sum_{a=2}^{3} J_a^2 \tag{2.19}$$

One finds that (H, J^2, J_z) form a complete commuting set. The states can be labelled by eigen values of H, J^2 and J_z. Denoting the states by $|E\,j\,m\rangle$, we have

$$H|Ejm\rangle = E|E\,j\,m\rangle$$

$$J^2|E_{jm}\rangle = j(j+1)|E\,j\,m\rangle$$

$$J_z|E\,j\,m\rangle = m|E\,j\,m\rangle, \quad -j \le m \le j \tag{2.20}$$

$2j = n =$ integer, (i.e., j integer or half integer). $j = 0, \frac{1}{2}, 1, \frac{3}{2}, 2, \ldots$. The energy levels are $2j+1 = n+1$ fold degenerate $(n = 0, 1, 2, \ldots)$

2.3 Lorentz Transformation

One expects that any physical theory must be consistent with the special theory of relativity and hence invariance under Lorentz transformation is required. In special theory of relativity, space and time are treated on an equal footing. Thus one talks about a 4-vector, whose zeroth component is the time and the other three components are the usual space coordinates. Under Lorentz transformation the quantity $t^2 - \vec{x}^2$ is invariant.

$$\text{Let, } x^\mu = (x^0, x^1, x^2, x^3),\ x^0 = t,\ \vec{x} = (x^1, x^2, x^3) \tag{2.21}$$

Under Lorentz transformation $x \to x' = \Lambda x$, i.e.,

$$x^\mu \to x'^{\,\mu} = \Lambda^\mu_\nu x^\nu \text{ (sum over } \nu) \tag{2.22}$$

such that,

$$t'^2 - \vec{x}'^2 = t^2 - \vec{x}^{\,2} \tag{2.23}$$

Define $g_{\mu\nu}$ such that

$$g_{\mu\nu} = 0,\ \mu \neq \nu, \quad \text{and} \quad g_{oo} = 1,\ g_{jj} = -1\ (1,2,3) \tag{2.24}$$

then we can write

$$t^2 - \vec{x}^2 = g_{\mu\nu}\, x^\mu\, x^\nu \tag{2.25}$$

Sum over repeated indices, one upper and one lower, is implied. From (2.23) and (2.25) we get

$$g_{\mu\nu}\Lambda^\mu_\alpha \Lambda^\nu_\beta = g_{\alpha\beta} \tag{2.26}$$

We can write this as $\tilde{\Lambda}^\mu_\alpha g_{\mu\nu}\Lambda^\nu_\beta = g_{\alpha\beta}$ (2.27)

consider $\left\{\dfrac{\partial}{\partial x^\mu}\right\} = \left(\dfrac{\partial}{\partial x^0}\dfrac{\partial}{\partial x^1}\dfrac{\partial}{\partial x^2}\dfrac{\partial}{\partial x^3}\right)$

$$\frac{\partial}{\partial x'^\mu} = \frac{\partial x^\nu}{\partial x^\mu}\frac{\partial}{\partial x^\nu} = (\Lambda^{-1})^\nu_\mu \frac{\partial}{\partial x^\nu} \tag{2.28}$$

Thus $\frac{\partial}{\partial x^\mu}$ and x^μ transform differently. A 4-vector which transforms like x^μ is called a contravariant 4-vector and is denoted by a^μ, whereas a 4-vector which transforms like $\partial_\mu = \frac{\partial}{\partial x^\mu}$ is called a covariant 4-vector and is denoted by a_μ. It can be easily seen that

$$a_\mu = g_{\mu\nu}a^\nu \quad \text{and} \quad a^\mu = g^{\mu\nu}a_\nu,$$

where $g^{\mu\nu} = g_{\mu\nu}$, $g^{\mu\alpha}g_{\alpha\nu} = g^\mu_\nu$, $g^\mu_\nu = 1$, $\mu = \nu$ and $g^\mu_\nu = 0$, $\mu \neq \nu$

$$a \cdot b = g_{\mu\nu}a^\mu b^\nu = a^\mu b_\nu = a_\mu b^\mu \tag{2.29}$$

$$a^\mu = (a^0, \vec{a}), \quad a_\mu = (a_0 = a^0, -\vec{a})$$

$$x^\mu = (x^0, x^1, x^2, x^3) = (t, \vec{x}) \quad x_\mu = (x_0, x_1, x_2, x_3) = (t, -\vec{x})$$

$$\partial_\mu = \left\{\frac{\partial}{\partial x^\mu}\right\} = \left(\frac{\partial}{\partial t}, \vec{\nabla}\right), \quad \vec{\nabla} = \left(\frac{\partial}{\partial x^1}, \frac{\partial}{\partial x^2}, \frac{\partial}{\partial x^3}\right)$$

$$\partial_\mu = \left\{\frac{\partial}{\partial x_\mu}\right\} = \left(\frac{\partial}{\partial t}, -\vec{\nabla}\right), \quad \partial \cdot \partial = \frac{\partial}{\partial t^2} - \vec{\nabla}^2$$

$$\partial \cdot a = \partial_\mu a^\mu = \frac{\partial a^0}{\partial t} + \vec{\nabla} \cdot \vec{a}$$

Another important 4-vector is the energy-momentum 4-vector

$$P^\mu = (P^0, P^1, P^2, P^3) = (E, \vec{P}), \quad P^0 = E, \quad \vec{P} = (P^1, P^2, P^3)$$

$$P_\mu = (P_0, P_1, P_2, P_3) = (E, -\vec{P})$$

where E is the energy of a particle and P^1, P^2, P^3 are the three cartesian components of the particle's 3-momentum.

$$P \cdot P = P^\mu P_\mu = E^2 - \vec{P}^2 = \text{invariant} = m^2$$

where m is the mass of the particle.

2.4 Relativistic Kinematics

In particle physics, one studies 2 kinds of processes, (a) collision process: $a+b \to 1+2+\cdots+n$, (b) decay process: $A \to 1+2+\cdots+n$ with $n \geq 2$. In the collision process, particles a and b collide to form a n-particle final state. If $n = 2$, it is called a scattering process and if $n > 2$ it is called a production process. In the decay process, a particle A decays into n-particles. In order that a process is physical, i.e., observable, three requirements have to be satisfied: (1) There has to be energy-momentum conservation, i.e., total energy and total 3-momentum of the initial state must be equal to those of the final state. (2) All the particles must be on-mass-shell, i.e., for the i-th particle $E_i^2 - \vec{p_i}^2 = m_i^2$. (3) The description of the process must be consistent with the special theory of relativity. In particular, the amplitude for the process must be Lorentz-invariant.

2 → 2 Process: Consider the process $a + b \to c + d$

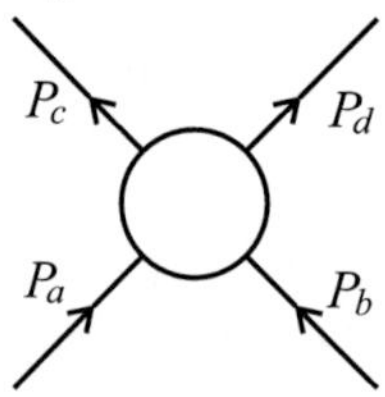

Fig. 2.1: 2 → 2 Process

with amplitude $A = A(p_a, p_b, p_c, p_d)$. We have

$$p_a + p_b = p_c + p_d$$

and $\quad p_i^2 = m_i^2 \quad (i = a, b, c, d)$ (2.30)

(2.30) implies that only 3 of the 4 momenta are independent. We take them to be p_a, p_b and p_c.

The invariant amplitude A can be a function of Lorentz invariant variables and if the particles are spinless only Lorentz invariant quantities are those constructed from the momenta, namely, $p_a \cdot p_b$, $p_b \cdot p_c$, $p_c \cdot p_a$, p_a^2, p_b^2, p_c^2. Because of the mass shell conditions, the last 3 are constants. Thus the only Lorentz invariant independent variables are $p_a \cdot p_b$, $p_b \cdot p_c$, $p_c \cdot p_a$.

Instead of these, it is more convenient to use the variables s, t, and u (called the Mandelstam Variables) which are related to them.

$$s = (p_a + p_b)^2 = (p_c + p_d)^2$$

$$t = (p_a - p_c)^2 = (p_d - p_b)^2$$

$$u = (p_b - p_c)^2 = (p_d - p_a)^2$$

$$\text{with} \quad s + t + u = \sum_i m_i^2 \tag{2.31}$$

Proof:
$$\begin{aligned} s + t + u &= (p_a + p_b)^2 + (p_a - p_c)^2 + (p_b - p_c)^2 \\ &= (p_a + p_b)^2 + (p_a - p_c)^2 + (p_d - p_a)^2 \\ &= 3p_a^2 + p_b^2 + p_c^2 + p_d^2 + 2p_a \cdot (p_b - p_c - p_d) \\ &= 3p_a^2 + p_b^2 + p_c^2 + p_d^2 + 2p_a \cdot (-p_a) \\ &= p_a^2 + p_b^2 + p_c^2 + p_d^2 \\ &= m_a^2 + m_b^2 + m_c^2 + m_d^2 \\ &= \sum_{t=a}^{d} m_i^2 = \text{constant} \end{aligned}$$

Thus of the three Mandelstam variables only 2 are independent. However, one writes the amplitude for the process as $A = A(s, t, u)$ and the fact $s + t + u = \text{constant}$ is understood.

Lab frame and CM frame

Since the amplitude is Lorentz-invariant one can work in any frame that is convenient. However, one uses either the Lab frame or the CM frame defined as follows.

Lab frame: In this frame one of the initial state particles, say the particle b, is taken to be at rest. Thus $p_b = (m_b, \vec{o})$ and $p_a = (E_a^L, p_a^L)$.

CM frame: In this frame, the total 3-momentum of the initial state particles is taken to be zero, $\vec{p}_a + \vec{p}_b = 0$.

Thus, $\vec{p}_a = -\vec{p}_b = p$, $p_a = (E_a^c, \vec{p})$, $p_b = (E_b^c, -\vec{p})$.

In terms of Lab frame variables

$$s = (p_a + p_b)^2 = p_a^2 + p_b^2 + 2p_a \cdot p_b = m_a^2 + m_b^2 + 2m_b \cdot E_a^L \tag{2.32}$$

and in terms of CM frame variables

$$W^2 = s = (p_a + p_b)^2 = (E_a + E_b)^2 - (\vec{p}_a + \vec{p}_b)^2 = (E_a^c + E_b^c)^2 \tag{2.33}$$

$W = E_a^c + E_b^c =$ total CM energy, E_a^c, E_b^c and $|\vec{p}|$ can be expressed in terms of s, m_a^2, m_b^2.

$$E_a^c = \frac{1}{2\sqrt{s}}(s + m_a^2 - m_b^2),\ E_b^c = \frac{1}{2\sqrt{s}}(s + m_b^2 - m_a^2) \tag{2.34}$$

$$|\vec{p}| = \frac{1}{2\sqrt{s}}\lambda(s, m_a^2, m_b^2) \tag{2.35}$$

where $\lambda(x, y, z) = (x^2 + y^2 + z^2 - 2xy - 2yz - 2zx)^{\frac{1}{2}}$ (2.36)

For the process, $a + b \to c + d$, one has $\vec{p}_c = -\vec{p}_d = \vec{q}$ in the CM frame. Therefore

$$E_c^a = \frac{1}{2\sqrt{s}}(s + m_c^2 - m_d^2),\ E_c^d = \frac{1}{2\sqrt{s}}(s + m_d^2 - m_c^2)$$

$$|\vec{q}| = \frac{1}{2\sqrt{s}}\lambda(s, m_c^2, m_d^2) \tag{2.37}$$

$$\begin{aligned} t = (p_a - p_c)^2 &= p_a^2 + p_c^2 - 2p_a \cdot p_c \\ &= m_a^2 + m_c^2 - 2(E_a E_c - |\vec{p}||\vec{q}| \cos\theta_c) \end{aligned} \tag{2.38}$$

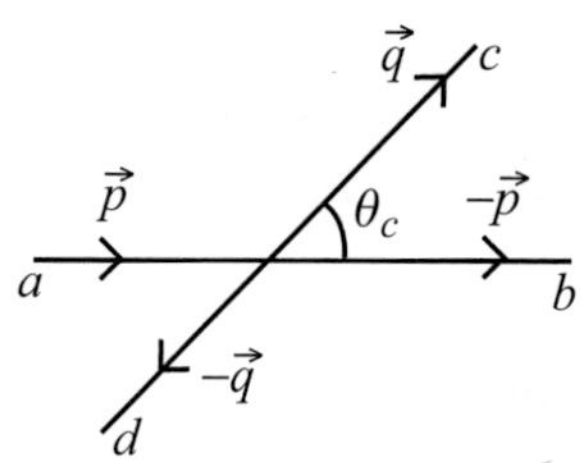

Fig. 2.2: CM frame

This gives $\cos\theta_c = \dfrac{2s(t - m_a^2 - m_c^2) + (s + m_a^2 - m_b^2)(s + m_c^2 - m_d^2)}{\lambda(sm_a^2m_b^2)\lambda(sm_c^2m_d^2)}$ (2.39)

The kinematically allowed region for the process $a + b \to c + d$ is $s \geq$ larger of

$\{(m_a + m_b)^2, (m_c + m_d)^2\}$, $\quad -1 \leq \cos\theta_c \leq 1$ (2.40)

For elastic scattering $m_a = m_c$, $m_d = m_b$. This means $E_a = E_c$, $E_b = E_d$ and $|\vec{p}| = |\vec{q}|$ and therefore

$$-4|\vec{p}|^2 \leq t \leq 0 \tag{2.41}$$

In the equal-mass case ($m_a = m_b = m_c = m_d = m$).

$$E_i = \frac{\sqrt{s}}{2}(i = a, b, c, d) \cdot |\vec{p}\,| = \left(\frac{s}{4} - m^2\right)^{\frac{1}{2}}$$

$$t = (p_a - p_c)^2 = -2|\,\vec{p}\,|^2(1 - \cos\theta_c)$$

$$u = (p_b - p_c)^2 = -2|\,\vec{p}\,|^2(1 + \cos\theta_c)$$

The allowed region is $s \geq 4m^2$, $t \leq 0$, $u \leq 0$ (2.42)

2.5 Crossing and Crossing symmetry

The amplitude $A(s, t, u)$ for the process $a + b \rightarrow c + d$ is related to the amplitudes for 5 other processes as a consequence of crossing symmetry and the TCP-theorem but for different allowed kinematic regions for the variables s, t, u. According to the crossing principle, the amplitude remains the same if a particle in the final state is replaced by the antiparticle corresponding to a particle in the initial state and vice versa with the sign of 4-momentum reversed. Processes are classified as s-channel, t-channel and u-channel processes depending on which Mandelstam variable is the energy variable.

Channel	**Process**	**Amplitude**	
s-channel:	$a + b \rightarrow c + d;\ \bar{c} + \bar{d} \rightarrow \bar{a} + \bar{b}$	$A(s, t, u)$	
t-channel:	$a + \bar{c} \rightarrow \bar{b} + d;\ b + \bar{d} \rightarrow \bar{a} + c$	$A(t, s, u)$	
u-channel:	$a + \bar{d} \rightarrow c + \bar{b};\ \bar{c} + \bar{b} \rightarrow \bar{a} + \bar{d}$	$A(u, t, s)$	(2.43)

For equal-mass case, the allowed kinematic regions are

$$\begin{aligned} &s\text{-channel:} \quad s \geq 4m^2,\ t \leq 0,\ u \leq 0 \\ &t\text{-channel:} \quad t \geq 4m^2,\ s \leq 0,\ u \leq 0 \\ &u\text{-channel:} \quad u \geq 4m^2,\ t \leq 0,\ s \leq 0 \end{aligned} \tag{2.44}$$

These can be depicted in a plot known as the Mandelstam plot. The inside of the central triangle is excluded for all channels. Examples of crossing:

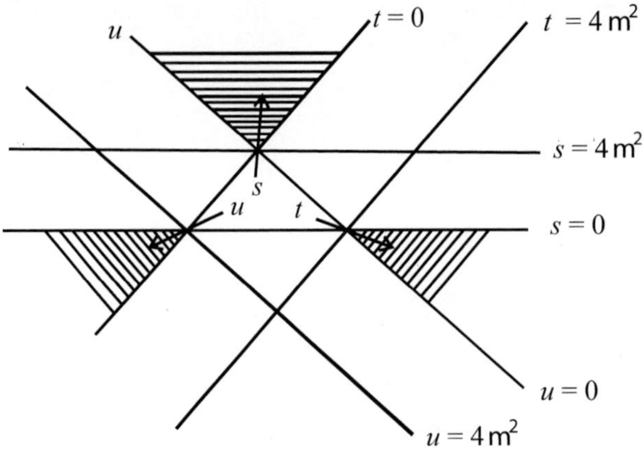

Fig. 2.3: Mandelstam plot

$$\begin{aligned}
&s\text{-channel:} \quad \pi^+ + p \to \pi^+ + p \\
&t\text{-channel:} \quad \bar{p} + p \to \pi^+ + \pi^- \\
&u\text{-channel:} \quad \pi^- + p \to \pi^- + p
\end{aligned} \tag{2.45}$$

2.6 Invariant Cross Section Formula

Consider the collision process $a + b \to f$. The probability of a colliding with b and producing the final state f is clearly inversely proportioned to area A of confinement of b. $P \propto \frac{1}{A}$ or $P = \frac{\sigma}{A}$, $\sigma =$ constant. Clearly if $A = \sigma$, $P = 1$. The cross section has the dimension of area. In practice, one does experiment with a beam of projectiles (particle a) and a volume of target particles (particle b) and measures event rate/volume which is given by

$$R = \frac{N_a N_b P}{t \cdot V} = \frac{N_a N_b \sigma}{t \cdot V \cdot A} = F_a n_b \sigma = L\sigma \tag{2.46}$$

where $F_a = \frac{N_a}{tA}$ = projectile flux = no. of particles incident per unit time per unit area and $n_b = \frac{N_b}{V}$ = target particle density. $L = \frac{N_a N_b}{t \cdot V \cdot A} = F_a n_b$ is called luminosity. One can also write $F_a = n_a |\vec{v_a} - \vec{v_b}|$, $n_a =$ projectile particle density.

$$L = n_a n_b |\vec{v_a} - \vec{v_b}| \tag{2.47}$$

$\vec{v_a} - \vec{v_b}$ = relative velocity of projectile particle and target particle. One can theoretically calculate the rate R if one knows the dynamics.

$$R = \int |A|^2 \delta_4 (P_a + P_b - P_f) d\varphi_f \tag{2.48}$$

where A = amplitude for the process and P_a, P_b are projectile and target particle 4-momenta and P_f is the total momentum of the final state particles. $P_f = \sum_{t=1}^{n} p_i$ and $d\varphi_f$ is the final state phase space volume element. Using covariant normalisation of single particle states

$$\langle \vec{p}, \lambda | \vec{p'}, \lambda' \rangle = (2\pi)^3 \delta_3(\vec{p} - \vec{p'})(2Ep)\delta_{\lambda\lambda'} \tag{2.49}$$

where $E_p = \sqrt{\vec{p}^2 + m^2}$. One finds $n_i = 2E_i$,

$$\text{and} \quad d\varphi_f = \prod_{i=1}^{n} \frac{d^3\vec{p_i}}{(2\pi)^3(2E_i)} \tag{2.50}$$

A useful result is

$$\int \frac{d^3p}{2E_p} = \int d^4p\delta(p^2 - m^2)\theta(p^0), p^2 = (p^0)^2 - \vec{p}^2 \tag{2.51}$$

with p^0 an independent variable.

From (2.46), (2.47),(2.48) and (2.49) one gets the invariant cross section formula

$$\begin{aligned} d\sigma(a + b \to f) &= \frac{1}{n_a n_b |\vec{v_a} - \vec{v_b}|} |A|^2 (2\pi)^4 \delta_4(P_a + P_b - P_f) d\phi_f \\ &= \frac{1}{(2E_a)(2E_b)|\vec{v_a} - \vec{v_b}|} |A|^2 (2\pi)^4 \delta_4(P_a + P_b - P_f) d\phi_f \end{aligned} \tag{2.52}$$

For collinear motion $E_a E_b |\vec{v_a} - \vec{v_b}| = \sqrt{(P_a \cdot P_b)^2 - m_a^2 m_b^2}$. Each factor on the rhs of Eq (2.52) is Lorentz invariant.

The process $a + b \to c + d$. Phase Space integration (CM frame). We have

$$\sigma(a + b \to c + d) = \frac{1}{L_{ab}} \int |A|^2 \cdot (2\pi)^4 (p_a + p_b - p_c - p_d) \frac{d^3p_c d^3p_d}{4E_c E_d (2\pi)^6} \tag{2.53}$$

where, $A = A_{ab \to cd} = A(p_a, p_b, p_c, p_d)$

$$\text{and} \quad L_{ab} = (2E_a)(2E_b)|\vec{v_a} - \vec{v_b}| \tag{2.54}$$

In the CM frame,

$$\begin{aligned} L_{ab} &= 4E_a E_b \left| \frac{\vec{p_a}}{E_a} - \frac{\vec{p_b}}{E_b} \right| = 4E_a E_b \left| \frac{\vec{p}}{E_a} + \frac{\vec{p}}{E_b} \right| = 4E_a E_b \left| \frac{1}{E_a} + \frac{1}{E_b} \right| |\vec{p}| \\ &= 4|\vec{p}|\sqrt{s} \end{aligned} \tag{2.55}$$

Let $I = \int |A|^2 \cdot (2\pi)^4 \delta_4(p_a + p_b - p_c - p_d) \frac{d^3p_c}{(2\pi)^3(2E_c)} \frac{d^3p_d}{(2\pi)^3(2E_d)}$

In the CM frame

$$
\begin{aligned}
I &= \frac{1}{4(2\pi)^2}\int |A|^2\delta(E_c+E_d-\sqrt{s})\delta_3(\vec{p}_c+\vec{p}_d)\frac{d^3p_c\,d^3p_d}{E_cE_d} \\
&= \frac{1}{4(2\pi)^2}\int |A|^2\delta(E_c+E_d-\sqrt{s})\frac{d^3p_c}{E_cE_d} \\
&= \frac{1}{4(2\pi)^2}\int |A|^2\delta(E_c+E_d-\sqrt{s})\frac{|\vec{p}_c|^2d|p_c|d\Omega}{E_cE_d} \\
&= \frac{1}{4(2\pi)^2}\int |A|^2\delta(E-\sqrt{s})dE\frac{d|p_c|}{dE}\cdot\frac{|\vec{p}_c|^2}{E_cE_d}d\Omega \qquad (2.56)
\end{aligned}
$$

Now $\dfrac{dE}{d|\vec{p}_c|} = \dfrac{d}{d|\vec{p}_c|}(E_c+E_d) = \dfrac{dE_c}{d|\vec{p}_c|}+\dfrac{dE_d}{d|\vec{p}_d|}\cdot\dfrac{d|\vec{p}_d|}{d|\vec{p}_c|} = \dfrac{|\vec{p}_c|}{E_c}+\dfrac{|\vec{p}_d|}{E_d}\cdot\dfrac{d|\vec{p}_d|}{d|\vec{p}_c|}$

$$
\begin{aligned}
&= |\vec{p}\,'|\left(\frac{1}{E_c}+\frac{1}{E_d}\right) = |\vec{p}\,'|\frac{E_c+E_d}{E_c\,E_d} \\
&= |\vec{p}\,'|\frac{\sqrt{s}}{E_c\,E_d} \quad (|p_c| = |p_d| = |\vec{p}\,'|)
\end{aligned}
$$

Using the result in (2.56) we get

$$
I = \frac{1}{4(2\pi)^2}\int |A|^2\frac{E_c\,E_d}{|\vec{p}\,'|\sqrt{s}}\cdot\frac{|\vec{p}\,'|^2}{E_cE_d}d\Omega = \frac{1}{4(2\pi)^2}\int |A|^2\frac{|\vec{p}\,'|}{\sqrt{s}}d\Omega \qquad (2.57)
$$

Substituting (2.55) and (2.57) in (2.53)we get

$$
\left(\frac{d\sigma}{d\Omega}\right)^{ab\to cd}_{CM} = \frac{1}{4|\vec{p}|\sqrt{s}}\cdot|A|^2\cdot\frac{1}{4(2\pi)^2}\cdot\frac{|\vec{p}\,'|}{\sqrt{s}}
$$

On simplification we get

$$
\left(\frac{d\sigma}{d\Omega}\right)^{ab\to cd}_{CM} = \frac{1}{64\pi^2\sqrt{s}}\cdot\frac{|\vec{p}\,'|}{|\vec{p}|}\cdot|A|^2 \qquad (2.58)
$$

Often, one is interested in spin-averaged cross section, i.e., one averages over all initial spin (polarization) states and sums over all final spin (polarization) states. In such cases,

$$
|A|^2 \to \overline{|A|^2} = \frac{1}{(2s_a+1)(2s_b+1)}\sum_{spins}|A|^2 \qquad (2.59)
$$

Particle Decay

Particles produced in high energy collisions are often unstable. Unstable particles decay into lighter-mass particles unless forbidden by conservation laws. The mean life time for decay of a particle A into n particles is given by

$$\frac{1}{\tau} = \Gamma = \frac{1}{2E_A} \int |A|^2 \cdot (2\pi)^4 \delta_4(P_A - P_f) d\phi_f \tag{2.60}$$

where Γ is the decay width, which is inverse of lifetime τ by Heisenberg uncertainty principle. As before,

$$P_f = \sum_{i=1}^{n} p_i \text{ and } d\phi_f = \prod_{i=1}^{n} \left[\frac{d^3 p_i}{(2\pi)^3 (2E_i)} \right]$$

Specializing to 2-body decay and CM frame (rest frame of A) one gets after carrying out the phase space integration

$$\Gamma(A \to 1+2) = \frac{1}{2M_A} \cdot |A|^2 \cdot \frac{1}{4(2\pi)^2} \cdot \frac{|\vec{p}\,'|}{\sqrt{s}} \cdot 4\pi$$

Simplification gives

$$\Gamma(A \to 1+2) = \frac{|\vec{p}\,'|}{8\pi M_A^2} |A|^2 \tag{2.61}$$

The 2 → 2 process. Phase space integration (Lab frame)

As an illustration of the phase space integration in the Lab frame we consider high energy elastic electron-proton scattering $e^- + p \to e^- + p$.

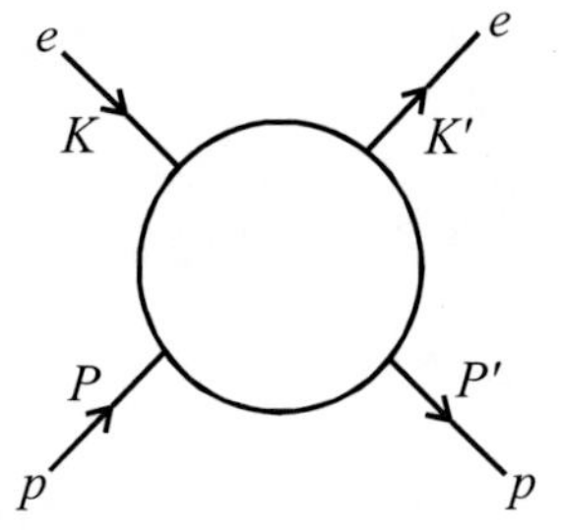

Fig. 2.4: Electron-proton elastic scattering

We have $K + P = K' + P'$, $K^2 = K'^2 = m^2$, $P^2 = P'^2 = M^2$, $P = (M, \vec{O})$, $K = (E, \vec{K})$, $K' = (E, \vec{K}\,')$, $|\vec{K}|, |\vec{K}\,'| \gg m$, $E \approx |\vec{K}|$, $E' \approx |\vec{K}\,'|$.

The spin averaged cross section is given by

$$\sigma = \frac{1}{L_{ab}} \int \overline{|A|^2} \cdot (2\pi)^4 \delta_4(P_i - P_f) d\phi_f \tag{2.62}$$

where $L_{ab} = (2E_a)(2E_p)|\vec{v}_e - \vec{v}_p|$

$$= 4\sqrt{(K \cdot P)^2 - m^2 M^2} \approx 4(KP) = 4ME \tag{2.63}$$

$$\delta_4(P_i - P_f) = \delta_4(K + P - K' - P') = \delta_4(P + q - P')$$

$$q = K - K' = \text{electron momentum transfer}$$

and $$d\phi_f = \frac{d^3K'}{(2\pi)^3(2E')} \cdot \frac{d^3P'}{(2\pi)^3(2E'_p)}$$

Using (2.51) we can carry out the integral in (2.62). We get

$$\sigma = \frac{1}{4ME}\int |\vec{A}|^2 \cdot \frac{(2\pi)^4}{2(2\pi)^6} \cdot \delta((P+q)^2 - M^2)\frac{d^3K'}{E'} \tag{2.64}$$

We have

$$\frac{d^3K'}{E'} = \frac{|\vec{K}'|^2 d|\vec{K}'|d\Omega}{E'} = \frac{|\vec{K}'|E'dE'd\Omega}{E'} \approx E'dE'd\Omega \tag{2.65}$$

Now $\delta((P+q)^2 - M^2) = \delta(P^2 + q^2 + 2P\cdot q - M^2) = \delta(q^2 + 2P\cdot q)$

$$q^2 = (K-K')^2 = K^2 + K'^2 - 2K\cdot K' = 2m^2 - 2(EE' - |\vec{K}||\vec{K}'|\cos\theta)$$

$$\approx -2EE'(1-\cos\theta)$$

θ = scatteromg angle in Lab frame. Therefore,

$$\delta(q^2 + 2P\cdot q) = \delta(-2EE'(1-\cos\theta) + 2M(E-E')) = \delta(f(E')) \tag{2.66}$$

where $f(E') = -2EE'(1-\cos\theta) + 2M(E-E')$

Now $$\int \delta(f(x))dx = \int \sum_i \frac{1}{|f'(x_i)|}\delta(x - x_i)dx$$

where $f(x_i) = 0,\ f'(x_i) = \frac{d}{dx}f(x)'$ at $x = x_i$ x_i = zero of $f(x)$

Using (2.65) and (2.66) we get from (2.64)

$$\left(\frac{d\sigma}{d\Omega}\right)^{ep\to ep}_{Lab} = \frac{1}{4ME}\cdot\frac{1}{2(2\pi)^2}\cdot\int |\vec{A}|^2\delta(f(E'))E'dE' \tag{2.67}$$

$f(E') = 0$ gives

$$-2EE'(1-\cos\theta) + 2M(E-E') = 0$$

or $$E' = \frac{E}{1 + \frac{E}{M}(1-\cos\theta)} \tag{2.68}$$

Also $$\frac{d}{dE'}f(E') = -2E(1-\cos\theta) - 2M$$

$$|f'(E')| = 2M\left[1 + \frac{E}{M}(1-\cos\theta)\right] = 2M\left(\frac{E}{E'}\right) \tag{2.69}$$

for E' given by (2.68). Using the δ function we can carry out the integral in (2.67) to get

$$\left(\frac{d\sigma}{d\Omega}\right)^{ep\to ep}_{Lab} = \frac{1}{4ME}\cdot\frac{1}{2(2\pi)^2}\cdot\overline{|A|^2}\cdot E'\cdot\frac{E'}{2ME} = \frac{1}{64\pi^2M^2}\left(\frac{E'}{E}\right)\overline{|A|^2} \quad (2.70)$$

where E' is given by (2.68.)

2.7 S-matrix and Transition Probability

The general form for the S-matrix element which gives the amplitude for transition from an initial state i to some final state f is

$$S_{fi} = \delta_{fi} - i(2\pi)^4\delta_4(P_i - P_f)A \quad (2.71)$$

where A=Lorentz invariant amplitude. For $f \neq i$ the transition probability is given by

$$\tilde{w}_{fi} = |(2\pi)^4\delta_4(P_i - p_f)A|^2 = (2\pi)^4\delta_4(0)\cdot(2\pi)^4\delta_4(P_i - p_f)|A|^2 \quad (2.72)$$

Now $\delta_4(K) = \delta_3(\vec{K})\delta(w)$. We have

$$\delta(w) = \frac{1}{2\pi}\int_{-\infty}^{\infty} e^{-iwt}dt = \frac{1}{2\pi}\lim_{T\to\infty}\int_{-T/2}^{T/2} e^{-iwt}dt$$

which gives

$$2\pi\delta(o) = \lim_{T\to\infty}\int_{-T/2}^{T/2} dt = T$$

$$\delta_3(\vec{K}) = \frac{1}{(2\pi)^3}\int_{-\infty}^{\infty} e^{i\vec{K}\cdot\vec{x}}d^3x.$$

This gives $(2\pi)^3\delta_3(0) = V(V\to\infty)$.

Combining the two results, we have $(2\pi)^4\delta_4(0) = VT$. This means $\tilde{w}_{fi}\to\infty$ as V and $T\to\infty$. However, $w_{fi} = (1/VT)\tilde{w}_{fi}$ is finite. Thus one can talk about transition probability per unit volume per unit time which is given by

$$w_{fi} = |A|^2(2\pi)^4\delta_4(P_i - P_f) \quad (2.73)$$

Event rate per unit volume per unit time is given by

$$R = \int w_{fi}d\phi_f \quad (2.74)$$

Substituting (2.73) in (2.74) we get

$$R = \int |A|^2 (2\pi)^4 \delta_4(P_i - P_f) d\phi_f \tag{2.75}$$

which is same as (2.48).

2.8 Quantum Numbers of Particles and their Determination

Particles are classified by quantum numbers. Some of these quantum numbers are related to the space-time transformation properties of the system while the others are related to the transformation properties of the system in some internal space. Examples of the space-time quantum numbers are mass, spin, intrinsic parity and C-parity. The quantum numbers like electric charge, i-spin, baryon number and strangeness are known as internal quantum numbers. Quantum numbers are determined by applying conservation laws and invariance principles to scattering and production processes involving the particles. As illustrations of how they are determined we shall consider the case of the pion which is the lightest hadron.

Spin of the Charged Pion: The spin of the charged pion is determined by studying the process $p+p \to \pi^+ + D$ and its inverse process $\pi^+ + D \to p+p$ (D = deuteron) and applying the principle of detailed balance which is a consequence of time-reversal invariance. In the CM frame, the process $a+b \to c+d$ is described by the left diagram (Fig. 2.5).

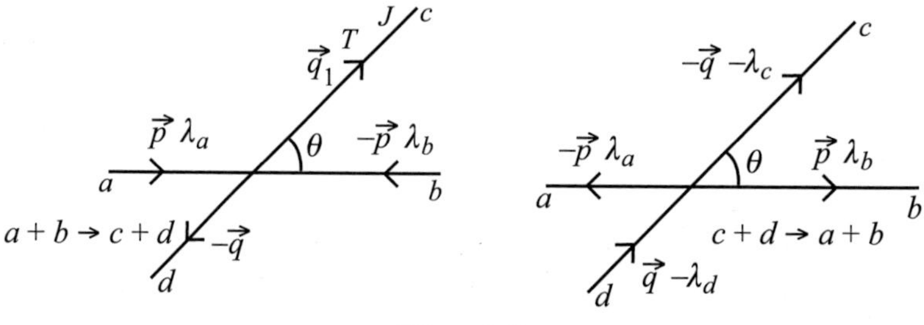

Fig. 2.5

Here λ's are the z-components of the spins the particles. $\vec{p}$, $\vec{q}$ are CM momentum of the initial and final particles. Under time reversal the process goes over to the process described by the right diagram which corresponds to the process $c+d \to a+b$. In terms of amplitude,

$$A_{ab\to cd}\left(\vec{p}, \vec{q}, \{\lambda_i\}\right) \xrightarrow{T} A_{cd\to ab}\left(-\vec{q}, -\vec{p}, \{-\lambda_i\}\right) \tag{2.76}$$

Time-reversal invariance implies

$$A_{ab\to cd}\left(\vec{p},\vec{q},\{\lambda_i\}\right)=A^*_{cd\to ab}\left(-\vec{q},-\vec{p},\{-\lambda_i\}\right) \tag{2.77}$$

This means

$$\begin{aligned}\sum_{\{\lambda_i\}}\left|A_{ab\to cd}\left(\vec{p},\vec{q},\{\lambda_i\}\right)\right|^2 &= \sum_{\{\lambda_i\}}\left|A_{cd\to ab}\left(-\vec{q},-\vec{p},\{-\lambda_i\}\right)\right|^2\\ &= \sum_{\{-\lambda_i\}}\left|A_{cd\to ab}\left(\vec{q},\vec{p},\{\lambda_i\}\right)\right|^2\\ &= \sum_{\{\lambda_i\}}\left|A_{cd\to ab}\left(\vec{q},\vec{p},\{\lambda_i\}\right)\right|^2\end{aligned} \tag{2.78}$$

The first equality results from.the fact that rotational invariance implies

$$A(\vec{p},\vec{q})=A(\vec{p}^{\,2},\vec{q}^{\,2},\vec{p},\vec{q})$$

and the second equality results from that fact $\Sigma_{\{\lambda_i\}}=\Sigma_{\{-\lambda_i\}}$ because $-s_i\leq\lambda_i\leq s_i$. Equation (2.78) is known as the principle of detailed balance. Now the spin-averaged cross sections for the processes $a+b\to c+d$ and $c+d\to a+b$ are, respectively, given by

$$\begin{aligned}\left(\frac{d\sigma}{d\Omega}\right)^{CM}_{ab\to cd} &= \frac{1}{64\pi^2 s}\frac{|\vec{q}|}{|\vec{p}|}\overline{|A_{ab\to cd}|^2}\\ &= \frac{1}{64\pi^2 s}\frac{|\vec{q}|}{|\vec{p}|}\frac{1}{(2s_a+1)(2s_b+1)}\sum_{\lambda_i}|A_{ab\to cd}(\vec{p},\vec{q},\{\lambda\}|^2\end{aligned} \tag{2.79}$$

and $$\left(\frac{d\sigma}{d\Omega}\right)^{CM}_{cd\to ab}=\frac{1}{64\pi^2 s}\frac{|\vec{p}|}{|\vec{q}|}\frac{1}{(2s_c+1)(2s_d+1)}\sum_{\lambda_i}|A_{cd\to ab}(\vec{q},\vec{p},\{\lambda\}|^2 \tag{2.80}$$

There is an additional factor of $\frac{1}{2}$ if the final state particles are identical (e.g., $\pi^+ + D\to p+p$). Therefore for same $s=W^2$ we have, because of eq. (2.78).

$$\frac{(d\sigma/d\Omega)^{CM}_{ab\to cd}}{(d\sigma/d\Omega)^{CM}_{cd\to ab}}=\frac{|\vec{q}|^2(2s_c+1)(2s_d+1)}{|\vec{p}|^2(2s_a+1)(2s_b+1)} \tag{2.81}$$

Applying (2.81) to the process $p+p\to\pi^+ + D$ and its inverse process $\pi^+ + D\to p+p$ we get

$$\frac{(d\sigma/d\pi)^{CM}_{pp\to\pi^+ + D}}{(d\sigma/d\pi)^{CM}_{\pi^+ + D\to pp}}=\frac{|\vec{q}|^2}{|\vec{p}|^2}\cdot\frac{3}{4}(2s_{\pi^+}+1)2 \tag{2.82}$$

$s_p = \frac{1}{2}$, $s_D = 1$, $|p| = (1/2\sqrt{s})\lambda\,(s, m_p^2, m_p^2)$ and $|q| = (1/2\sqrt{s})\lambda\,(s, m_\pi^2\; m_D^2)$.

Experimental data implies that $s_{\pi^+} = 0$.

Spin of π^0: Although π^0 is believed to be a member of an $I = 1$ multiplet along with π^+ and π^-, thus having spin 0 like π^+ (and π^-), a simple argument due to Young derived from symmetry principles shows that observation of the decay $\pi^0 \to 2\gamma$ would be inconsistent with $J_{\pi^0} = 1$.

Proof: If ε_1 and ε_2 are the photon helicities then in the π^0 rest frame

$$\vec{\varepsilon_1} \cdot \vec{k} = 0 = \varepsilon_2 \cdot \vec{k},\ \vec{k} = \frac{1}{2}(\vec{k}_1 - \vec{k}_2),\ \vec{k}_1 + \vec{k}_2 = 0 \tag{2.83}$$

If $j_{\pi^0} = 1$, then the invariant amplitude in this frame is of the form

$$A = \vec{\varepsilon_\pi} \cdot \vec{A} \tag{2.84}$$

where $\vec{A}$ is to be constructed from $\vec{\varepsilon_1}$, $\vec{\varepsilon_2}$ and $\vec{k}$. Possible forms for $\vec{A}$ are

$$\left.\begin{array}{l} 1)\ \vec{k}(\vec{\varepsilon_1} \cdot \vec{\varepsilon_2}) \\ 2)\ \vec{\varepsilon_1} \times \vec{\varepsilon_2} \\ 3)\ \vec{k} \times (\vec{\varepsilon_1} \times \vec{\varepsilon_2}) \end{array}\right\} \tag{2.85}$$

Since photons are bosons, $\vec{A}$ must be symmetric under exchange of photons Thus (1) and (2) are excluded since under $1 \leftrightarrow 2$, $\vec{k} \to -\vec{k}$ and $\vec{\varepsilon_1} \times \vec{\varepsilon_2} \to \vec{\varepsilon_2} \times \vec{\varepsilon_1} = -\vec{\varepsilon_1} \times \vec{\varepsilon_2}$. (1) is also excluded by parity invariance, and (3) is excluded because of gauge invariance $\vec{k} \times (\vec{\varepsilon_1} \times \vec{\varepsilon_2}) = \vec{\varepsilon_1}(\vec{k} \cdot \vec{\varepsilon_2}) - \vec{k} \cdot \vec{\varepsilon_1})\vec{\varepsilon_2} = 0$ since $\vec{\varepsilon_1} \cdot \vec{k} = \vec{\varepsilon_2} \cdot \vec{k} = 0$. Thus $j_{\pi^0} \neq 1$ if angular momentum is conserved in $\pi^0 \to 2\gamma$.

Intrinsic parity of the pion: Apart from orbital parity which arises due to relative motion of particles, particles are assigned intrinsic parities which appear naturally in a field theoretic description. Phenomenologically, assignment of intrinsic parities are done in order to ensure parity conservation in strong and electromagnetic processes.

Experimental data indicate that the capture of π^- in liquid H_2 and liquid D_2 occurs from the atomic s-states. When π^- is stopped in liquid deuterium the following reactions take place

$$\begin{aligned} \pi^- + \mathrm{D} &\to n + n \\ &\to n + n + \gamma \end{aligned} \tag{2.86}$$

Consider the process $\pi^- + \mathrm{D} \to n + n$. Angular momentum conservation gives $J_i = J_f$. $\vec{J_i} = \vec{s_i} + \vec{l_i}$, $\vec{s_i} = \vec{s_\pi} + \vec{s_d}$, $(s_\pi = 0,\ s_d = 1)$, $l_i = 0$, $s_i = 1$. Thus

$J_i = 1$. Now $\vec{J_f} = \vec{s}_f + \vec{l}_f$, $\vec{s}_f = \vec{s}_1 + \vec{s}_2$, $(\vec{s}_1 = \vec{s}_2 = \vec{s}_n = (1/2))$, $s_f = 0,\ 1$. $s_f = 1$ spin state is symmetric and $s_f = 0$ spin state is antisymmetric under exchange of particles. Since we have 2 identical fermions in the final state, ψ_f must be antisymmetric under particle exchange.

$\psi_f = \psi_{space}\psi_{spin}$

$P_{12}\psi_{space(1,2)} = (-)^{l_f}\psi_{space(1,2)}$

$P_{12}\psi_{spin(1,2)} = (-)^{s_f+1}\psi_{spin(1,2)}$

$P_{12}\psi_f = (-)^{l_f+s_f+1}\psi_f = -\psi_f$

Therefore, we have two choices, (a) $s_f = 0$, $l_f = \text{even} = 0, 2, \ldots$ and (b) $s_f = 1$, $l_f = \text{odd} = 1, 3, \ldots$. The allowed final states are ${}^1S_0$, ${}^1D_2$, ${}^3P_{0,1,2}$, ${}^3F_{2,3,4}$. Of these only ${}^3P_1$ corresponds to $J_f = 1$. Therefore, angular momentum conservation and spin-statistics connection requires the final state to be ${}^3P_1$.

Parity conservation requires $\eta_\pi\eta_D(-)^{l_i} = \eta_n^2(-)^{l_f}$ or $\eta_\pi\eta_D(-)^0 = \eta_n^2(-)^1$ or $\eta_\pi\eta_D = -\eta_n^2$. η's are the intrinsic parities. Now $\eta_D = \eta_p\eta_n = \eta_n^2$. Thus $\eta_\pi = -1$, i.e., π^- has odd (negative) intrinsic parity.

Intrinsic parity of the π^0 meson: Like spin, intrinsic parity of π^0 is also determined by the fact of occurrence of the process $\pi^0 \to 2\gamma$ in which parity is conserved. Since π^0 has zero spin we have 2 choices for the matrix element of this process in the π^0 rest frame.

$$\text{(i) } (\vec{\varepsilon_1}\cdot\vec{\varepsilon_2})f(k^2) \qquad \text{(ii) } \vec{k}\cdot(\vec{\varepsilon_1}\times\vec{\varepsilon_2})g(k^2) \tag{2.87}$$

The above choices are consistent with rotational invariance and spin statistics requirement for photons. To satisfy parity invariance one has to choose (1) if π^0 is a scalar particle (even or positive intrinsic parity); (2) is to be chosen if π^0 is a pseudoscalar particle (odd or negative intrinsic parity). One can choose between the two by determining the relative orientation of the polarization vectors $\vec{\varepsilon_1}$ and $\vec{\varepsilon_2}$ of the 2 photons. Experimentally, there is a way to determine the relative orientation of the polarisation planes first suggested by Kroll and Wada. In the rare decay $\pi^0 \to e^+e^-e^+e^-$, the memory of the polarisations of the 2 virtual photons is retained and the planes of the two e^+e^- pairs tend to be perpendicular for pseudoscalar π^0 and parallel for scalar π^0. The predicted angular distribution is of the form

$$P(\theta) = 1 + a\cos 2\theta \tag{2.88}$$

$\theta =$ angle between the two decay planes. $a > 0$ if π^0 has $J^P = 0^+$ and $a < 0$ if π^0 has $J^P = 0^-$.

$$a_{expt} = -0.8 \pm 0.2 \tag{2.89}$$

Thus π^0 has odd(negative) intrinsic parity like π^-.

C-Parity of π^0: The charge conjugation operator C converts a particle into its antiparticle and charge conjugation invariance corresponds to particle-antiparticle symmetry. Under charge conjugation operation electric charge changes sign since particle and antiparticle have opposite charges. A neutral (all "charges" zero) particle is its own antiparticle.

Thus $|a\rangle \xrightarrow{C} |\overline{a}\rangle, \quad Q \xrightarrow{C} -Q$

$$|\pi^0\rangle \xrightarrow{C} |\pi^0\rangle, \quad |\pi^+\rangle \xrightarrow{C} |\pi^-\rangle \tag{2.90}$$

Thus π^- is charge conjugate of π^+ (and vice versa) and π^0 is self conjugate. Thus a self-conjugate or a neutral particle is an eigenstate of C. The eigenvalue is known as C-parity. Therefore, C parity can be assigned to a neutral particle or particle-anti-particle system

$$C|\pi^0\rangle = \eta|\pi^0\rangle \tag{2.91}$$

$$C^2 = 1 \Rightarrow \eta^2 = 1 \text{ or } \eta = \pm 1 \tag{2.92}$$

The sign is determined by using C-invariance in processes induced by strong or electromagnetic interactions. C-invariance of electromagnetic interaction implies that the photon has odd (negative) C-parity.

$$L^{em}_{int} = j_\mu A^\mu, \quad j_\mu \xrightarrow{C} -j_\mu \tag{2.93}$$

This means $A^\mu \xrightarrow{C} -A^\mu$. Thus the photon has odd C-parity. C-parity of a system of n-photons is $(-)^n$. The occurrence of the em process $\pi^0 \to 2\gamma$ implies $\eta_{\pi^0} = (-)^2 = +1$. Thus π^0 has even(positive) C-parity.

C-parity of a particle-antiparticle pair: A system of a particle and its antiparticle becomes a system of identical particles if one includes charge as a degree of freedom. This means

$$\psi_{tot} = \psi_{space} \cdot \psi_{spin} \cdot \psi_{charge} \tag{2.94}$$

is symmetric for $b\vec{b}$ and anti-symmetric for $f\vec{f}$. (b=boson, f=femion)

$$P_{12}\psi_{tot}(1,2) = -\psi_{tot}(1,2) \text{ for } f\vec{f}$$
$$P_{12}\psi_{tot}(1,2) = +\psi_{tot}(1,2) \text{ for } b\vec{b}$$
$$P_{12}\psi_{space}(1,2) = (-)^l\psi_{space}(1,2)$$
$$P_{12}\psi_{spin}(1,2) = (-)^s\psi_{space}(1,2) \text{ for } b\vec{b}$$
$$P_{12}\psi_{spin}(1,2) = (-)^{s+1}\psi_{space}(1,2) \text{ for } f\vec{f}$$

Thus $C(-)^l(-)^s = +1$ for $b\vec{b}$

and $C(-)^l(-)^{s+1} = -1$ for $f\vec{f}$ (2.95)

where $C = C$-parity. Thus $C = (-)^{l+s}$ for both a $b\vec{b}$, and $f\vec{f}$.

Example

Positronium $(e^+e^-)^3S_1$ state has $C = (-)^{0+1} = -1$. $^1S_0$ state has

$$C = (-)^{0+0} = +1 \tag{2.96}$$

Therefore, the $^3S_1$ state of positronium cannot decay into two photons. Energy-momentum conservation and C-invariance require that the minimum number of photons in the decay of $^3S_1$ state of $(e^=e^-)$ is three and of $^1S_0$ state is two. This means that the $^3S_1$ state of positronium is longer-lived than the $^1S_0$ state. Thus the ground state of positronium must be $^3S_1$ because of C-invariance.

Intrinsic parity of spin $\frac{1}{2}$ fermions: The intrinsic parity of a spin $\frac{1}{2}$ fermion is not defined. However, it can be shown that a fermion-antifermion $(f\overline{f})$ pair has odd (negative) intrinsic parity. It the intrinsic parity of f is η, that of is $\overline{f}$ is $-\eta^*(\eta = e^{i\theta})$. Thus the intrinsic parity of $f\overline{f}$ pair is

$$-|\eta|^2 = -1 \tag{2.97}$$

Therefore, the total parity of a $f\overline{f}$ pair is $(-)^l(-1) = (-)^{l+1}$ where $(-)^l =$ orbital parity. It follows that the ground state $(^3S_1)$ of positronium (e^+e^-) has spin-parity $J^P = 1^-$.

S-matrix and Interaction Dynamics

Physical quantities like cross sections and decay rates are proportional to $|A|^2$, where A is the invariant amplitude that appears in the definition of the S-matrix element. In order to compute the S-matrix we require an explicit knowledge of the dynamics of particle interactions. In non-relativistic quantum mechanics (NRQM) this is given by the potential energy term in the Hamiltonian. In particle physics one studies processes which involve creation and annihilation of particles and antiparticles. Moreover, most of the phenomena of interest occur only when one probes short distances for which high energy projectiles are needed. The appropriate theoretical framework is therefore the relativistic quantum field theory (RQFT). In a field theoretical description, one associates a field (= a system with infinitely many degrees of freedom) with each particle. The interaction between two particles is usually assumed to be mediated by a third particle which corresponds to the quantum of another field. In the non-relativistic limit, the range of the force (or the potential) is given by the Compton wavelength of the particle mediating the interaction. Thus the infinite range of the electromagnetic force and the masslessness of the photon are consistent. As first suggested by Yukawa, the short but finite range of the nuclear force implies that the mass of the mediating particle involved (the pion) is about 300 m_e. Similarly, the extremely short range of the weak force suggests that the masses of the weak vector bosons (which mediate the weak interactions) is about 100M_p.

The basic quantity in RQFT is the Lagrangian density which is a function of the fields and their space-time derivatives. Its form is determined by assumed and expected symmetry requirements. The interaction dynamics is contained in that part of the Lagrangian density which involves the product of three or more fields. The best studied and most successful quantum field theory is the Quantum Electrodynamics (QED). The interaction Lagrangian density L_I in QED is obtained from that in classical electrodynamics, namely,

$$L_I = eJ^{\mu}_{em}(x)A_{\mu}(x) \tag{2.98}$$

with the electro magnetic current $J^{\mu}_{em}(x)$ given in terms of the electron field $\psi(x)$, i.e.,

$$J^{\mu}_{em}(x) = \overline{\psi}(x)\gamma^{\mu}\psi(x) \tag{2.99}$$

Here $A_{\mu}(x)$ is the electromagnetic field which satisfies Maxwell equations and $\psi(x)$ satisfies the Dirac equation. γ^{μ} ($\mu = 0, 1, 2, 3$) are 4×4 Dirac matrices and $\overline{\psi}(x) = \psi^{\dagger}(x)\gamma^0$, (the minus sign in (2.99) indicates that the electron charge is $-e$).

The S-operator is an exponential function of L_I. Given L_I, one can use the perturbation theory (provided its use is justified) to make a series expansion in powers of the coupling constant, and thus can compute the S-matrix element upto any given order in the perturbation theory. It terms out that given L_I, one can derive a set of rules and draw diagrams, called the Feyman diagrams, which contribute to the S-matrix element corresponding to a particular order in the perturbation theory. An nth order diagram is obtained by joining n first order diagrams in a particular way. Some first and second order diagrams corresponding to different interactions are given below.

1) Electromagnetic interactions (QED)

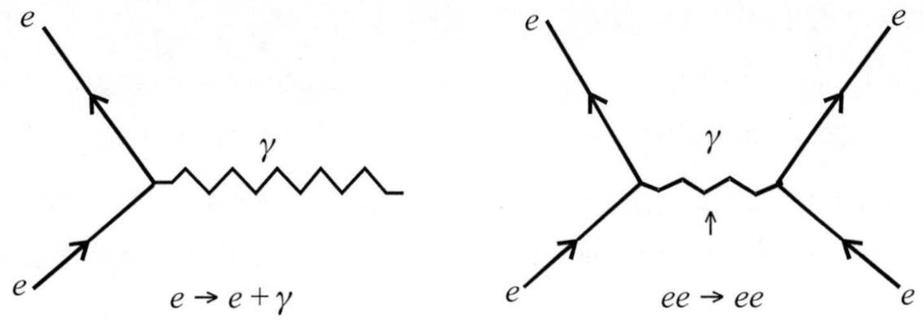

2) Strong Interactions

a) Pion-nucleon interactions (Yukawa)

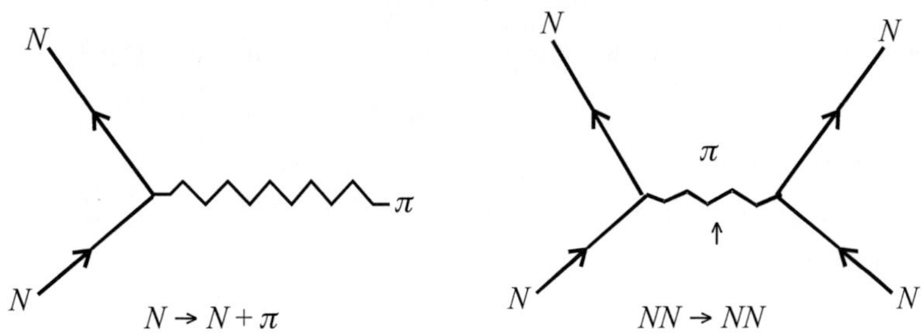

b) Quark-gluon interactrion (QCD)

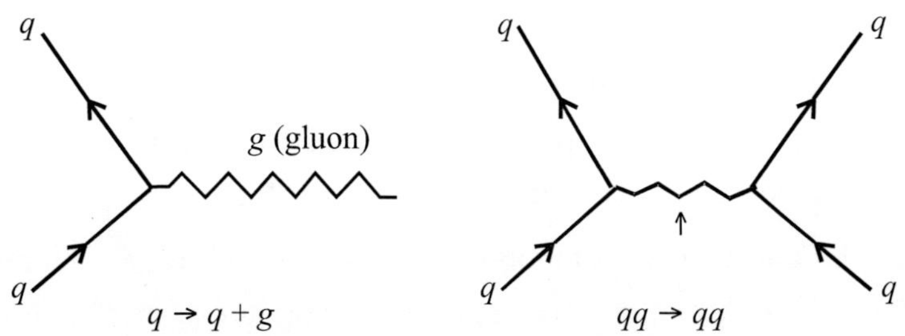

3) Weak interactions (GSW model)

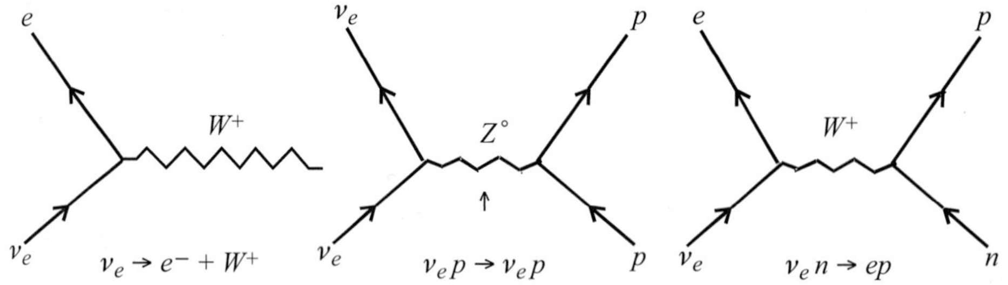

Problems

1. Calculate the maximum kinetic energy of the electron emitted in the muon decay at rest. ($\mu^- \rightarrow e^- + \vec{\nu}_e + \nu_\mu$)
2. Obtain the threshold energy for the production of antiprotions in proton-proton collisions in a fixed-target experiment.
3. Prove that the minimum opening angle θ between the two photons emitted in the decay of the neutral pion in flight ($\pi^0 \rightarrow \gamma + \gamma$) is given by
$$\cos\frac{\theta}{2} = \beta = \frac{v}{c}$$
where v is the velocity of the pion in the rest frame of the observer and c is the velocity of light in vacuum.
4. Derive an expression for the energy of γ-ray produced in the decay of the neutral pion in terms of the energy(E) and velocity β (in unit of c) of the pion and the angle θ of its emission with respect to the pion direction of motion in the Lab frame. Hence show that the energy of the photon lies in the range given by
$$\frac{E}{2}(1-\beta) \leq E_\gamma \leq \frac{E}{2}(1+\beta).$$
5. A 32 GeV electron is made to collide head-on with an 800GeV proton in a colliding-beam experiment. Calculate the total available CM energy of the ep system. How much energy the incident electron must have in a fixed-target experiment in order to make available the same total CM energy for the ep system as calculated above?
6. Find the energy of the muon emitted in the decay of a charged pion ($\pi^+ \rightarrow \mu^+ + \nu_\mu$) at rest assuming the neutrino to be massless. For a

neutrino with a small but finite mass show that, compared to the massless neutrino case, the magnitude of the muon 3-momentum emitted in the same decay would be reduced by the fraction

$$\left|\frac{\Delta p}{p}\right| = \frac{m_\nu^2(m_\pi^2 + m_\mu^2)}{(m_\pi^2 + m_\mu^2)^2}$$

where p is the magnitude of the muon 3-momentum in the massless neutrino case.

7. The solar constant (the rate at which the solar energy is incident on the earth's surface) is about 2 cals cm^{-2} min^{-1}. The fusion of hydrogen into helium in the solar core produces 26 MeV energy for every helium atom formed plus 2 ν_es carrying only a small fraction of this energy. Calculate the expected solar neutrino flux incident on the earth.

Chapter 3

Lie Group and Lie Algebra

3.1 Lie Group

Consider a set $G = \{g(\theta) = g(\theta_1 \cdots \theta_r)\}$. Here $g(\theta)$ is a function of r real parameters $\theta_1 \cdots \theta_r (\alpha_a \leq \theta_a \leq \beta_a)$. The set is said to form a group if an operation called group multiplication is defined among the elements, which has the following four properties:

1. Closure

$$g(\theta)g(\phi) = g(\xi)$$

where $\xi_a = f_a(\theta;\phi) = f_a(\theta_1 \cdots \theta_r;\ \phi_1 \cdots \phi_r)$ i.e., the product of any 2 elements of the set is also an element of the set.

2. Associativity

The group multiplication satisfies the property

$g(\theta)\{g(\theta')g(\theta'')\} = \{g(\theta)g(\theta')\}g(\theta'')$

3. Existence of Identity

There exists an element $g(\theta_0)$ in the set with the property

$g(\theta)g(\theta_0) = g(\theta_0)g(\theta) = g(\theta)$

$\theta_0 = 0(\theta_a = 0, a = 1, \dots r)$ and $g(0) = 1$

4. Existence of an inverse

For each element $g(\theta)$ of the set there exists an element $g(\overline{\theta})$ such that

$g(\theta)g(\overline{\theta}) = g(\overline{\theta})g(\theta) = g(0) = 1$

$\overline{\theta} = -\theta$ and $g(\overline{\theta}) = g(-\theta) = g^{-1}(\theta)$ is the inverse element of $g(\theta)$

The group is said to constitute a **continuous group** if all functions involved are continuous functions. A Lie group is a special class of continuous groups if one is dealing with only analytic functions (i.e., functions for which all derivatives exist).

Generators of a Lie group: Since $g(\theta)$ is an analytic function, we can make a Taylor series expansion for $g(\theta)$. For θ_a's to be infinitesimal we can write

$$g(\theta) = 1 - i\sum_{a=1}^{r} \theta_a F_a + O(\theta_a^2) \tag{3.1}$$

where $$F_a = \left.\frac{\partial}{\partial\theta_a} g(\theta)\right|_{\theta_1=\cdots=\theta_r=0} \tag{3.2}$$

F_a's are known as the generators of the Lie group. Clearly, the number of generators = number of parameters of the group.

3.2 Lie Algebra

A set of elements $S = \{A, B, C, \cdots\}$ is said to form a Lie Algebra if two operations called "addition" $(A+B)$ and "multiplication" $(A*B)$ are defined for elements of the set with the following properties:

1. $A + B$ and αA belong to the set (α = a scalar) i.e., the set forms a linear vector space under "addition".
2. "Multiplication" has the two properties

   (i) $A * B = -B * A$ (multiplication is non-cumulative)

   (ii) $(A * B) * C + (B * C) * A + (C * A) * B = 0$ (Jacobi identity is satisfied) (3.3)

Examples of Lie Algebras: Two familiar examples of Lie algebras are

(1) a set of ordinary 3-vectors $\{\vec{A}, \vec{B}, \ldots\}$ with

$\vec{A} * \vec{B} = \vec{A} \times \vec{B}$ (vector product)

and (2) a set of $n \times n$ matrices $\{A, B, \ldots\}$ with

$A * B = [A, B] = AB - BA$ (commutator)

Th. 1: The generators of a Lie group form a Lie Algebra.
Proof. Let $G = \{g(\theta) = g(\theta_1, \ldots \theta_r)\}$ be a Lie group. Consider the "commutator element"

$$g(\xi) = g(\theta)g(\phi)g^{-1}(\theta)g^{-1}(\phi)$$

where $$\xi_c = f_c(\theta, \varphi) = f_c(\theta_1 \cdots \theta_r;\ \varphi_1 \cdots \varphi_r),\ (c = 1, \ldots r) \tag{3.4}$$

Also let θ and ϕ (hence ξ) to be infinitesimal; we have Taylor series expansions

$$\left.\begin{aligned} g(\theta) &= 1 - i\sum_{a=1}^{r} \theta_a F_a + 0(\theta_a^2) \\ g(\phi) &= 1 - i\sum_{b=1}^{r} \phi_b F_b + 0(\phi_b^2) \\ \text{and} \quad g(\xi) &= 1 - i\sum_{b=1}^{r} \xi_c F_c + 0(\xi_c^2) \end{aligned}\right\} \tag{3.5}$$

Furthermore, since $\xi_c = f_c(\theta, \phi)$ is analytic function of (θ, ϕ), we have also Taylor series expansion

$$\xi_c = f_c(\theta, \phi) = A_c + \sum_{a=1}^{r} \theta_a B_{ac} + \sum_{b=1}^{r} \phi_b C_{bc} + \sum_{a,b} \theta_a \theta_b D_{abc} + \sum_{a,b} \phi_a \phi_b E_{abc} + \sum_{a,b} \theta_a \phi_b f_{abc} + 0(\theta_a^3, \phi_b^3, \theta_a^2 \phi_b \theta_a \phi_b^2) \tag{3.6}$$

Now if either θ or ϕ is zero, $g(\xi) = 1$ which means $\xi_c = 0, \ \forall c$

This implies $f_c(0, \phi) = f_c(\theta, 0) = 0$

This means $A_c, B_{ac}, C_{bc}, D_{abc}, E_{abc}$ are all zero and

$$\xi_c = f_c(\theta, \phi) = \sum_{a,b}^{r} \theta_a \phi_b f_{abc} + 0(\theta_a^2 \phi_b, \theta_a \varphi_b^2)$$

From, Eq. (3.4), (3.5) and (3.7) we get

$$\left(1 - i\sum_a \theta_a F_a + 0(\theta_a^2)\right)\left(1 - i\sum_b \phi_b F_b + 0(\phi_b^2)\right) \left(1 + i\sum_a \theta_a F_a + 0(\theta_a^2)\right)\left(1 + i\sum_b \phi_b F_b + 0(\phi_b^2)\right)$$

$$= 1 - i\sum_c \left(\sum_{a,b} \theta_a \phi_b f_{abc}\right) F_c + 0(\theta_a^2 \phi_b, \theta_a \phi_b^2)$$

$$\text{or} \quad 1 + (-i)^2 \sum_a \theta_a \varphi_b [F_a, F_b] + \cdots = 1 - i\sum_c \sum_{a,b} \theta_a \varphi_b f_{abc} F_c + \cdots$$

$$\text{or} \quad [F_a, F_b] = i\sum_c f_{abc} F_c \tag{3.8}$$

Thus $S = \{F_a\}$ forms a Lie algebra, the dimension of the Lie algebra (no. of elements) being equal to the no of generators of the associated Lie group. $f_{abc'}$s are called the "structure constants" of the Lie algebra. They are characteristic of the Lie algebra and, in fact, define the Lie algebra. Since F_a's form a Lie algebra, they satisfy the properties

$$[F_a, F_b] = -[F_b, F_a] \tag{3.9a}$$

and $$[[F_a, F_b], F_c] + [[F_b, F_c], F_a] + [[F_c, F_a], F_b] = 0 \tag{3.9b}$$

Eqns. (3.9a) and (3.9b) lead to the following properties of the structure constants:

$$f_{abc} = -f_{bac} \tag{3.10a}$$

and $$\sum_{d=1}^{r} \{f_{abd} f_{dce} + f_{bcd} f_{dae} + f_{cad} f_{dbe}\} = 0 \tag{3.10b}$$

$(a, b, c, = 1, \ldots, r)$

Simple and Semi-simple Lie Algebra

Consider a subset X of a Lie algebra S. The subset X is called a subalgebra of the Lie algebra, if X itself forms a Lie algebra, i.e. $[X, X] \epsilon X$ (This means the commutator of two elements belonging to X is a linear combination of elements of X only).

A subalgebra X of a Lie algebra S is said to be an ideal of S if $[S, X] \epsilon X$, i.e. all commutators involving X are in that subalgebra. An invariant subgroup is generated by an ideal.

A Lie algebra is said to be simple, if it is not abelian and if it does not have an ideal.

A Lie algebra is said to be semi-simple if none of its ideals are abelian.

A Lie group is called simple or semi-simple accordingly as the associated Lie algebra is simple or semi-simple. A general Lie group can be written as a direct sum of simple Lie groups, semi-simple Lie groups and a number of abelian invariant subgroups.

A Lie group which can be written as the direct product of two simple Lie groups is semi-simple.

3.3 Representations

By representation of a group $G = \{g\}$, one means a mapping of the group elements onto a set of operators in a vector space, $g \to D(g)$, which preserves the group properties:

$$D(g)D(g') = D(gg')$$
$$D(1) = 1 \qquad (3.11)$$
$$\text{and} \quad D(g^{-1})[D(g)]^{-1}$$

One is often interested in **matrix representations**, i.e. where $D(g)'s$ are a set of $n \times n$ non-singular matrices. n is called the **dimension** of the representation.

Similarly, a representation of a Lie algebra $S = \{F_a\}$ corresponds to a mapping of the elements of the algebra onto operators in a vector space, $F_a \to D(F_a)$ which preserves the properties of the algebra, i.e.

$$[D(F_a), D(F_b)] = i\sum_c f_{abc}D(F_c) \qquad (3.12)$$

Now, an infinitesimal element of a Lie group $G = \{g(\theta)\}$ can be written as

$$g = 1 - i\sum_{a=1}^{r} \theta_a F_a + 0(\theta_a^2)$$

where the generators F_a form a Lie algebra $S = \{F_a\}$

We have

$$D(g) = 1 - i\sum_{a=1}^{r} \theta_a D(F_a) + 0(\theta_a^2) \qquad (3.13)$$

Thus the representation of a Lie algebra determines the representation of the associated Lie group uniquely in the neighbourhood of the identity element. (For a matrix representation 1 is the $n \times n$ unit matrix).

Th. 2: There always exists a representation of a Lie algebra of dimension equal to the dimension of the Lie algebra.

Proof. Consider a representation $D = \{D(F_a)\}$ of a r-dimensional Lie algebra $S = \{F_a\}$, where $\{D(F_a)\}'s(a = 1, \ldots r)$ are a set of $r \times r$ matrices whose elements are given by

$$[D(F_a)]_{bc} = if_{bac} = -if_{abc}(a, b, c = 1, \ldots r,) \qquad (3.14)$$

We have

$$\begin{aligned}[D(F_a), D(F_b)]_{de} &= \sum_c \{[D(F_a)]_{dc}[D(F_b)]_{ce} - [D(F_b)]_{dc}[D(F_a)]_{ce}\} \\ &= i^2\sum_c \{f_{dac}f_{cbe} - f_{dbc}f_{cae}\} = i^2\sum_c \{f_{dac}f_{cbe} + f_{bdc}f_{cae}\}\end{aligned}$$

$$= -i^2 \sum_c f_{abc} f_{cde} = +i^2 \sum_c f_{abc} f_{dce}$$

$$= i \sum_c f_{abc} [D(F_c)]_{de}$$

This means

$$[D(F_a), D(F_b)] = i f_{abc} D(F_c)$$

Thus the matrices $D(F_a)$ whose elements are given by (3.14) constitute a r-dimensional representation of the Lie algebra, r being the number of generators of the associated Lie group. This representation is called the **adjoint** representation of the Lie algebra.

Th. 3: The structure constants of a simple Lie algebra are completely anti-symmetric in its indices.

Proof. Consider the adjoint representation $D\{D_a = D(F_a)\}$ of a Lie algebra $S = \{F_a\}$ with $[D(b)]_{ac} = i f_{abc}$

Let, $g_{bb'} = \mathrm{Tr}(D_b D_{b'}) = \sum_{a,c} (D_b)_{ac} (D_{b'})_{ca} = (i)^2 \sum_{a,c} f_{abc} f_{cb'a}$

$$= \mathrm{Tr}(D_{b'} D_b) = g_{b'b}$$

Thus $g_{bb'} = g_{b'b}$ and also $g_{bb'}$ is real.

This means that $g_{bb'}$ is a real symmetric matrix and therefore can be diagonalized by an orthogonal transformation. Thus we can always write

$$g_{bb'} = \lambda \delta_{bb'} \tag{3.16}$$

We get

$$\mathrm{Tr}\left([D_a, D_b] D_c\right) = i \sum_{c'} f_{abc'} \mathrm{Tr}\left(D_{c'} D_c\right) = i \sum_{c'} f_{abc'} \lambda \delta_{c'c} = i \lambda f_{abc} \tag{3.17}$$

Now $\mathrm{Tr}\left([D_a, D_b] D_c\right) = \mathrm{Tr}\left(D_a D_b D_c - D_b D_a D_c\right) = \mathrm{Tr}\left(D_c D_a D_b - D_a D_c D_b\right)$

$$= -\mathrm{Tr}\left([D_a, D_c] D_b\right) = -i \lambda f_{acb} \tag{3.18}$$

Comparing eqns. (3.17) and (3.18) we get

$$f_{abc} = -f_{acb} \tag{3.19}$$

Also from the fundamental property of the commutator,

$$[D_a, D_b] = -[D_b, D_a]$$

We have

$$f_{abc} = -f_{bac} \tag{3.20}$$

Eqs. (3.19) and (3.20) together give

$$f_{abc} = -f_{bac} = -f_{acb} = -f_{cba} \tag{3.21a}$$

This means that the structure constants f'_{abc}s are completely anti symmetric in its indices.

Because of the cyclic property of the trace we have the identities

$$\mathrm{Tr}\,([D_a, D_b]D_c) = \mathrm{Tr}\,([D_b, D_c]D_a) = \mathrm{Tr}\,([D_c, D_a]D_b)$$

Hence $f_{abc} = f_{bca} = f_{cab} = -f_{cba} = -f_{acb} = -f_{bac}$ (3.21b)

Schur's Lemma: Let O be an irreducible representation of a group G in a linear vector space V, and A be a linear application of $V(A : V \to V)$

If for $\forall g \epsilon G, [D(g), A] = 0$, then A is a multiple of the identity sperator, i.e. $A = \lambda \cdot 1$

Consequences of Schur's Lemma

1. Any irreducible representation of an Abelian group is 1-dimensional.
2. All Casimir operators (group invariants) are multiples of the identity operator.

3.4 Transformation Groups

From the standpoint of Physics, Lie groups are mostly realised as transformations in a vector space and groups are classified according to the nature of the transformation and the dimension of the defining vector space.

The groups $O(n)$ and $SO(n)$: Consider transformations in a n-dimensional real vector space R_n.

$$x \xrightarrow{R} x' = Rx \text{ such that } (x, x) = (x', x') \tag{3.22}$$

In terms of components,

$$x_i \xrightarrow{R} x'_i = \sum_{j-1}^{n} R_{ij} x_j \quad (i - 1, \ldots n)$$

and $(x, x) = \sum_{i=1}^{n} x_i^2$

Here R_{ij}s are elements of a $n \times n$ real, non singular matrix.

Now $(x', x') = \sum_{i=1}^{n} x_i'^2 = \sum_i \sum_j \sum_k R_{ij} x_j R_{ik} x_k$

and $(x, x) = \sum_{j=1}^{n} x_j^2 = \sum_j \sum_k x_j x_k \delta_{jk}$

Using (3.22) we get

$$\sum_i R_{ij}R_{ik} = \delta_{jk} \quad \text{or,} \quad \tilde{R}R = 1, \quad \tilde{R} = \text{transpose of } R$$

$$\text{and} \quad (\det R)^2 = 1 \tag{3.23}$$

A matrix for which $R^{-1} = \tilde{R}$ is called an orthogonal matrix and the transformations (3.22) is known to form a group, called the $O(n)$ group. Moreover, if one chooses $\det R = +1$, the group is called the $SO(n)$ group. One can easily see $O(n) = SO(n)(\times)I$, where I is a constant matrix with determinant ± 1

Considering infinitesimal transformations, one can write

$$R = 1 + \epsilon\Omega + 0(\epsilon^2)$$

$$\tilde{R}R = 1 \Rightarrow (1 + \epsilon\tilde{\Omega})(1 + \epsilon\Omega) = 1$$

Retaining terms upto $0(\epsilon)$ we get $\Omega + \tilde{\Omega} = 0$

i.e. $\Omega_{ij} = -\Omega_{ji}$

$$\Omega_{ii} = 0, \Omega_{ij} = -\Omega_{ji}, i \neq j$$

Thus one can write a general orthogonal transformation as

$$R = 1 + \sum_{a=1}^{r} \epsilon_a \Omega_a$$

Here Ω_a's are $n \times n$ linearly independent anti symmetric matrices with

$$r = n^2 - \left\{\frac{1}{2}n(n-1) + n\right\} = \frac{1}{2}n(n-1) \tag{3.25}$$

The groups $U(n)$ and $SU(n)$: Consider transformations in a n-dimensional complex vector space C_n

$$x \xrightarrow{U} x' = Ux \text{ such that } (x, x) = (x', x') \tag{3.26}$$

In terms of components,

$$x_i \xrightarrow{U} x_i' = \sum_{j=1}^{n} U_{ij}x_j, \; (x, x) = \sum_i |x_i|^2 = \sum_i x_i x_i^*$$

$$\text{Now} \quad (x', x') = \sum_i x_i' x_i'^* = \sum_i \sum_j \sum_k U_{ij}x_j U_{ik}^* x_k^*$$

$$= (x, x) = \sum_j x_j x_j^* = \sum_j \sum_k x_j x_k^* \delta_{jk}$$

$$\Rightarrow \quad \sum_i U_{ij}U^*_{ik} = \delta_{jk} \quad \text{or} \quad U^+U = 1, U^+_{ik} = U^*_{ki}$$

$$\text{and} \quad |\det U|^2 = 1 \tag{3.27}$$

$U^+ =$ hermitean adjoint of U

A matrix which satisfies (3.27) is called a unitary matrix and the transformation is known as unitary transformation.

Clearly, the set of transformations in a n-dim. complex vector space forms a group, called the $U(n)$ group. If one also chooses $\det U = +1$, then the group is called the $SU(n)$ group.

$$U(n) = SU(n) \times U(1), U(1) = e^{i\theta}1.$$

Considering infinitesimal transformations, one can write

$$U = 1 - i\epsilon H + o(\epsilon^2)$$

$$U^+U = 1 \;\Rightarrow\; (1 + i\epsilon H^+ \cdots)(1 - i\epsilon H + \cdots) = 1$$

or, $\quad i\epsilon(H^+ - H) + o(\epsilon^2) = 0 \quad$ or, $H^+ = H$, i.e. H is hermitean.

Furthermore,

$$\det U = 1 \;\Rightarrow\; \det|1 - i\epsilon H| = 1$$

or, $\quad 1 - i\epsilon(\mathrm{Tr}\, H) = 1 \quad$ or, $\mathrm{Tr}\, H = 0$

Thus H is a $n \times n$ traceless, hermitean matrix,

$$\sum_i H_{ii} = 0, \quad H_{ji} = H^*_{ij}$$

number of independent parameters

$$r = 2n^2 - \left\{2 \cdot \frac{1}{2}n(n-1) + n + 1\right\} = n^2 - 1 \tag{3.28}$$

$$\text{Thus } U = 1 - i\sum_{a=1}^{r} \epsilon_a H_a \tag{3.29}$$

where H_a's are a set of $n \times n$ linearly independent, traceless hermitean matrices.

The Symplectic group $Sp(n)$: Consider transformation in an even dimensional ($n = 2n'$) real vector space

$$x \xrightarrow{S} x' = Sx \qquad y \xrightarrow{S} y' = Sy$$

$$\text{such that } (x, gy) = (x', gy') \tag{3.30}$$

where g is a matrix whose crossdiagonal elements are nonzero with the elements in the upper half and elements in the lower half being +1 and −1, respectively. For example, for $n = 2$,

$$g = \begin{pmatrix} 0 & 1 \\ -1 & 0 \end{pmatrix} \tag{3.31}$$

and $$(x, gy) = (x_1, x_2) \begin{pmatrix} 0 & 1 \\ -1 & 0 \end{pmatrix} \begin{pmatrix} y_1 \\ y_2 \end{pmatrix} = x_1 y_2 - x_2 y_1 \tag{3.32}$$

Thus under $Sp(2)$ transformation, $x_1 y_2 - x_2 y_1$ is invariant. An infinitesimal transformation can be written in the form

$$S = 1 + \epsilon A + 0(\epsilon^2) \tag{3.33}$$

$$(x', gy') = (Sx, gSy) = (x, gy) \Rightarrow \tilde{S} g S = g \tag{3.34a}$$

or $$(1 + \epsilon \tilde{A} + \cdots) g (1 + \epsilon A + \cdots) = g$$

or $$g + \epsilon(\tilde{A} g + g A) + O(\epsilon^2) = g$$

or $$\tilde{A} g + g A = 0 \tag{3.34b}$$

$$S = 1 + \sum_{a=1}^{r} \epsilon_a A_a \tag{3.35}$$

$$r = n^2 - \frac{1}{2} n(n-1) = \frac{1}{2} n(n+1) \tag{3.36}$$

The group $SO(3)$: Defining vector space and hence representation is 3-dimensional ($n = 3$)

$$x \xrightarrow{R} x' = Rx, \quad x = \begin{pmatrix} x_1 \\ x_2 \\ x_3 \end{pmatrix}, \quad R = 3 \times 3 \text{ matrix}$$

$$R = 1 + \sum_{a=1}^{r} \theta_a \Omega_a, \; r = \frac{1}{2} n(n-1) = 3, \tag{3.37}$$

Ω_a's are a set of three 3×3 linearly independent antisymmetric matrices. *A* suitable choice is

$$\Omega_1 = \begin{pmatrix} 0 & 0 & 0 \\ 0 & 0 & -1 \\ 0 & 1 & 0 \end{pmatrix}, \; \Omega_2 = \begin{pmatrix} 0 & 0 & 1 \\ 0 & 0 & 0 \\ -1 & 0 & 0 \end{pmatrix}, \; \Omega_3 = \begin{pmatrix} 0 & -1 & 0 \\ 1 & 0 & 0 \\ 0 & 0 & 0 \end{pmatrix} \tag{3.38}$$

Ω_a's satisfy the CR's

$$[\Omega_1, \Omega_2] = \Omega_3, \ [\Omega_2, \Omega_3] = \Omega_1, \ [\Omega_3, \Omega_1] = \Omega_2 \tag{3.39}$$

Writing

$$L_a = i\Omega_1, \quad [L_a, L_b] = i \sum_c \epsilon_{abc} L_c$$

$$\text{and} \quad R = 1 - i \sum_{a=1}^{3} \theta_a L_a \tag{3.40}$$

The group $SU(2)$: Defining vector space and hence representation is 2-dimensional ($n = 2$)

$$x \xrightarrow{U} x' = Ux, \ x = \begin{pmatrix} x_1 \\ x_2 \end{pmatrix}, U = 2 \times 2 \text{ matrix}$$

$$U = 1 - i \sum_{a=1}^{r} \theta_a \frac{\sigma_a}{2}, \ r = n^2 - 1 = 3,$$

σ_a's are a set of three 2×2 traceless hermitean matrices. A suitable choice is (Pauli matrices)

$$\sigma_1 = \begin{pmatrix} 0 & 1 \\ 1 & 0 \end{pmatrix}, \quad \sigma_2 = \begin{pmatrix} 0 & -i \\ i & 0 \end{pmatrix}, \quad \sigma_3 = \begin{pmatrix} 1 & 0 \\ 0 & -1 \end{pmatrix} \tag{3.42}$$

σ_a's satisfy the CRs

$$[\sigma_1, \sigma_2] = 2i\sigma_3, \quad [\sigma_2, \sigma_3] = 2i\sigma_1, \quad [\sigma_3, \sigma_1] = 2i\sigma_1 \tag{3.43}$$

$$\text{Furthermore,} \ \sigma_a \sigma_b + \sigma_b \sigma_a = 2\delta_{ab}, 1 \tag{3.44}$$

$SO(3)$ and $SU(2)$ have the same Lie algebra.

Rank of a Lie Group

An important quantity in the study of Lie groups and associated Lie algebras is the rank. The rank of a Lie group is the maximum number of generators that can be diagonal in any given representation and thus corresponds to the maximum number of commuting generators.

Group	Rank (l)	No. of generators (r)
$SU(n)$	$n-1$	n^2-1
$SO(n)$	$\frac{n}{2}$, $n=$ even; $\frac{n-1}{2}$, $n=$ odd	$\frac{1}{2}n(n-1)$
$Sp(n)$	$\frac{n}{2}$ ($n=$ even)	$\frac{1}{2}n(n+1)$

Problems

1. Consider a set of three 3×3 matrices whose elements are given by

$$(t_a)_{bc} = -i\epsilon_{abc}$$

where $a, b, c = 1, 2, 3$. Show that they satisfy the $SU(2)$ commutation relations $[t_a, t_b] = i\epsilon_{abc}t_c$

Find the eigen values and eigen states of the matrix t_3. Hence show that t_1, t_2, t_3 correspond to the adjoint (vector) representation of $SU(2)$.

2. Consider a $SU(2)$ doublet $\psi = \begin{pmatrix} \psi_1 \\ \psi_2 \end{pmatrix}$

Show that the composite operators $S = \psi^+\psi$ and $\vec{V} = \psi^+\frac{\vec{\tau}}{2}\psi, \vec{\tau} = (\tau_1, \tau_2, \tau_3)$ being the Pauli matrices, transform as a $SU(2)$ scalar and a $SU(2)$ vector, respectively.

3. Show that for every $SU(2)$ matrix U, there exists a matrix S which connects U to its complex conjugate matrix U^* through the similarity transformation $S^{-1}US = U^*$

If $\psi = \begin{pmatrix} \psi_1 \\ \psi_2 \end{pmatrix}$ is an $SU(2)$ doublet, then show that $\psi_c = S\psi*$ is also an $SU(2)$ doublet, that is, ψ and ψ_c have the same properties under $SU(2)$ transformations.

4. Prove that $e^{i\frac{\vec{\theta}}{2}\cdot\vec{\tau}} = 1\cdot\cos\frac{\theta}{2} + i(\vec{\tau}\cdot\hat{n})\sin\frac{\theta}{2}$

where $\vec{\tau} = (\tau_1, \tau_2, \tau_3)$ are the Pauli matrices and $\vec{\theta} = \theta\hat{n}(\theta$ is the magnitude and is the direction of $\vec{\theta})$.

5. Given a set of composite quark field operators $F^a = \int q^+(x)\frac{\lambda^a}{2}q(x)d^3x$ where the quark field operators $q(x) = \begin{pmatrix} q_1(x) \\ q_2(x) \\ q_3(x) \end{pmatrix}$ satisfy the anti commutation relations $\{q_a(x), q_b^+(y)\} = \delta_{ab}\delta_3(\vec{x} - \vec{y})$ and the λ_a's $(a = 1, \ldots, 8)$ are the 3×3 Gell-Mann matrices which satisfy the $SU(3)$ commutation relations, $\left[\frac{\lambda^a}{2}, \frac{\lambda^b}{2}\right] = if_{abc}\frac{\lambda^c}{2}$ then show that $\{F^a\}$, if assumed to be time-independent, generate the $SU(3)$ Lie algebra

$$[F^a, F^b] = if_{abc}F^c$$

6. If $[A, B] = B$, calculate $e^{i\alpha A}Be^{-i\alpha A}$

Chapter 4

I-Spin and the $SU(2)_I$ Symmetry

The hadron spectrum is characterized by the appearance of "charge multiplets", i.e. particles with same spin-party (J^P) and nearly same mass but with different electric charges. This mass degeneracy is attributed to the existence of some internal symmetry. For light hadrons, this symmetry is known as the I-spin symmetry.

The concept of I-spin symmetry was introduced by Heisenberg in 1932 in order to explain the charge independence of nuclear forces. Low energy neutron-proton and proton-proton scattering data suggest $V_{np} \approx V_{pp}$. On the other hand, binding energy data for mirror nuclei, after subtracting electromagnetic effects suggest $V_{pp} \approx V_{nn}$. Combining the two equalities one finds $V_{np} \approx V_{pp} \approx V_{nn}$ in the same angular momentum state. Moreover, $M_n - M_p = 1.3$ MeV.

Heisenberg proposed that charge independence of nuclear forces can be understood if the proton and the neutron are not regarded as two different particles, but two different charge states of the same particle, called the "nucleon", analogous to two different spin states (spin up $\uparrow$ and spin down $\downarrow$ states) of a spin $\frac{1}{2}$ particle. Thus the proton and the neutron states form the basis states of a 2-dim complex vector space and the general state of a nucleon is a linear combination of these basis states.

$$|N\rangle = \alpha|p\rangle + \beta|n\rangle \tag{4.1}$$

writing $|p\rangle$ and $|n\rangle$ as 2-dim column vectors $|p\rangle = \begin{pmatrix} 1 \\ 0 \end{pmatrix}$, $|n\rangle = \begin{pmatrix} 0 \\ 1 \end{pmatrix}$

one has $\quad N = \begin{pmatrix} \alpha \\ \beta \end{pmatrix} \qquad (4.2)$

Consider transfomations

$$|N\rangle \xrightarrow{U} |N'\rangle = U|N\rangle \tag{4.3}$$

such that $\langle N|N\rangle = \langle N'|N'\rangle$

One therefore has

$$U^+U = 1 \quad \text{and} \quad |\det U|^2 = 1 \tag{4.4}$$

The set of such transformations form a group. Restricting to $\det U = +1$, the group, according to standard notation is $SU(2)$. We name it $SU(2)_I$. Thus spin and i-spin have same group structure except that the two groups act on different vector spaces.

One can write an infinitesimal $SU(2)_I$ transformation as

$$U(\delta\theta) = 1 - i \sum_{a=1} \delta\theta_a \frac{\tau_a}{2} + 0(\delta\theta^2)$$

τ_as are 2×2 linearly independent traceless, hermitean matrices. The standard choice is the Pauli matrices.

$$\tau_1 = \begin{pmatrix} 0 & 1 \\ 1 & 0 \end{pmatrix}, \ \tau_2 = \begin{pmatrix} 0 & -i \\ i & 0 \end{pmatrix}, \ \tau_3 = \begin{pmatrix} 1 & 0 \\ 0 & 1 \end{pmatrix} \tag{4.5}$$

(we name the Pauli matrices above τ_a's instead of σ_a's in order to distinguish i-spin from spin). Direct computation gives

$$[\tau_a, \tau_b] = 2i \sum_c \epsilon_{abc}\tau_c, \{\tau_a, \tau_b\} = 2\delta_{ab} \cdot 1 \tag{4.6}$$

Regarding τ_a's as 2-dim representation of some abstract operators, we can write the abstract $SU(2)_I$ (finite) group clement as

$$U(\theta) = U(\theta_1, \theta_2, \theta_3) = \exp\left(-i \sum_{a=1}^{3} \theta_a I_a\right) \tag{4.7}$$

$$\text{with} \quad [I_a, I_b] = i \sum_{c=1}^{3} \epsilon_{abc} I_c \tag{4.8}$$

Equation (4.8) defines the $SU(2)_I$ Lie algebra.

Under I-spin transformation,

$$H \to H' = UHU^{-1}$$

I-spin invariance implies $H' = H$ i.e.

$$UHU^{-1} = H \quad \text{or} \quad [U, H] = 0$$

$$\text{which implies } [I_a, H] = 0 \tag{4.9}$$

The irreducible representations of $SU(2)_I$ (i.e. possible i-spin multiplets) are found in the same way as for $SU(2)_I$, Since no two I_a's commute, only one of them can be diagonalized at a time. We choose it to be I_3.

Define the operator

$$I^2 = \sum_{a=1}^{3} I_a^2 \tag{4.10}$$

called the 'Casimir operator' of $SU(2)_I$. Clearly

$$[I^2, I_a] = 0, \quad [I^2, H] = 0 \tag{4.11}$$

Thus (H, I^2, I_3) form a complete commuting set. Defining operators

$$I_+ = I_1 + iI_2, \; I_- = I_1 - iI_2 \tag{4.12}$$

and using (4.8) one easily obtains

$$[I_3, I_+] = I_+, \; [I_3, I_-] = -I_-, \; [I_+, I_-] = 2I_3$$

Denoting the states by $|II_3\rangle$, one finds

$$\begin{aligned} I^2|II_3\rangle &= I(I+1)|II_3\rangle \\ I_3|II_3\rangle &= I_3|II_3\rangle \\ I_\pm|II_3\rangle &\sqrt{(I \mp I_3)(I \pm I_3 + I)}|I, I_3 \pm 1|\rangle \end{aligned} \tag{4.14}$$

$I_3 = I$ is the highest weight state. $-I \le I_3 \le I$ with $2I = n =$ integer $= 0, 1, 2, 3, \ldots$ The multiplets are labelled by I and the states of a multiplet by I_3. The multiplicity of an I-spin multiplet is

$$2I + 1 = n + 1.$$

$I = \frac{1}{2}$ is the lowest non-trivial multiplet and the nucleon belongs to this multiplet.

$$|p\rangle = \left|\frac{1}{2}, \frac{1}{2}\right\rangle, \quad |n\rangle = \left|\frac{1}{2}, \frac{1}{2}\right\rangle$$

Define $\tau_\pm = \dfrac{1}{2}(\tau_1 \pm i\tau_2)$

$$\tau_+|n\rangle = |p\rangle, \quad \tau_-|p\rangle = |n\rangle \tag{4.15}$$

***I*-spin of pions:** Pions were introduced by Yukawa (1935) in order to explain the short range nature of nuclear forces. Nucleons interact via exchange of pions, just as charged particles interact via exchange of photons. Finite range of nuclear force implies nonzero mass for pions (in the same way as infinite range of coulomb force is due to zero mass of the photon).

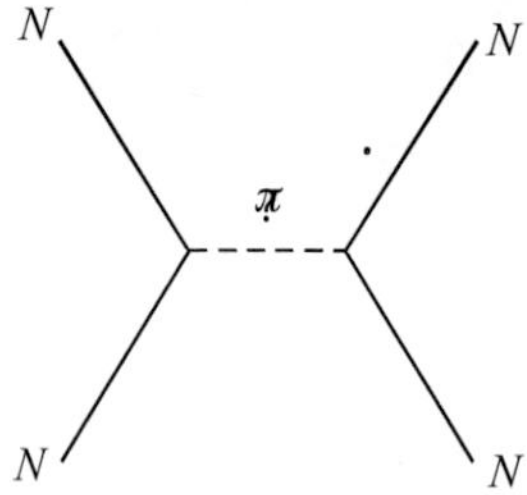

Fig. 4.1

Charge conservation and *i*-spin conservation implies that there should be 3 pions π^+, π^-, π^0. They could thus be assigned to an $I = 1$ multiplet, Moreover, they should be degenerate in mass. It turns out,

$m_{\pi^+} = m_{\pi^-} = 139.6$ MeV, $m_{\pi^0} = 135.0$ MeV

$|\pi^+\rangle = |1,\, +1\rangle$, $|\pi^0\rangle = |1,\, 0\rangle$, $|\pi^-\rangle = |1,\, -1\rangle$

$I_+|\pi^-\rangle = \sqrt{2}|\pi^0\rangle$, $I_+|\pi^0\rangle = \sqrt{2}|\pi^+\rangle$, $I_-|\pi^+\rangle = \sqrt{2}|\pi^0\rangle$, $I_-|\pi^0\rangle = \sqrt{2}|\pi^-\rangle$

$\boldsymbol{I}$-spin of antinucleons

$$\left[\frac{\tau_a}{2}, \frac{\tau_b}{2}\right] = i\epsilon_{abc}\frac{\tau_c}{2}$$

implies $$\left[\frac{-\tau_a^*}{2}, \frac{-\tau_b^*}{2}\right] = i\epsilon_{abc}\left(\frac{-\tau_c^*}{2}\right) \tag{4.16}$$

This means $\hat{\tau}_a = -\tau_a^*$ also provide a 2-dim repr. of $SU(2)_I$.

But one is interested only in inequivalent representations. Two representations of the same dimension are said to be equivalent if the two sets of matrices are connected by a similarity transformation. Thus $\{\tau_a\}$ and $\{\hat{\tau}_a\}$ are equivalent if there exists a non-singular matrix S such that

$$S\hat{\tau}_a S^{-1} = \tau_a (a = 1, 2, 3) \tag{4.17}$$

It can be easily seen that such a matrix exists.

since $\hat{\tau}_a = -\tau_a^*$, (4.17) gives

$$S\tau_a^* S^{-1} = -\tau_a \tag{4.18}$$

Now $\tau_1^* = \tau_1$, $\tau_2^* = -\tau_2$, $\tau_3^* = \tau_3$

Therefore, $$S\tau_1 = -\tau_1 S, \quad S\tau_2 = \tau_2 S, \quad S\tau_3 = -\tau_3 S \tag{4.19}$$

i.e. S commutes with τ_2 and anticommutes with τ_1 and τ_3

Therefore, $S = a\tau_2$

$$S^+S = 1 \text{ implies, } |a| = 1$$

we choose $S = -i\tau_2 = \begin{pmatrix} 0 & -1 \\ 1 & 0 \end{pmatrix}$ (4.20)

Now $N \xrightarrow{U} N' = UN,\ U = e^{-i\Sigma_a \theta_a \frac{\tau_a}{2}}$

and $N^* \xrightarrow{U} N^{*\prime} = U^*N^*,\ U^* = e^{+i\sum_a \theta_a \frac{\tau_a^*}{2}},\ SU^*S^1 = U$

Define $N_c = SN^*,\ N_c \xrightarrow{U} N_c' = SN^{*\prime} = SU^*N^* = SU^*S^{-1} \cdot SN^*$

$$= (SU^*S^{-1})N_c = UN_c \tag{4.21}$$

Thus N_c transforms in the same way as N

For $N = \begin{pmatrix} p \\ n \end{pmatrix},\ N^* = \begin{pmatrix} \overline{p} \\ \overline{n} \end{pmatrix}$,

$$N_c = SN^* = \begin{pmatrix} 0 & -1 \\ 1 & 0 \end{pmatrix} \begin{pmatrix} \overline{p} \\ \overline{n} \end{pmatrix} = \begin{pmatrix} -\overline{n} \\ \overline{p} \end{pmatrix} \tag{4.22}$$

Thus N_c forms a $I = \frac{1}{2}$ multiplet. We choose

$$|\overline{n}\rangle = -\left|\frac{1}{2}, +\frac{1}{2}\right\rangle \text{ and } |\overline{p}\rangle = \left|\frac{1}{2}, -\frac{1}{2}\right\rangle \tag{4.23}$$

One can write the relation between electric charge Q (in unit of e) and the third component of i-spin I_3 for $N, \overline{N}$ and π in the form

$$Q = I_3 + \frac{B}{2} \tag{4.24}$$

where B is the baryon number.

Coupling of i-spins: rules for coupling of two or more i-spins are the same as ordinary spins. Let

$$\vec{I} = \vec{I}^{(1)} + \vec{I}^{(2)}$$

then $\left|I^{(1)} - I^{(2)}\right| \le I \le I^{(1)} + I^{(2)},\ I_3 = I_3^{(1)} + I_3^{(2)}$

and $\sum_1 (2I + I) = \left(2I^{(1)} + 1\right)\left(2I^{(2)} + 1\right)$ (4.25)

***NN* system**

$$I^{(1)} = I^{(2)} = \frac{1}{2}$$

Therefore $I = 0, 1$

The total i-spin states in terms of component states are as follows:

$$|1,1\rangle = |pp\rangle$$

$$\left|1,0\right\rangle = \frac{1}{\sqrt{2}}\left(|pn\rangle + |np\rangle\right)$$

$$|1,-1\rangle = |nn\rangle$$

$$\left|0,0\right\rangle = \frac{1}{\sqrt{2}}\left(|pn\rangle - |np\rangle\right)$$

$N\overline{N}$ system—Similarly, we get for the $N\overline{N}$ states

$$|1,1\rangle \equiv |p\overline{n}\rangle$$

$$\left|1,0\right\rangle = \frac{1}{2}\,|p\overline{p} - |n\overline{n}\rangle, \left|0,0\right\rangle = \frac{1}{\sqrt{2}}\left(p\overline{p}\rangle + |n\overline{n}\rangle\right)$$

$$|1,-1\rangle = |n\overline{p}\rangle \tag{4.27}$$

Generalized Symmetry Principle: When one includes i-spin as an additional degree of freedom, particles belonging to the same i-spin multiplet may be regarded as identical particles and the total wave function of a system of such particles, is given by the product

$$\psi_{tot} = \psi_{space}, \psi_{spin}, \psi_{i-spin} \tag{4.28}$$

ψ_{tot} will be symmetre or antisymmetre depending on whether the particles are bosons or fermions. This is easily checked for NN system.

***I*-spin and πN Scattering:** consider the πN system. The total i-spin

$$\vec{I} = \vec{I}^{(N)} + \vec{I}^{(\pi)}, \quad \text{with } I^{(N)} = \frac{1}{2}, \; I^{(\pi)} = 1$$

$$I \text{ can have values } I = \frac{3}{2}, \frac{1}{2} \tag{4.29}$$

One can easily write the total i-spin states in terms of particle states by using raising and lowing operators $I_{\pm} = I_1^{(1)} \pm I_1^{(2)}$

$$|II_3\rangle$$

$$\left|\frac{3}{2} + \frac{3}{2}\right\rangle = |\pi^+ p\rangle$$

$$\left|\frac{3}{2} + \frac{1}{2}\right\rangle = \sqrt{\frac{2}{3}}|\pi^0 p\rangle + \sqrt{\frac{1}{3}}|\pi^+ n\rangle$$

$$\left|\frac{3}{2}, -\frac{1}{2}\right\rangle = \sqrt{\frac{2}{3}}|\pi^0 n\rangle + \sqrt{\frac{1}{3}}|\pi^- p\rangle$$

$$\left|\frac{3}{2},-\frac{3}{2}\right\rangle = |\pi^- n\rangle$$

$$\left|\frac{1}{2},+\frac{1}{2}\right\rangle = -\sqrt{\frac{1}{3}}|\pi^0 p\rangle + \sqrt{\frac{2}{3}}|\pi^+ n\rangle$$

$$\left|\frac{1}{2},-\frac{1}{2}\right\rangle = -\sqrt{\frac{1}{3}}|\pi^0 n\rangle + \sqrt{\frac{2}{3}}|\pi^- p\rangle \tag{4.30}$$

There are six possible πN scattering processes that can be studied experimentally. We denote the corresponding scattering amplitudes as follows:

$$\begin{aligned} &a_+ : \pi^+ p \to \pi^+ p \qquad && b_- : \pi^- n \to \pi^- n \\ &a_- : \pi^- p \to \pi^- p && b_+ : \pi^+ n \to \pi^+ n \\ &a_0 : \pi^- p \to \pi^0 n && b_0 : \pi^+ n \to \pi^0 p \end{aligned} \tag{4.31}$$

where $a_+ = \langle \pi^+ p|S|\pi^+ p\rangle, \quad \sigma_+ = \lambda |a_+|^2$

$\lambda =$ kinematic factor $= f(s, m_N^2, m_\pi^2)$

and similarly for other processes. Of the six possible amplitudes only two are independent because of i-spin invariance.

I-spin invariance of strong interaction implies

$$\langle II_3|S|I'I_3'\rangle = a_I \delta'_{II'} \delta_{I_3 I_{3'}} \tag{4.34}$$

$$\langle f|S|i\rangle = \sum_{\substack{I,I_3\\ I',I_3}} \langle f|II_3\rangle\langle II_3|S|I'I_3'\rangle\langle I'I_3'|i\rangle = \sum_I a_I \langle I_3|f\rangle^* \langle II_3|i\rangle \tag{4.35}$$

Charge conservation ensures no sum over I_3.

Eq. (4.35) means that any scattering amplitude can be written in terms of i-spin amplitudes. For πN system, there are only two such independent amplitudes $a_{\frac{3}{2}}$ and $a_{\frac{1}{2}}$.

Consider the three amplitudes

$$a_+ = \langle \pi^+ p|S|\pi^+ p\rangle \quad a_- = \langle \pi^- p|S|\pi^- p\rangle \quad a_0 = \langle \pi^0 n|S|\pi^- p\rangle \tag{4.36}$$

$|\pi^+ p\rangle$ is a pure $I = \frac{3}{2}$ state, whereas $|\pi^- p\rangle$ and $|\pi^0 n\rangle$ are mixtures of $I = \frac{3}{2}$ and $I = \frac{1}{2}$ states. In particular,

$$\begin{pmatrix} \left|\frac{3}{2},-\frac{1}{2}\right\rangle \\ \left|\frac{1}{2},-\frac{1}{2}\right\rangle \end{pmatrix} = \begin{pmatrix} \sqrt{2/3} & \sqrt{1/3} \\ -\sqrt{1/3} & \sqrt{2/3} \end{pmatrix} \begin{pmatrix} |\pi^0 n\rangle \\ |\pi^- p\rangle \end{pmatrix}$$

Inverting the relation we get,

$$\begin{pmatrix} |\pi^0 n\rangle \\ |\pi^- p\rangle \end{pmatrix} = \begin{pmatrix} \sqrt{2/3} & -\sqrt{1/3} \\ \sqrt{1/3} & \sqrt{2/3} \end{pmatrix} \begin{pmatrix} \left|\frac{3}{2}, -\frac{1}{2}\right\rangle \\ \left|\frac{1}{2}, -\frac{1}{2}\right\rangle \end{pmatrix}$$

Thus we get,

$$\left.\begin{aligned} a_+ &= \langle\pi^+ p|S|\pi^+ p\rangle = a_{3/2} \\ a_- &= \langle\pi^- p|S|\pi^- p\rangle = \frac{1}{3}(a_{3/2} + 2a_{1/2}) \\ a_0 &= \langle\pi^0 n|S|\pi^- p\rangle = \frac{\sqrt{2}}{3}(a_{3/2} - a_{1/2}) \end{aligned}\right\} \tag{4.37}$$

The three amplitudes α_+, α_- and α_0 are related by

$$a_- + \sqrt{2}a_0 = a_+ \tag{4.38}$$

which means $|a_+| \leq |a_-| + \sqrt{2}|a_0|$

$$\begin{aligned} \sigma_+ &= \sigma(\pi^+ p \to \pi^+ p) = \lambda|a_+|^2 = \lambda\left|a_{\frac{3}{2}}\right|^2 \\ \sigma_- &= \sigma(\pi^- p \to \pi^- p) = \lambda|a_-|^2 = \frac{\lambda}{9}\left|a_{\frac{3}{2}} + 2a_{\frac{1}{2}}\right|^2 \\ \sigma_0 &= \sigma(\pi^- p \to \pi^- n) = \lambda|a_0|^2 = \frac{2\lambda}{9}\left|a_{\frac{3}{2}} - a_{\frac{1}{2}}\right|^2 \end{aligned} \tag{4.39}$$

$$\sigma_{tot}(\pi^+ p) = \sigma_+ = \lambda|a_{\frac{3}{2}}|^2$$

$$\sigma_{tot}(\pi^- p) = \sigma_- + \sigma_0 = \frac{\lambda}{3}(|a_{3/2}|^2 + 2|a_{1/2}|^2) \tag{4.40}$$

One can have three special cases:

(1) $|a_{3/2}| \gg |a_{1/2}|$

In this case

(a) $\sigma_+ : \sigma_- : \sigma_0 \approx 9 : 1 : 2$

(b) $\dfrac{\sigma_{tot}(\pi^- p)}{\sigma_{tot}(\pi^+ p)} \approx \dfrac{1}{3}$ (4.41)

(2) $|a_{1/2}| \gg |a_{3/2}|$

In this case

(a) $\sigma_- : \sigma_0 \approx \left|\dfrac{2}{3}a_{1/2}\right|^2 : \left|\dfrac{\sqrt{2}}{3}a_{1/2}\right|^2 \approx 2 : 1$

$$\text{(b)} \quad \frac{\sigma_{tot}(\pi^+ p)}{\sigma_{tot}(\pi^- p)} \ll 1 \tag{4.42}$$

(3) $|a_{3/2}| \approx |a_{1/2}|$

In this case

$$\sigma_0 \approx 0,\ \sigma_+ \approx \sigma_-,\ \sigma_{tot}(\pi^+ p) \approx \sigma_{tot}(\pi^- p) \tag{4.43}$$

The first two cases occur at low energies, and the third case at high energies.

Baryon Resonances

If one plots σ_{tot} against $\sqrt{s}$ one sees a peak in both $\sigma_{tot}(\pi^+ p)$ and $\sigma_{tot}(\pi^- p)$ at $\sqrt{s} = 1236$ MeV($T_\pi^{lab} \approx 195$ MeV) with $\sigma_{tot}(\pi^+ p) \approx 3\ \sigma_{tot}(\pi^- p)$.

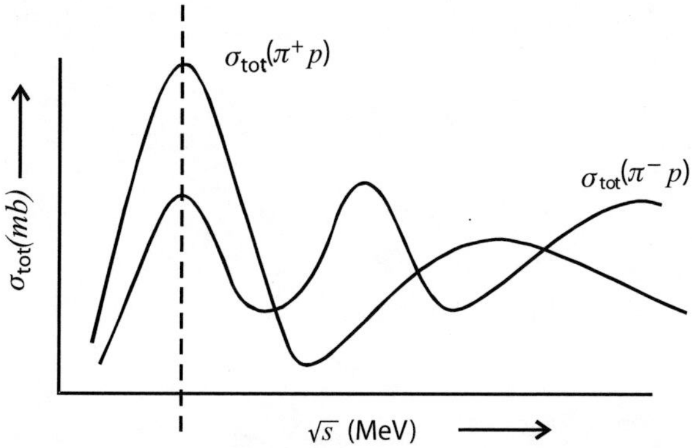

Fig. 4.2

This peak in σ_{tot} is interpreted due to the formation of a resonance (intermediate state) $\Delta(1236)$ with $\Gamma \approx 120$ McV ($\pi N \to \Delta \to \pi N$). This resonance has quantum numbers $J^P = \frac{3}{2}^+$, $I = \frac{3}{2}$, and is known as the (3,3) resonance. Many such resonances have been seen at higher energies:

$$N(1470),\ J^P = \frac{1}{2}^+;\ N(1520),\ J^P = \frac{3}{2}^-;\ \Delta(1650),\ J^P = \frac{1}{2}^- \text{ and so on.}$$

Resonances in the πN channel have quantum numbers

$$B = 1,\ I = \frac{1}{2}, \frac{3}{2},\ J = \ell \pm \frac{1}{2},\ P = (-)^{\ell+1}\ (\ell = 0, 1, 2, \ldots)$$

J is determined by measuring angular distributions and parity is determined by measuring the polarization of the scattered nucleon.

Two Pion System and Vector Mesons

$$\vec{I} = \vec{I}^{(1)} + \vec{I}^{(2)}$$

$$I^{(1)} = I^{(2)} = I^{(\pi)} = 1 \Rightarrow I = 0, 1, 2 \tag{4.44}$$

There are 9 possible states. The total *i*-spin eigenstates are given by $|II_3\rangle$

$$|2, +2\rangle = |\pi^+\pi^+\rangle$$

$$|2, +1\rangle = \frac{1}{\sqrt{2}}\left(|\pi^+\pi^0\rangle + |\pi^0\pi^+\rangle\right)$$

$$|2, 0\rangle = \frac{1}{\sqrt{6}}\left(|\pi^+\pi^-\rangle + |\pi^-\pi^+\rangle + 2|\pi^0\pi^0\rangle\right)$$

$$|2, -1\rangle = \frac{1}{\sqrt{2}}\left(|\pi^0\pi^-\rangle + |\pi^-\pi^0\rangle\right)$$

$$|2, -2\rangle = |\pi^-\pi^-\rangle$$

$$|1.1\rangle = \frac{1}{\sqrt{2}}\left(|\pi^+\pi^0\rangle - |\pi^0\pi^+\rangle\right)$$

$$|1, 0\rangle = \frac{1}{\sqrt{2}}\left(|\pi^+\pi^-\rangle - |\pi^-\pi^+\rangle\right)$$

$$|1, -1\rangle = \frac{1}{\sqrt{2}}\left(|\pi^0\pi^-\rangle - |\pi^-\pi^0\rangle\right)$$

$$|0, 0\rangle = \frac{1}{\sqrt{3}}\left(|\pi^+\pi^-\rangle + |\pi^-\pi^+\rangle - |\pi^0\pi^0\rangle\right) \tag{4.45}$$

The $I = 0, 2$ states are symmetric, while the $I = 1$ state is antisymmetric under exchange of particles.

According to the generalized symmetry principle, the total wave-function.

$$\psi_{tot} = \psi_{space} \cdot \psi_{i-spin} \tag{4.46}$$

must be symmetric for a system of pions. This means that ψ_{space} and ψ_{i-spin} must be either both symmetric or both antisymmetric. Thus for a two-pion system $L + I$ = even, where I = total *i*-spin and L is the orbital angular momentum for the relative motion of the two pions. The generalized symmetry principle requires that the system of two π^0s cannot be in an $I = 1$ state.

Meson Resonances

At high energies, multipion production takes place in πN, NN and $N\overline{N}$ collisions. For example,

$$\pi + N \to N + \pi + \pi$$

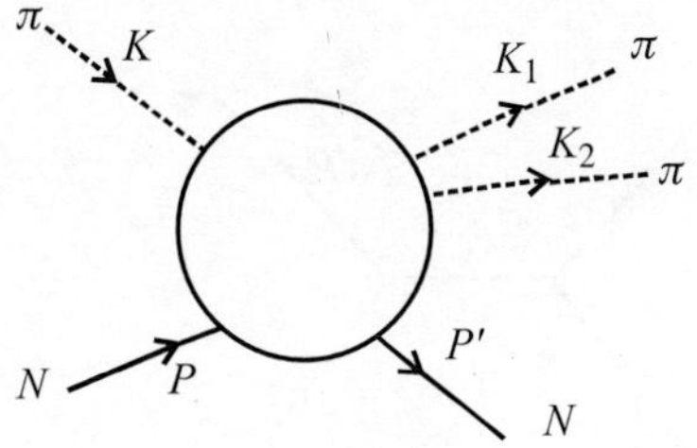

Fig. 4.3

Momentum conservation gives $K + P = P' + K_1 + K_2$

If one plots the number of two pions ($N_{\pi\pi}$) against the invariant mass m^* of the system, where

$$m^{*2} = (K_1 + K_2)^2 = (E_1 + E_2)^2 - (\vec{K}_1 + \vec{K}_2)^2$$

one sees a peak (resonance) at $m^* = 773$ MeV with $\Gamma = 150$ MeV

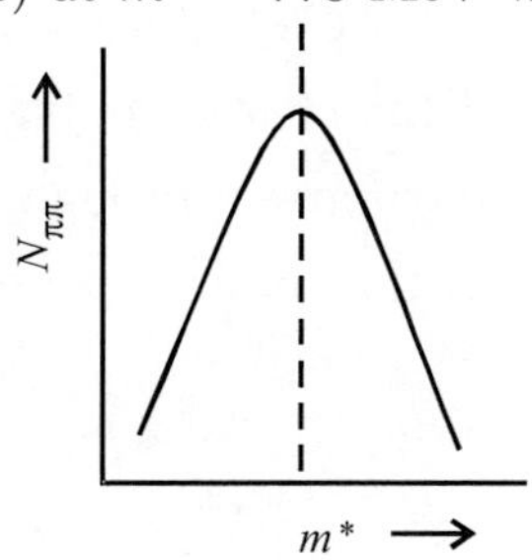

Fig. 4.4

The peak occurs for $\pi^+\pi^0$, $\pi^+\pi^-$ and $\pi^0\pi^-$ but not for other charge combinations ($\pi^+\pi^+$, $\pi^0\pi^0$, $\pi^-\pi^-$). This is interpreted as due to the production of a (short-lived) meson resonance which subsequently decays into 2 pions.Thus we can look upon the process $\pi + N \to N + \pi + \pi$ as a two-step process.

$$\pi + N \to N + \rho, \rho \to \pi\pi$$

Thus ρ is an intermediate state in $\pi\pi$ setting ($\pi\pi \to \rho \to \pi\pi$) with one of the initial pions being off mass shell (virtual pion).

The possible quantum numbers of the 2-pion system are

$$I = 0, 1, 2,\ J = L,\ P = (-)^L(-)^2 = (-)^L,\ C = (-)^L$$

$\rho(773)$ has quantum numbers $J^{PC} = 1^{--}$, $I = 1$, $B = 0$. A hadron with $J^{PC} = 1^{--}$(same as the photon) is called a vector meson. $\rho(773)$ is the lightest vector meson.

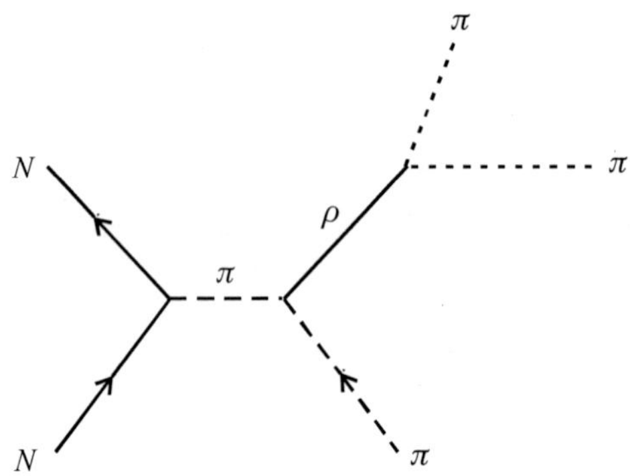

Fig. 4.5

***G*-Parity:** Important selection rules for strong interaction processes involving certain systems of hadrons originate from the introduction of a quantum number called the G-parity. The corresponding operator is known as G-conjugation.

$$G = Ce^{i\pi I_2} \tag{4.47}$$

Thus G is the product of the charge conjugation operator C and $R = e^{i\pi I_2}$, the rotation operator for rotation through 180° about Y-axis in I-spin space. In analogy to action of the similar operator for rotation in ordinary space, we have

$$e^{i\pi I_2}|II_3\rangle = (-)^I |I, -I_3\rangle, \quad I = \text{integer}.$$

Therefore,

$$G\begin{pmatrix}\pi^+\\ \pi^0\\ \pi^-\end{pmatrix} = Ce^{i\pi I_2}\begin{pmatrix}\pi^+\\ \pi^0\\ \pi^-\end{pmatrix} = C\begin{pmatrix}-\pi^-\\ -\pi^0\\ -\pi^+\end{pmatrix} = \begin{pmatrix}-\pi^+\\ -\pi^0\\ -\pi^-\end{pmatrix} = -\begin{pmatrix}\pi^+\\ \pi^0\\ \pi^-\end{pmatrix}$$

$$\text{or} \quad G|\vec{\pi}\rangle = -|\vec{\pi}\rangle \tag{4.48}$$

Thus the pion has negative (odd) G-parity, Under G-conjugation.

$$I_3 \xrightarrow{G} I_3, \; B \xrightarrow{G} -B, \; Q \xrightarrow{G} Q \tag{4.49}$$

Therefore one can have simultaneous eigen states of G and Q if $B = 0$, e.g. ordinary mesons. Similarly the $N\overline{N}$ system is an eigen state of G-parity.

The G-parity of $N\overline{N}$ system is $G = (-)^{L+S+I}$. In the process

$N + \overline{N} \to n\pi$

$$G_i = G_f \text{ gives } (-)^{L+S+I} = (-)^n \tag{4.99}$$

n is even or odd according as $L + S + I$ is even or odd.

(S = total spin $= 0, 1$; I = total i-spin $= 0, 1$; L = relative angular momentum $= 0, 1, 2, \ldots$)

Problems

1. State, giving reasons, which of the following 2-particle states can or cannot exist in an $I = 1$ state

   (a) $\pi^0\pi^0$ (b) $\pi^+\pi^-$ (c) $\pi^+\pi^+$

2. Show that i-spin invariance implies the following cross section ratio

   (a) $\dfrac{\sigma(p + p \to {}^2H + \pi^+)}{\sigma(p + n \to {}^2H + \pi^0)} = 2;$ (b) $\dfrac{\sigma(p + {}^2H \to {}^3He + \pi^0)}{\sigma(p + {}^2H \to {}^3H + \pi^+)} = \dfrac{1}{2};$

   (c) $\dfrac{\sigma(\pi^+ + p \to \Delta^{++})}{\sigma(\pi^- + p \to \Delta^0)} = 3$

3. Show that i-spin invariance predicts that at CM energy corresponding to the $N^*(1470)$ resonance

   $\sigma(\pi^+p \to \pi^+p) : \sigma(\pi^-p \to \pi^-p) : \sigma(\pi^-p \to \pi^0 n) \approx 0 : 2 : 1$

4. A resonance named ρ is seen in the $\pi\pi$ channel in the production process $\pi + N \to N + \pi + \pi$ at $\sqrt{s} = 773$ MeV with width $\Gamma =$ 150 MeV. The resonance is seen when the 2 pions have different charges $(\pi^+\pi^0, \pi^+\pi^-, \pi^0\pi^-)$ but not when they have the same charges $(\pi^+\pi^+, \pi^-\pi^-, \pi^0\pi^0)$. Obtain the possible spin-parity, C-parity, i-spin and G-parity of the resonance.

5. A neutral meson is known whose mass is 549 MeV and width 2.6 KeV. Its principal decay modes are $\eta \to \gamma\gamma$, $\eta \to \pi^0\pi^0\pi^0$ and $\eta \to \pi^+\pi^-\pi^0|$.

   Obtain its possible quantum numbers (spin, parity, C-parity, i-spin and G-parity), giving reasons. How does one explain its small width?

6. Suppose that the nucleon appears in the isodoublet representation $\Psi = \begin{pmatrix} \psi_p \\ \psi_n \end{pmatrix}$ and the pion in the isotriplet representation $\phi = \begin{pmatrix} \phi_1 \\ \phi_2 \\ \phi_3 \end{pmatrix}$

   with $\phi_{\pi^0} = \phi_3$ and $\phi_{\pi^\pm} = \frac{1}{\sqrt{2}}(\phi, \mp i\phi_2)$

   Show that the isospin-invariant, Lorentz-invariant pion-nucleon interaction Lagrangian is given by

$$L_{\text{int}} = g\overline{\psi}\gamma_5(\overrightarrow{\tau} \cdot \overrightarrow{\phi})\psi$$

   where $\overline{\psi} = \psi^{+}\gamma^0$. ($\gamma^0$ and γ_5 are the 4×4 Dirac matrices) and thus the relations among the various pion-nucleon coupling constants are

$$g_{pp\pi^0} = -g_{nn\pi^0} = \frac{1}{\sqrt{2}}g_{pn\pi^+} = \frac{1}{\sqrt{2}}g_{np\pi^-} = g$$

Chapter 5

Strangeness (S) Quantum Number and the $SU(3)_f$ Symmetry

5.1 Strangeness

The "strangeness" quantum number S was introduced in 1953 independently by Gell-Mann and Nishijima in order to explain the properties of the K-mesons and hyperons $(\wedge, \Sigma, \equiv, \Omega)$ whose behaviour appeared strange at the time they were first seen in cosmic rays. These particles were later very copiously produced in high-energy accelerator experiments. It was soon pointed out by Pais that K-mesons and hyperons were always produced in association via strong interactions (that is, whenever a hyperon is produced, one or more K-mesons are produced simultaneously), but they decay singly via weak interactions.

As first suggested by Gell-Mann and Nishijima, this apparent puzzle is resolved, if one introduces a new quantum number which they named "strangeness" (S) with the following properties:

1. $S = 0$ for π and N; $S \neq 0$ for K-mesons and hyperons.
2. S is conserved ($\Delta S = 0$) in strong and electromagnetic processes, but is not conserved in weak processes involving K-mesons and hyperons. $\Delta S = 0, \pm 1$ for first order weak processes.
3. K-mesons and hyperons can be put in i-spin multiplets, just like pions and nucleons, each i-spin multiplet being characterized by a definite value of S. This means that the strangeness operator commutes with i-spin operator ($[S, I_a] = 0$)
4. The relation between electric charge Q (measured in unit of $|e|$) and the third component of i-spin (I_3) is modified to include K-mesons and hyperons. The modified relation, known as the Gell-Mann-Nishijima relation is

$$Q = I_3 + \frac{1}{2}(B + S) = I_3 + \frac{Y}{2}, \; Y = B + S \tag{5.1}$$

Y is called "hypercharge". It turns out that Y is a better quantum number than S for particle classification.

5. Under G-Conjugation, S changes sign like the baryon number $B(S \xrightarrow{G} -S)$. Thus G-parity can be defined only for hadrons or hadron systems for which $B = S = 0$. Thus the symmetry group of hadrons is at least $SU(2)_I \times U(I)_Y$ with generators $I_a (a = 1, 2, 3)$ and Y. The Lie algebra is given by

$$[I_a, I_b] = i\epsilon_{abc} I_c, \; [I_a, Y] = 0 \tag{5.2}$$

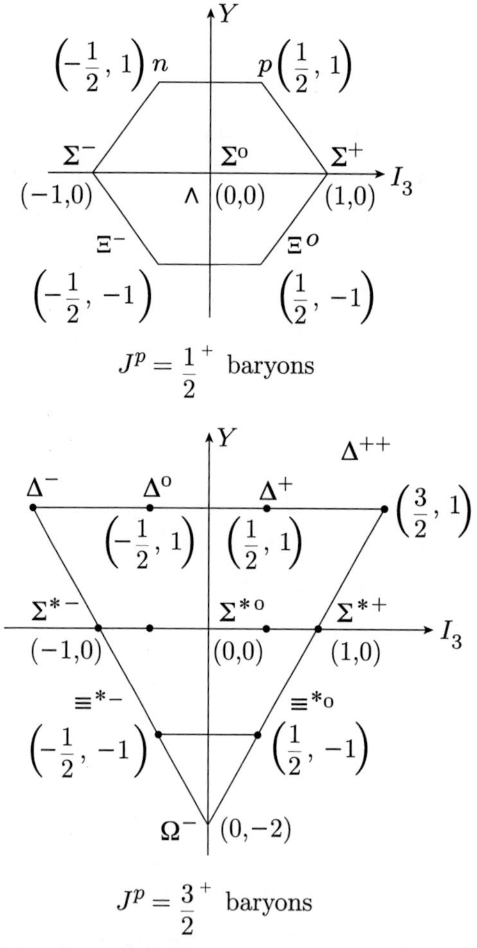

Fig. 5.1. Lowest-lying hadron multiplets

A close look at the hadron spectra shows that a number of i-spin multiplets with the same spin-parity J^P seem to cluster together in mass. Light hadrons appear in supermultiplets with multiplicity 8, that is in octets (eight fold way). This indicates that the symmetry group G of hadrons must be higher than $SU(2)_I \times U(1)_Y$.

Desired properties of G : The symmetry group G of hadrons is expected to have the following properties:

1. G must be a Lie group of rank $\ell \geq 2$. Two of the commuting generators are to be identified with I_3 and Y.
2. G must have a representation of dimension 8.
3. G must have an SU(2) subgroup.

The smallest simple Lie group with the above properties is the group $SU(3)$.

5.2 The Group $SU(3)$

The $SU(3)$ group is defined by considering proper unitary transformations in a 3-dimensional complex vector space.

$$x = \begin{pmatrix} x_1 \\ x_2 \\ x_3 \end{pmatrix} \xrightarrow{U} x' = \begin{pmatrix} x_1' \\ x_2' \\ x_3' \end{pmatrix}, \quad x_i' = \sum_{j=1}^{3} U_{ij} x_j \tag{5.3}$$

U is a 3×3 matrix.

$$\sum_i |x_i'|^2 = \sum_i \sum_j \sum_k U_{ij} x_j U_{ik}^* x_k^* = \sum_j |x_j|^2 = \sum_j \sum_k x_j x_k^* \delta_{jk}$$

$$\text{gives } \sum_{i=1}^{3} U_{ij} U_{ik}^* = \delta_{jk} \quad \text{or} \quad U^+ U = 1 \tag{5.4}$$

Thus U is a unitary matrix (5.4) gives $|\det\ U|^2 = 1$

We choose $\det U = 1$. The set of 3×3 unitary matrices with determinant $+1$ forms a group, called the $SU(3)$ group.

One can write an infinitesimal element of the group as

$$U(\delta\theta) = 1 - i \sum_{a=1}^{r} \delta\theta_a \frac{\lambda_a}{2} \tag{5.5}$$

U unitary $\Rightarrow \lambda_a'$s are hermitean

$\det U = +1 \Rightarrow \lambda_a'$s are traceless.

There can be $r = n^2 - 1 = 3^2 - 1 = 8$ traceless, hermitian, linearly independent matrices. $\mathrm{Tr}\lambda_a = 0 \Rightarrow$ only 2 of the diagonal elements are independent. There can be thus atmost 2 simultaneously diagonal matrices which means the rank of $SU(3)$ is 2

Gell-Mann matrices: λ_a matrices may be chosen as 3-dimensional generalisation of 2×2 Pauli matrices.

$$\lambda_1 = \begin{pmatrix} 0 & 1 & 0 \\ 1 & 0 & 0 \\ 0 & 0 & 0 \end{pmatrix}, \quad \lambda_2 = \begin{pmatrix} 0 & -i & 0 \\ i & 0 & 0 \\ 0 & 0 & 0 \end{pmatrix}, \quad \lambda_3 = \begin{pmatrix} 1 & 0 & 0 \\ 0 & -1 & 0 \\ 0 & 0 & 0 \end{pmatrix}$$

$$\lambda_4 = \begin{pmatrix} 0 & 0 & 1 \\ 0 & 0 & 0 \\ 1 & 0 & 0 \end{pmatrix}, \quad \lambda_5 = \begin{pmatrix} 0 & 0 & -i \\ 0 & 0 & 0 \\ i & 0 & 0 \end{pmatrix}, \quad \lambda_6 = \begin{pmatrix} 1 & 0 & 0 \\ 0 & 0 & 1 \\ 0 & 1 & 0 \end{pmatrix}$$

$$\lambda_7 = \begin{pmatrix} 0 & 0 & 0 \\ 0 & 0 & -i \\ i & i & 0 \end{pmatrix}, \quad \lambda_8 = \frac{1}{\sqrt{3}} \begin{pmatrix} 1 & 0 & 0 \\ 0 & 1 & 0 \\ 0 & 0 & -2 \end{pmatrix} \tag{5.6}$$

The matrices are also normalized like the Pauli matrices, i.e.,

$$\mathrm{Tr}(\lambda_a \lambda_b) = 2\delta_{ab} \tag{5.7}$$

The above matrices are known as the Gell-Mann matrices. Explicit computation gives

$$[\lambda_a, \lambda_b] = 2if_{abc}\lambda_c \quad \text{(sum over } c \text{ is implied)} \tag{5.8a}$$

$$\{\lambda_a, \lambda_b\} = \frac{4}{3}\delta_{ab} \cdot 1 + 2d_{abc}\lambda_c \tag{5.8b}$$

f_{abc}'s and d_{abc}'s are real constants whose values are known from explicit computation of the commutators and anticommutators. f_{abc}'s are known as the structure constants of the $SU(3)$ Lie algebra.

From Eqs. (5.7) and (5.8) one gets

$$\mathrm{Tr}\left([\lambda_a, \lambda_b]\lambda_c\right) = 4if_{abc} \tag{5.9a}$$

$$\mathrm{Tr}\left(\{\lambda_a, \lambda_b\}\lambda_c\right) = 4d_{abc} \tag{5.9b}$$

Properties of structure constants:

1. $[\lambda_a, \lambda_b] = -[\lambda_b, \lambda_a] \;\Rightarrow\; f_{abc} = -f_{bac}$

2. $[[\lambda_a, \lambda_b], \lambda_c] + [[\lambda_b, \lambda_c], \lambda_a] + [[\lambda_c, \lambda_a], \lambda_b] = 0$

$$\Rightarrow \sum_d (f_{abd} f_{dce} + f_{bcd} f_{dae} + f_{cad} f_{dce}) = 0$$

Using the cyclic property of trace, $\mathrm{Tr}\,(AB) = \mathrm{Tr}\,(BA)$ one can see that f_{abc}'s are completely antisymmetric and d_{abc}'s are completely symmetric in a, b, c.

$$f_{abc} = f_{bca} = f cab = -f_{bac} = -f_{cba} = -f_{acb} \tag{5.10}$$

$$d_{abc} = d_{bca} = d_{cab} = d_{bac} = d_{cba} = d_{acb} \tag{5.11}$$

The last 3 equalities in (5.10) and (5.11) follow from the properties of commutators and anti-commutators, respectively:

$$([A, B] = -[B, A] \quad \text{and} \quad \{A, B\} = \{B, A\})$$

The independent f_{abc}'s are

$$\begin{aligned} &f_{123} = 1 \\ &f_{147} = f_{246} = f_{257} = f_{345} = \frac{1}{2} \\ &f_{156} = f_{367} = -\frac{1}{2} \\ &f_{458} = f_{678} = \sqrt{\frac{3}{2}} \end{aligned} \tag{5.12}$$

The commutation relation (5.8a) can be rewritten as

$$\left[\frac{\lambda_a}{2}, \frac{\lambda_b}{2}\right] = i f_{abc} \frac{\lambda_c}{2} \tag{5.13}$$

Writing $D(F_a) = \frac{\lambda_a}{2}$ we see that the Lie Algebra of $SU(3)$ is given by

$$[F_a, F_b] = i f_{abc} F_c \tag{5.14}$$

where F_a's are the abstract generators of the $SU(3)$ group and λ_a's are 3-dimensional representation of F_a's. They are known as the fundamental representations of $SU(3)$. The abstract elements of the $SU(3)$ group are then given by

$$U(\theta) = e^{-i \sum_{a=1}^{8} \theta_a F_a} \tag{5.15}$$

with the F_a's satisfying (5.14).

Conjugate representation: From(5.13) we see that $\hat{\lambda}_a = -\lambda_a^*$ satisfies the Lie algebra of $SU(3)$

$$\text{Thus}\quad D(F_a) = \frac{\hat{\lambda}_a}{2} = \frac{-\lambda_a^*}{2}, \tag{5.16}$$

provides another 3-dimensional representation of $SU(3)$ Lie algebra. The 2 sets of representations are inequivalent.

Proof. We have

$$\lambda_8 = \frac{1}{\sqrt{3}}\begin{pmatrix} 1 & 0 & 0 \\ 0 & 1 & 0 \\ 0 & 0 & -2 \end{pmatrix}, \quad \hat{\lambda}_8 = -\lambda_8^* = \frac{1}{\sqrt{3}}\begin{pmatrix} -1 & 0 & 0 \\ 0 & -1 & 0 \\ 0 & 0 & +2 \end{pmatrix}$$

$$\det \lambda_8 = \frac{-2}{\sqrt{3}} \quad \text{and} \quad \det \hat{\lambda}_8 = \frac{+2}{\sqrt{3}}, \quad \det \lambda_8 \neq \det \hat{\lambda}_8$$

Therefore, λ_8 and $\hat{\lambda_8}$ cannot be connected by a similarity transformation and hence are inequivalent. $\hat{\lambda}_a$'s constitute the conjugate representation. Thus $SU(3)$ has two inequivalent fundamental representations.

Adjoint Representation: Let us consider eight 8×8 matrices whose elements are given by.

$$(D_a)_{bc} = i f_{bac}(a, b, c = 1, \cdots 8) \tag{5.17}$$

$$\begin{aligned} \text{Now}\quad [D_a, D_b]_{de} &= \sum_c \{D_a\}_{dc}(D_b)_{ce} - (D_b)_{dc}(D_a)_{ce} \\ &= i^2 \sum_c \{f_{dac}f_{cbe} - f_{dbc}f_{cae}\} \\ &= i^2 \sum_c \{f_{dac}f_{cbe} + f_{bdc}f_{cae}\} \\ &= i^2 \sum_c f_{abc}f_{dcc} \\ &= i \sum_c f_{abc}(D_c)_{de} \end{aligned}$$

This gives $[D_a, D_b] = i \sum_c f_{abc} D_c$

Therefore, the matrices D_a's $(a = 1, 8)$ satisfy (5.14), the Lie algebra of $SU(3)$. Thus $\{D(F_a) = D_a; a = 1, 8\}$ with elements given by (5.17) forms a 8-dimensional representation of $SU(3)$ Lie algebra which is called the adjoint representation of $SU(3)$, since the dimension of the representation is equal to the number of generators of $SU(3)$, $(N = r = 8)$.

In order to find all possible representations (multiplets) of $SU(3)$ Lie algebra, one can proceed in the same way as in the case of $SU(2)$, namely the angular momentum Lie algebra. One first constructs pairs of step-up and step-down operators out of the non-diagonal generators. Since the two diagonal generators are chosen to be F_3 and F_8, there are 3-pairs of such operators. They are

$$I\pm = F_1 \pm iF_2,\ V\pm = F_4 \pm iF_5 \ \text{ and } \ U\pm = F_6 \pm iF_7 \tag{5.18}$$

Given the commutation relations of F_a's (hermitean generators) one can find the CR's involving F_3, F_8 and $I\pm, V\pm$ and $U\pm$ (non hermitean generators). Next one constructs the Casimir operators of $SU(3)$, which are defined by $[C_i, F_a] = 0$. One finds that there are 2 Casimir operators of $SU(3)$:

$$C_2 = \sum_{a-1}^{8} F_a^2 = -\frac{2}{3} \sum_{a,b,c=1}^{8} f_{abc} F_a F_b F_c \tag{5.19}$$

and $$C_3 = \sum_{a,b,c=1}^{8} d_{abc} F_a F_b F_c \tag{5.20}$$

One finds that $(C_2, C_3, C_1, F_3, F_8)$ form a complete set of commuting operators. Here $C_1 = \sum_{a=1}^{3} F_a^2 = I^2$ is the Casimir operator of one of the $SU(2)$ sub-algebras of $SU(3)$.

The states of an $SU(3)$ multiplet are simultaneous eigen states of the above set of operators. The eigenvalues of C_2, C_3 (given in terms of 2 integers p, q) characterize a multiplet and eigenvalues of C_1, F_3, F_8 distinguish the individual states of the multiplet. It turns out that the procedure described above which is a generalization of $SU(2)$ case is also the general procedure for finding the representation of any simple Lie algebra.This will be discussed in Ch. 6

Problems

1. Show that i-spin invariance implies the following cross section ratio

(a) $\dfrac{\sigma(K^- + n \to \Lambda + \pi^-)}{\sigma(K^- + p \to \Lambda + \pi^0)} = 2$ (b) $\dfrac{\sigma(\pi^+ + n \to \Lambda + K^+)}{\sigma(\pi^- + p \to \Lambda + K^0)} = 1$

2. Find a relation between the total cross sections (at a given CM energy) for the reactions

$$\pi^- p \to \Sigma^o K^o,\ \pi^- p \to \Sigma^- K^+ \ \text{ and } \ \pi^+ p \to \Sigma^+ K^+$$

3. State through which i-spin channels the reactions $K^- + p \to \Sigma^o + \pi^o$ and $K^- + p \to \Sigma^+ + \pi^-$ may proceed. Find the ratio of their cross sections assuming that one of the i-spin channels dominate.

4. The A_2^- meson has mass 1310 MeV and width 80 MeV. It is known to decay into $\pi^+\pi^-\pi^o$ and K^oK^-. What are its possible spin-parity, i-spin, strangeness and G-parity assignments?

5. The capture of negative kaons in helium sometimes leads to the formation of a hypernucleus (one in which a neutron is replaced by a $\wedge$-hyperon) according to the reaction

$$K^- + 4_{\text{He}} \to {}^4H_\wedge + \pi^o$$

Study of the branching ratio of ${}^4H_\wedge$ and isotropy of its decay products establish the fact that this hypernucleus has spin zero. Show that this implies negative parity for K^- independently of the orbiral angular momentum state from which the K^- is captured.

6. Define the operators

$$U_+ = F_6 + iF_7,\ U_- = F_6 - iF_7 \ \text{ and } \ U_3 = \frac{\sqrt{3}}{2}F_8 - \frac{1}{2}F_3$$

where F_a's $(a = 1, \ldots, 8)$ are the generators of the $SU(3)$ group, $[F_a, F_b] = if_{abc}F_c$

(a) Show that U_+, U_- and U_3 form a $SU(2)$ Lie algebra,

$$[U_+, U_-] = 2U_3,\ [U_3, U_+] = U_+,\ [U_3, U_-] = -U_-$$

(b) Show that $Y = U_3 + \dfrac{Q}{2}$

where Y and Q are the hypercharge and the electric charge operators, respectively.

(c) Draw the $U_3 - Q$ diagram for the baryon octet to which the nucleon belongs.

(d) Show that Q commutes with the generators of the U-spin (U_+, U_-, U_3). What does this imply?

(e) Assuming that hadrons belonging to the same *i*-spin multiplet have the same mass, while hadrons belonging to the same *U*-spin multiplet have masses given by

$$m = a + bU_3$$

where a and b are constants, show that the baryon octet masses satisfy the relation

$$\frac{1}{2}(m_N + m_\Xi) = \frac{1}{4}(3m_\Lambda + m_\Sigma)$$

Chapter 6

General Treatment of Simple Lie Algebras

6.1 Introduction

The general treatment of Lie algebra due to Cartan, Weyl and Dynkin is an extension of the method used for finding the irreducible representations of $SU(2)$ by means of raising and lowering operators $J_{\pm}$. The generators are split into a set $\{H_i\}$ which commute with one another, and which accordingly can be assigned simultaneous eigenvalues, and another set $\{E_\alpha\}$ which are generalised raising and lowering operators.

The different Lie groups can be classified systematically in terms of this splitting, and of the nature of the nonvanishing commutation relations between $\{H_i\}$ and $\{E_\alpha\}$ and those between E_α and E_β's. This classification amounts to specifying a number of (simple) roots, whose lengths and scalar products are severely restricted. These properties can be summarized by a certain matrix or equivalently by a Dynkin diagram. It turns out that there are four regular infinite series of Lie algebras, the so-called classical Lie algebras, along with five exceptional Lie algebras which do not belong to any such series.

The representation theory of Lie algebras is also modelled on the treatment of $SU(2)$ by the raising and lowering operators. In particular, one might recall that the states of an irreducible representation $|jm\rangle$ can be obtained by applying lowering operators to the highest m-value state $|jj\rangle$. In the Cartan language, this is the state of highest weight and the concept generalises to more complicated Lie algebras with more than one commuting generator.

6.2 The Lie Algebra in the Cartan Basis

The structure constants determine the Lie algebra, but the converse is clearly not true, because one has the freedom of choosing new basis vectors which are linear combination of the old ones, thus changing the form of the commutation relations. As a prototype, consider the Lie algebra,

$$[J_a, J_b] = i\epsilon_{abc} J_c \tag{6.1}$$

By using J_3 and the complex combination $J_\pm = J_1 \pm iJ_2$, these can be recast in the form

$$\begin{aligned} [J_3, J_\pm] &= \pm J_\pm \\ [J_+, J_-] &= 2J_3 \end{aligned} \tag{6.2}$$

In this form the commutation relations look rather different, but of course, they are completely equivalent to the first form. The Cartan way of presenting the commutation relations is a generalization of this second form.

One starts by looking for a maximum set of commuting generators. For this, one considers solutions of the eigen value problem

$$[A, X] = \rho X \tag{6.3}$$

where $$A = \sum_{a=1}^{r} \alpha_a F_a, \quad X = \sum_{a=1}^{r} x_a F_a \tag{6.4}$$

and $$[F_a, F_b] = i \sum_c f_{abc} F_c \tag{6.5}$$

If we substitute (6.4) in (6.3) and use (6.5), we get

$$(if_{abc}\alpha_a x_b - \rho x_c)F_c = 0 \tag{6.6}$$

Since F_c's are linearly independent, (6.6) gives

$$\sum_{a,b} if_{abc}\alpha_a x_b - \rho x_c = 0 \tag{6.7}$$

We can write (6.7) in the form

$$\sum_b (\Lambda_{cb} - \rho\delta_{cb})x_b = 0 \tag{6.8}$$

$$\Lambda_{cb} = i \sum_a f_{abc}\alpha_a$$

Eqn. (6.8) is a set of homogeneous linear equations. For nontrivial solution the determinant of coefficients must be zero.

$$\det \|\Lambda - \rho \cdot 1\| = 0 \tag{6.9}$$

Eqn. (6.9) is the eigen value problem for the matrix Λ. Since the l.h.s. in (6.9) is a rth degree polynomial, it has r roots, which are the eigen values of the matrix Λ. Since Λ is a hermitian matrix, all its eigen values are real. Cartan further proved that,

(1) Only the root $\rho = 0$ is degenerate with multiplicity l. Clearly $l \geq 1$ (since A commutes with A). l is called the **rank** of the Lie algebra.

(2) All the nonzero roots ($\rho \neq 0$) are nondegenerate.

(3) For each nondegenerate root ρ, there exists another root $\rho' = -\rho$ (Λ is an antisymmetric matrix). Thus the nonzero roots occur in pairs. This means

$$p = \frac{r-l}{2} = \text{integer} \tag{6.10}$$

We denote the eigenvectors corresponding to the zero root ($\rho = 0$) by H_i's ($i = 1, 2, \ldots, l$) and the eigenvectors corresponding to the nonzero roots ($\rho \neq 0$) by E_α's.

We can then write the CR's in the form

$$[H_i, H_j] = 0 \ (i, j = 1, \ldots l) \tag{6.11}$$

$$[H_i, E_\alpha] = \rho^i_\alpha E_\alpha \ (\alpha = \ldots \pm p) \tag{6.12}$$

with $\rho^i_{-\alpha} = -\rho^i_\alpha$ (6.13)

The set of numbers $\rho_\alpha = (\rho^1_\alpha, \ldots, \rho^l_\alpha)$ ($\alpha = \pm 1, \ldots, \pm p$) can be regarded as the components of a vector in a l-dim. vector space. They are called **root vectors** or simply **roots**. The set of generators $\{H_i\}$ ($i = 1, \ldots, l$) are said to form the Cartan subalgebra.

$[H_i, E_\alpha] = \rho^i_\alpha E_\alpha \ \Rightarrow \ [H_j, E_\alpha] = \rho^j_\alpha E_\alpha$

This can be proved by using Jacobi identity for H_i, H_j, E_α and the fact that the non-zero roots are nondegenerate which means there is only one linearly independent eigenvector for each nondegenerate root.

$$[[H_i, H_j],\ E_\alpha] + [[H_j, E_\alpha],\ H_i] + [E_\alpha, H_i],\ H_j\} = 0$$

gives $[H_i, [H_j, E_\alpha]] = \rho^i_\alpha [H_j, E_\alpha]$

This means that $[H_j, E_\alpha]$ is an eigenvector of H_i with the same eigenvalue ρ^i_α. But since ρ^i_α is non degenerate,

$$[H_j, E_\alpha] \propto E_\alpha,$$

which means $[H_j, E_\alpha] = \rho^j_\alpha \ E_\alpha$. Using the Jacobi identity involving H_i, E_α and E_β,

$$[H_i, [E_\alpha, E_\beta]] + [E_\alpha, [E_\beta, H_i]] + [E_\beta, [H_i, E_\alpha]] = 0$$

we get $[H_i, [E_\alpha,\ E_\beta]] = (\rho^i_\alpha + \rho^i_\beta)[E_\alpha,\ E_\beta]$ (6.14)

One has 3 possibilities,

(i) $\beta = -\alpha$

In this case since $\rho^i_{-\alpha} = -\rho^i_\alpha$ one has

$$[H_i, [E_\alpha,\ E_{-\alpha}]] = 0 \Rightarrow [E_\alpha,\ E_{-\alpha}] = \sum_i C^i_\alpha H_i \quad (6.15)$$

(ii) $\beta \neq -\alpha,\ \rho^i_\alpha + \rho^i_\beta = \rho^i_\gamma$ is a root

In this case since the non-zero roots are non-degenerate

$$[E_\alpha,\ E_\beta] = N_{\alpha\beta}\ E_\gamma \quad (6.16)$$

where E_γ is the generator corresponding to the root ρ_γ and $N_{\alpha\beta}$ is a constant.

(iii) $\beta \neq -\alpha,\ \rho^i_\alpha + \rho^i_\beta$ is not a root.

In this case,

$$[E_\alpha,\ E_\beta] = 0 \quad (6.17)$$

It can be easily seen that the constants. C^i_α in Eq. (6.15) are proportional to the roots ρ^i_α.

From (6.15)

$$\mathrm{Tr}\,([E_\alpha,\ E_{-\alpha}]H_i) = \sum_j C^j_\alpha \mathrm{Tr}\,(H_j H_i) = \sum_j C^j_\alpha \lambda \delta_{ji} = \lambda C^i_\alpha$$

Using cyclic property of trace

$$\mathrm{Trace}([E_\alpha,\ E_{-\alpha}]H_i) = \mathrm{Trace}([H_i,\ E_\alpha]E_{-\alpha}) = \rho^i_\alpha \mathrm{Trace}(E_\alpha\ E_{-\alpha}) = \lambda' \rho^i_\alpha$$

Since λ and λ' are constants, $C^i_\alpha \ \propto\ \rho^i_\alpha$

This means that, with proper choice of normalisation, we can write the commutation relation (6.15) in the from

$$[E_\alpha,\ E_{-\alpha}] = 2\Sigma_\alpha \quad (6.18)$$

where, $\Sigma_\alpha = \dfrac{\Sigma {\rho_\alpha}^i H_i}{|\rho_\alpha|^2},\ |\rho_\alpha|^2 = \sum_i |{\rho_\alpha}^i|^2$ (6.19)

Further more, $[\Sigma_\alpha,\ E_{\pm\alpha}] = \frac{1}{|\rho_\alpha|^2}\sum_i \rho^i_\alpha[H_i,\ E_{\pm\alpha}] = \frac{1}{|\rho_\alpha|^2}\sum_i {\rho_\alpha}^i \rho^i_{\pm\alpha}\ E_{\pm\alpha}$

$$= \pm E_{\pm\alpha}, (\rho^i_{-\alpha} = -\rho^i_\alpha) \tag{6.20}$$

Thus E_α, $E_{-\alpha}$ and Σ_α form a $SU(2)$ subalgebra ($\alpha = 1, \cdots p$).

Thus a simple Lie algebra has $p = \frac{1}{2}(r-l)$, $SU(2)$ subalgebras. However, not all of them are linearly independent. For $SU(3)$, there are 3 $SU(2)$ subalgebras, two of which are independent.

Theorem: If ρ_α and ρ_β are two roots of a simple Lie algebra, then

$$m = \frac{2\rho_\alpha \cdot \rho_\beta}{|\rho_\alpha|^2} \quad \text{and} \quad n = \frac{2\rho_\beta \cdot \rho_\alpha}{|\rho_\beta|^2} \tag{6.21}$$

are integers and $\rho_\beta - m\rho_\alpha$ and $\rho_\alpha - n\rho_\beta$ are roots.

Proof. $[\Sigma_\alpha, E_\beta] = \frac{1}{|\rho_\alpha|^2}\sum_i {\rho_\alpha}^i[H_i, E_\beta] = \frac{1}{|\rho_\alpha|^2}\sum_i {\rho_\alpha}^i {\rho_\beta}^i E_\beta$

$$= \frac{m}{2}E_\beta, \quad m = \frac{2}{|\rho_\alpha|^2}\sum_i {\rho_\alpha}^i \rho^i_\beta$$

This means

$$\Sigma_\alpha\,(E_\beta|m_\alpha\rangle) = ([\Sigma_\alpha,\ E_\beta] + E_\beta\ \Sigma_\alpha)\,|m_\alpha\rangle$$
$$= \left(m_\alpha + \frac{m}{2}\right)(E_\beta|m_\alpha\rangle)$$

$\Rightarrow$ $E_\beta|m_\alpha\rangle$ is an eigen vector of $\sum_\alpha$ with eigen value $m'_\alpha = m_\alpha + \frac{m}{2}$.

Since Σ_α is a generator of $SU(2)$, eigenvalues of Σ_α are integers, or half integers. Therefore, $m = 2m'_\alpha - 2m_\alpha =$ integer.

Similarly, it can be proved that n is an integer.

Schwarz inequality, $|\rho_\alpha \cdot \rho_\beta|^2 \le |\rho_\alpha|^2|\rho_\beta|^2$ gives

$$\frac{2(\rho_\alpha \cdot \rho_\beta)}{|\rho_\alpha|^2} \times \frac{2(\rho_\beta \cdot \rho_\alpha)}{|\rho_\beta|^2} \le 4$$

This means $mn \le 4$ (6.22)

If φ is the angle between the two root vectors ρ_α and ρ_β then

$$mn = 4\cos^2\varphi \quad \text{or,} \quad \cos^2\varphi = \frac{mn}{4} \le 1 \tag{6.23}$$

Therefore, allowed values of $\cos^2\varphi$ are 0, $\frac{1}{4}$, $\frac{1}{2}$, $\frac{3}{4}$ and 1. We assume that $0° \le \varphi \le 90°$

(m, n)	$\cos^2\varphi$	$0 \leq \varphi \leq 90°$	$\frac{\lvert\rho_p\rvert}{\lvert\rho_\alpha\rvert} = \sqrt{m/n}$
(0, 0)	0	90°	undefined
(1, 1)	1/4	60°	1
(1, 2)	1/2	45°	$1/\sqrt{2}$
(1, 3)	$\sqrt{3/2}$	30°	$1/\sqrt{3}$

Thus the angle between roots can be multiples of 90°, 60°, 45° and 30°.

For $l = 1$ the root diagram is 1-dimensional. There are only two non-zero roots $\rho_{\pm 1}$. There is only one root diagram.

$$\longleftarrow \bullet \longrightarrow$$

$$\rho_1 = -\rho_{-1} = 1 \tag{6.24}$$

total number of generators $r = l + 2p = 3$.

The generators are H_1, E_1 and E_{-1}. The $CR's$ are

$$[H_1,\ E_{\pm 1}] = \pm E_{\pm 1}$$

$$[E_1,\ E_{-1}] = 2H_1$$

This is just the angular momentum algebra with

$$H_1 = J_3,\ E_{\pm 1} = J_1 \pm\ iJ_2$$

Thus there is only one independent Lie algebra of rank-1, namely the angular momentum algebra $SU(2)$. All rank-1 Lie algebra are isomorphic to it.

For $\ell = 2$ the root diagram is 2-dimensional. The number of non-zero roots depends on the choice of angle φ. (Angle between roots are multiples of φ). There are four root diagrams for four possible choices of φ, namely, $\varphi = 90°$, 60°, 45° and 30°.

For $\varphi = 90°$

$$r = \ell + 2p = 2 + 2 \cdot 2 = 6 \tag{6.26}$$

$\rho_1 = (1,\ 0),\ \rho_{-1} = (-1,\ 0) \qquad \rho_2 = (0,\ 1),\ \rho_{-2} = (0,\ -1)$

The root diagram corresponds to the $SO(4)$ Lie algebra with 6 generators.

$[r = \frac{1}{2}n(n-1) = \frac{1}{2} \cdot 4 \cdot 3 = 6]$,

$SO(4) \approx SU(2) \times SU(2)$, is not simple; it is semisimple.

For $\varphi = 60°$,

$$r = l + 2p = 2 + 6 = 8, \tag{6.27}$$

The roots are

$$\rho_1 = (1, 0),\ \rho_{-1} = (-1, 0),\ \rho_2 = \left(\frac{1}{2}, \frac{\sqrt{3}}{2}\right),\ \rho_{-2} = \left(\frac{-1}{2}, \frac{-\sqrt{3}}{2}\right),$$

$$\rho_3 = \left(\frac{1}{2}, \frac{-\sqrt{3}}{2}\right),\ \rho_{-3} = \left(\frac{-1}{2}, \frac{\sqrt{3}}{2}\right)$$

The root diagram corresponds to $SU(3)$ Lie algebra.

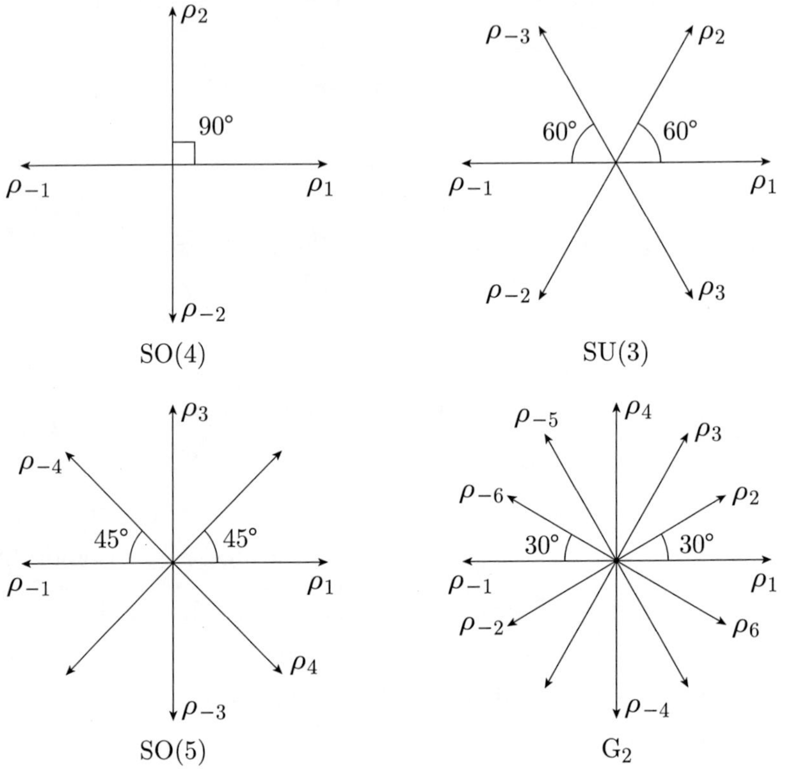

Fig. 6.1. The root diagrams

For $\varphi = 45°$,

$$r = l + 2p = 2 + 8 = 10. \tag{6.28}$$

The roots are

$$\rho_1 = (1, 0),\ \rho_{-1} = (-1, 0),\ \rho_2 = \left(\frac{1}{2}, \frac{1}{2}\right),\ \rho_{-2} = \left(-\frac{1}{2}, -\frac{1}{2}\right)$$

$$\rho_3 = (0, 1), \rho_3 = (0, -1),\ \rho_4 = \left(\frac{1}{2}, \frac{-1}{2}\right),\ \rho_{-4} = \left(\frac{-1}{2}, \frac{1}{2}\right)$$

The root diagram corresponds to $SO(5)$ or $Sp(4)$ Lie algebra.

$\varphi = 30°$

$\rho_1 = (1,0)$, $\rho_{-1} = (-1,0)$, $\rho_2 = \left(\frac{1}{2}, \frac{1}{2\sqrt{3}}\right)$, $\rho_{-2} = \left(-\frac{1}{2}, -\frac{1}{2\sqrt{3}}\right)$

$\rho_3 = \left(\frac{1}{2}, \frac{\sqrt{3}}{2}\right)$, $\rho_{-3} = \left(\frac{-1}{2}, -\frac{\sqrt{3}}{2}\right)$, $\rho_4 = \left(0, \frac{1}{\sqrt{3}}\right)$, $\rho_{-4} = \left(0, -\frac{1}{\sqrt{3}}\right)$

$\rho_5 = \left(\frac{1}{2}, \frac{-\sqrt{3}}{2}\right)$, $\rho_{-5} = \left(-\frac{1}{2}, \frac{\sqrt{3}}{2}\right)$, $\rho_6 = \left(\frac{1}{2}, \frac{-1}{2\sqrt{3}}\right)$, $\rho_{-6} = \left(-\frac{1}{2}, \frac{1}{2\sqrt{3}}\right)$

and $r = l + 2p = 2 + 12 = 14$ (6.29)

The root diagram corresponds to the exceptional Lie group G_2.

Proceeding in this way one can enumerate all possible Lie algebras. It turns out that they consist of four infinite series of classical Lie algebras.

labelled $\begin{matrix} SO(2l) & SO(2l+1) & SU(l+1) & \text{and} & Sp(2l) \\ D_1 & B_1 & A_i & \text{and} & C_l \end{matrix}$, respectively.

together with 5 exceptional Lie algebras G_2, F_4, E_6, E_7 and E_8. (the subscripts indicate the rank). They are called exceptional because unlike the classical Lie groups they cannot be realised as transformations on a linear vector space.

6.3 *SU*(3) Lie Algebra in Cartan Basis

There are 2 hermitean diagonal generators H_1 and H_2 and 3 pairs of non-hermitean generators. $E_\pm\alpha(\alpha = 1,2,3)$. $E_{-\alpha} = E_\alpha^+$. The CR's are

$$[H_1, H_2] = 0$$

$$[H_i, E_{\pm\alpha}] = \pm\rho_\alpha^i E_{\pm\alpha} \begin{pmatrix} i = 1,2 \\ \alpha = 1,2,3 \end{pmatrix}$$

$$[E_\alpha, E_{-\alpha}] = 2\Sigma_\alpha$$

$$[E_2, E_3] = E_1$$

$$[E_1, E_{-3}] = E_2$$

$$[E_{-2}, E_1] = E_3 \tag{6.30}$$

In the above $\Sigma_\alpha = \frac{1}{\rho_\alpha^2} \sum_i \rho_\alpha^i H_i$.

All other commutators are either zero or related to these.

From (6.27) and (6.19) we get,

$$\Sigma_1 = H_1 \Sigma_2 = \frac{1}{2}H_1 + \frac{\sqrt{3}}{2}H_2,\ \Sigma_3 = \frac{1}{2}H_1 - \frac{\sqrt{3}}{2}H_2 \tag{6.31}$$

(6.31) gives $\sum_2 + \sum_3 = \sum_1$

This means that only 2 of the 3 $SU(3)$ subalgebras are independent.

The generators in the Cartan basis are linear combinations of those in the canonical (hermitean) basis. For the $SU(3)$, the relations are as follows:

$H_1 = F_3,\ H_2 = F_8$

$E_{\pm 1} = F_1 \pm iF_2, E_{\pm 2} = F_4 \pm iF_5, = E_{\pm 3} = F_6 \mp iF_7$

$$\Sigma_1 = F_3,\ \Sigma_2 = \frac{1}{2}F_3 + \frac{\sqrt{3}}{2}F_8,\ \Sigma_3 = \frac{1}{2}F_3 - \sqrt{\frac{3}{2}}F_8 \tag{6.32a}$$

For $SU(3)_f$, the flavor group of hadrons, they were renamed

$E_{\pm 1} = I_\pm,\ E_{\pm 2} = V_\pm,\ E_{\pm 3} = U_\pm,$

$$\Sigma_1 = I_3,\ \Sigma_2 = V_3,\ \Sigma_3 = -U_3, \tag{6.32b}$$

The subgroups generated by $(I_+, I_-, I_3), (U_+, U_-, U_3)$ and (V_+, V_-, V_3) are known as I-spin, U-spin and V-spin, respectively.

6.4 Simple Roots and Dynkin Diagrams

Clearly, not all the roots are linearly independent. In fact, since they are l-dim.vectors (l = rank of the group), at most l of them can be linearly independent.

An extremely useful way of choosing such a basis is by first introducing the concept of positive roots.

One can define a positive root as one whose first non-zero entry is positive. A simple root is a positive root that cannot be expressed as the sum of two other positive roots. It turns out that the number of simple roots equals the rank l of the group and they form a basis in the root vector space, Thus any root can be expressed as an integer linear combination of simple roots and the non-zero coefficients are either all positive integers or all negative integers. It turns out that the scalar product of 2 simple roots $S_i \cdot S_j < 0$ and therefore the angle between them is obtuse.

For $SU(3)$ there are 3 positive roots ρ_1, ρ_2, ρ_3. Since $\rho_1 = \rho_2 + \rho_3, \rho_1$ is not a simple root. The simple roots are therefore

$$S_1 = \rho_2,\ S_2 = \rho_3 \quad \text{and} \quad S_1 \cdot S_2 = \rho_2 \cdot \rho_3 = -\frac{1}{2}.$$

An essential information about the group algebra is encoded in these scaler products. One way of presenting this information is by means of the Cartan matrix defined by

$$K_{ij} = 2S_i \cdot S_j / S_j^2 \tag{6.33}$$

The diagonal elements are superfluous since $K_{ii} = 2$. If all the roots are of same length (as in the case of $SU(3)$), K is symmetric. Since the simple roots are linearly independent, K is non-singular, ie det $(K) \neq 0$. The Cartan matrices for $SU(3)$, $SO(5)$ and G_2 (rank 2 Lie groups) are

$$\begin{aligned} SU(3) : K &= \begin{pmatrix} 2 & -1 \\ -1 & 2 \end{pmatrix} \\ SO(5) : K &= \begin{pmatrix} 2 & -2 \\ -1 & 2 \end{pmatrix} \\ G_2 : K &= \begin{pmatrix} 2 & -1 \\ -3 & 2 \end{pmatrix} \end{aligned} \tag{6.34}$$

The Cartan matrix can be represented in a graphical form known, as a Dynkin diagram. In such a diagram, the l simple roots $S_1, \ldots S_l$ are represented by small circles C_j. Any two circles C_i, C_j are then joined together by n_{ij} lines, where $n_{ij} = K_{ij}K_{ji}$ (no summation). The only possibilities for n_{ij} are 0, 1, 2 and 3.

If, $n_{ij} = 0$, the roots are not connected

$n_{ij} = 1$, the roots have the same length

$n_{ij} = 2, 3$ the roots have unequal lengths.

When the lengths are unequal we put an arrow pointing towards the root of smaller length.

The Dynkin diagrams corresponding to rank-2 Lie algebra are shown below.

6.5 Representations and Weights

The representation theory of compact Lie groups is based on the approach used for $SU(2)$, with the difference, however, that in general we will have several elements H_i of the Cartan subalgebra analogous to J_3 and several step operators E_α analogous to $J_\pm$.

Since the H_i's commute they can have simultaneous eigenstates which are labelled by their eigen values.

$$H_i|\omega\rangle = \omega^i|\omega\rangle \quad (i = 1, \ldots, l) \tag{6.35}$$

The set of numbers w^i may be regarded as the components of a vector in a l-dim. vector space, called a weight vector or simply a weight and the vector space is called the weight space.

Properties of weights

(1) We have $H_i(E_\alpha|\omega\rangle) = ([H_i, E_\alpha] + E_\alpha H_i)|\omega\rangle$

$$= (\rho^i_\alpha E_\alpha + E_\alpha \omega^i)|\omega\rangle \quad \text{[using (6.12) and (6.35)]}$$

$$= (\omega^i + \rho^i_\alpha)(E_\alpha|\omega\rangle) \tag{6.36}$$

Thus $E_\alpha|\omega\rangle$ is an eigenstate of H_i with eigen value $\omega^i + \rho^i_\alpha$.

This means that if $\omega = (\omega^1 \cdots \omega^l)$ is a weight then $\omega + \rho_\alpha = (\omega^1 + \rho^1_\alpha, \ldots \omega^l + \rho^l_\alpha)$ is a weight.

In general, we have more than one step operator and therefore can shift the weight in different directions. The weights therefore lie on a lattice, the weight lattice and they are displaced by ρ_α's in the same way as roots themselves.

(2) The root lattice is a subset of weight lattice obtained by displacement from the origin. The roots together with l zero roots are in fact the weights of the adjoint representation.

(3) $$\frac{2\rho_\alpha \cdot \omega}{\rho^2_\alpha} = p = \text{integer} \tag{6.37}$$

Proof. Since $\Sigma_\alpha = \frac{1}{\rho^2_\alpha}\Sigma_i \rho^i_\alpha \cdot H_i$ forms the third member of the $SU(2)$ subalgebra, $\{E_\alpha, E_{-\alpha}, \sum_\alpha\}, 2\Sigma_\alpha$ have integer eigen values. Now

$$2\Sigma_\alpha|\omega\rangle = \frac{2}{\rho^2_\alpha}\sum_i \rho^i_\alpha \cdot H_i|\omega\rangle = \frac{2}{\rho^2_\alpha}\sum_i \rho^i_\alpha \omega_i|\omega\rangle = \left(\frac{2}{\rho^2_\alpha}\rho_\alpha \cdot \omega\right)|\omega\rangle \tag{6.38}$$

Therefore, $|\omega\rangle$ is an eigenvector of $2\Sigma_\alpha$ with eigenvalue $\frac{2}{\rho^2_\alpha}\rho_\alpha \cdot \omega$.

Hence $\frac{2}{\rho^2_\alpha}\rho_\alpha \cdot \omega = p = \text{integer}$.

Fundamental weights: Any weight can be written as an integral linear combination of what are known as the fundamental weights. The number of fundamental weights is equal to the rank of the group.

The fundamental weights (f_i) play the same role for weight space as the simple roots (S_j) do for the root space.

$$\omega = \sum_{i=1}^{l} p_i f_i \tag{6.39}$$

where $p_i s$ are integers. The fundamental weights f_j's have the following properties.

(1) f_j's form a basis in the weight space.

(2) f_j's are dual to the simple roots.

$$\frac{2s_i \cdot f_j}{s_i^2} = \delta_{ij} \quad (i, j = 1, \ldots l) \tag{6.40}$$

The fact that p_i's in (6.39) are integers follows from (6.37) and (6.40).

Highest weight: A set of weights $\{\omega_\alpha = (\omega_\alpha^1 \cdots \omega_\alpha^1)\}$ is said to form an irreducible set if the set is closed under group transformation. They form an invariant subspace of the full weight space. Each irreducible set is characterized by the existence of what is known as the highest weight ω_h which has the property

$$E_\alpha|\omega_h\rangle = 0 \quad (\alpha = 1, \ldots p) \tag{6.41}$$

The notation "highest weight" arises from eq. (6.39) where the weights are expressed as integral linear combination of the fundamental weights. The weights can therefore be ordered by their heights, the first nonvanishing coefficient p_i. Dynkin has shown that the necessary and sufficient condition for a weight to be the highest weight of an irreducible representation is that all the integers $p_i \geq 0$.

The fundamental weights (f_i) themselves are the highest weights of some IR's called the fundamental representations. The number of fundamental representations is equal to the rank of the group. They are denoted by $D(1, 0, \ldots 0)$, $D(0, 1, 0 \ldots 0)$, $D(0, \ldots 1)$. An IR is then denoted by a set of integers $D = D\ (p_1, \ldots p_l)$ with the highest weight given by

$$\omega_h = \sum_{i=1}^{l} p_i f_i \tag{6.42}$$

Fundamental roots of $SU(3)$ are

$$S_1 = \left(\frac{1}{2}, \frac{\sqrt{3}}{2}\right) \quad \text{and} \quad S_2 = \left(\frac{1}{2}, -\frac{\sqrt{3}}{2}\right).$$

From eq. (6.40) we see that the fundamental weights of $SU(3)$ are

$$f_1 = \left(\frac{1}{2}, \frac{1}{2\sqrt{3}}\right), \; f_2 = \left(\frac{1}{2}, -\frac{1}{2\sqrt{3}}\right) \tag{6.43}$$

There are two fundamental representations of the group $SU(3)$, denoted by $D(1,0)$ and $D(0,1)$.

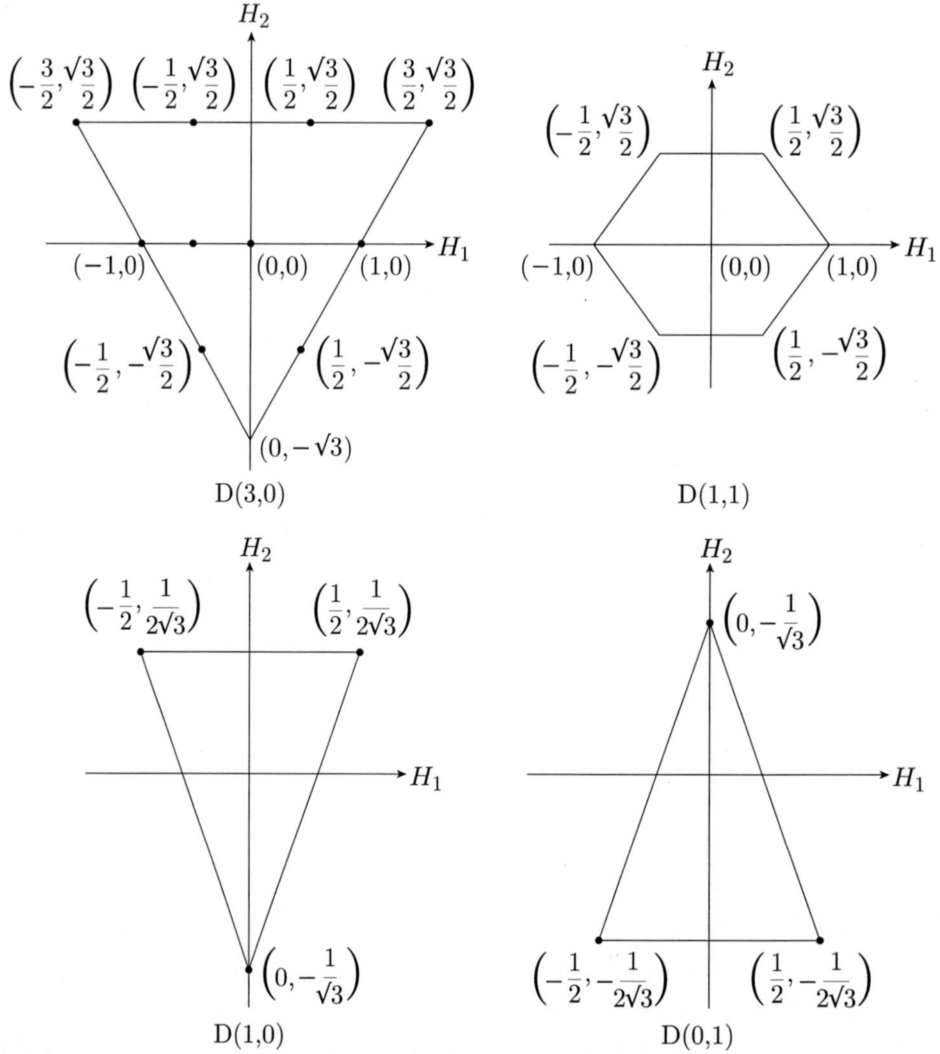

Fig. 6.3. The Weight diagrams of SU(3)

An IR of $SU(3)$ is denoted by a set of 2 positive integers $D(p,q)$ with the highest weight given by

$$\omega_h(p,q) = pf_1 + qf_2 = \left(\frac{p+q}{2}, \frac{p-q}{2\sqrt{3}}\right) \tag{6.44}$$

The adjoint representation is denoted by $D(1,1)$ with the highest weight

$$\omega_h(1,1) = f_1 + f_2 = (1,0) \tag{6.45}$$

The weight diagrams for the two fundamental representations, the adjoint representation and the decuplet $D(3,0)$ are shown in fig. 6.3.

The weight diagram of the adjoint representation of $SU(3)$ is the same as the $I_3 - Y$ plot for the $J^P = \frac{1^+}{2}$ baryon octet and $J^P = 0^-$ meson octet if we identify

$$H_1 = F_3 = I_3 \quad \text{and} \quad H_2 = F_8 = \frac{\sqrt{3}}{2}Y.$$

6.6 Eigenvalues of the Quadratic Casimir Operator of $SU(3)$

We have,

$$C_2 = F^2 = \sum_{a=1}^{8} F_a^2 = H_1^2 + H_2^2 + \frac{1}{2}\sum_{\alpha=1}^{3}(E_\alpha E_{-\alpha} + E_{-\alpha}E_\alpha)$$

$$= H_1^2 + H_2^2 + \sum_{\alpha=1}^{3}(E_{-\alpha}E_\alpha + 2H_1) \tag{6.45}$$

Using (6.45),

$$C_2|\omega_h\rangle = (H_1^2 + H_2^2 + 2H_1 + \sum_\alpha E_{-\alpha}E_\alpha)|\omega_h\rangle$$

$$= \left\{\left(\frac{p+q}{2}\right)^2 + \left(\frac{p-q}{2\sqrt{3}}\right)^2 + 2\left(\frac{p+q}{2}\right)\right\}|\omega_h\rangle.$$

[using (6.44) and (6.41)]

$$= \left\{\frac{1}{3}(p^2+q^2+pq) + (p+q)\right\}|\omega_h\rangle \tag{6.46}$$

Since C_2 has the same eigenvalue for all members of a multiplet, the eigenvalue $C_2(p,q)$ for an IR $D(p,q)$ of $SU(3)$ is given by

$$C_2(p,q) = \frac{1}{3}(p^2+q^2+pq) + (p+q).$$

$$C_2|\omega\rangle = C_2(p,q)|\omega\rangle, \omega \,\epsilon D(p,q) \tag{6.47}$$

6.7 Young Tableau

A Young Tableau (YT) is a set of boxes arranged in a number of rows such that

(i) rows are aligned on the left

(ii) rows are counted from top to bottom

(iii) If λ_i = no. of boxes in the i-th row, then $\lambda_{i+1} \leq \lambda_i$

(iv) boxes are numbered in the dictionary order

Young Tableaus were invented to provide a pictorial description of permutation symmetries of a set of n objects. Thus they provide IR's of the permutation group S_n. S_n is a discrete finite group with no. of elements $= n!$ S_2 has 2 elements, 1 and P_{12}. Instead of them, one can consider $S_{12} = \frac{1}{2}(1 + P_{12})$ and $A_{12} = \frac{1}{2}(1 - P_{12})$, the symmetrization and antisymmetrization operators. They produce symmetric and antisymmetric arrangement of 2-particle states. S_{12} is represented by boxes in the same row and A_{12} by boxes in the same column.

$S_2 =$ 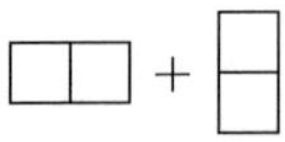 (6.48)

S_{12} A_{12}

For $n \geq 3$ more possibilities appear apart from the completely symmetric and antisymmetric arrangements. For example, for S_3 we have 3 possibilities.

$S_3 =$ 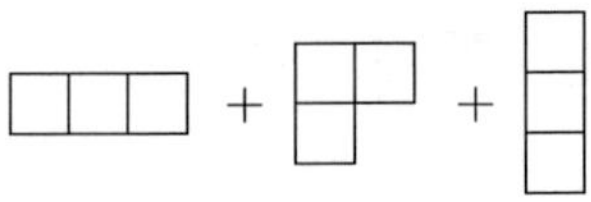 (6.49)

S M A

Symmetric, Mixed symmetry and Antisymmetric.

Weight of a Tableau

The weight ω_i of a tableau i of S_n is defined as the no. of ways in which the tableau can be obtained from tableaus of S_{n-1}. The weights have the property

$$\sum_i \omega_i^2 = n! \tag{6.50}$$

The weights of allowed tableaus can be obtained by adding boxes corresponding to particles one by one with the restrictions stated above.

$$S_2 = \boxed{1} \times \boxed{2} = \underset{S}{\begin{array}{|c|c|}\hline 1 & 2 \\ \hline \end{array}} + \underset{A}{\begin{array}{|c|}\hline 1 \\ \hline 2 \\ \hline \end{array}} \tag{6.51}$$

$$S_3 = \boxed{1} \times \boxed{2} \times \boxed{3} = \underset{S}{\begin{array}{|c|c|c|}\hline 1 & 2 & 3 \\ \hline \end{array}} + \underset{M_\alpha}{\begin{array}{|c|c|}\hline 1 & 2 \\ \hline 3 & \\ \cline{1-1} \end{array}} + \underset{M_\beta}{\begin{array}{|c|c|}\hline 1 & 3 \\ \hline 2 & \\ \cline{1-1} \end{array}} + \underset{A}{\begin{array}{|c|}\hline 1 \\ \hline 2 \\ \hline 3 \\ \hline \end{array}} \tag{6.52}$$

M_α = symmetric in 1, 2 but no definite symmetry w.r.t. other pairs.

M_β = antisymmetric in 1, 2 but no definite symmetry w.r.t. other pairs.

$$S_2 = \boxed{1} \times \boxed{2} = \underset{S}{\begin{array}{|c|c|}\hline 1 & 2 \\ \hline \end{array}} + \underset{A}{\begin{array}{|c|}\hline 1 \\ \hline 2 \\ \hline \end{array}} \tag{6.53}$$

$$S_3 = \boxed{1} \times \boxed{2} \times \boxed{3} = \underset{S}{\begin{array}{|c|c|c|}\hline 1 & 2 & 3 \\ \hline \end{array}} + \underset{M_\alpha}{\begin{array}{|c|c|}\hline 1 & 2 \\ \hline 3 & \\ \cline{1-1} \end{array}} + \underset{M_\beta}{\begin{array}{|c|c|}\hline 1 & 3 \\ \hline 2 & \\ \cline{1-1} \end{array}} + \underset{A}{\begin{array}{|c|}\hline 1 \\ \hline 2 \\ \hline 3 \\ \hline \end{array}} \tag{6.54}$$

$$\begin{aligned} S_3 &: \sum_i \omega_i^2 = 1^2 + 2^2 + 1^2 = 6 = 3! \\ S_4 &: \sum_i \omega_i^2 = 1^2 + 3^2 + 2^2 + 3^2 + 1^2 = 24 = 4! \end{aligned} \tag{6.55}$$

6.8 Young Tableaus and Irreducible Representations of $SU(n)$

YT's provide a powerful tool for handling the IR's of $SU(n)$, particularly for $n \geq 3$. This arises from the fact that any IR if $SU(n)$ can be generated from the C.G series decomposition of the tensorial or direct product of appropriate number of its fundamental representations. The elements in the representation space of the tensors thus obtained are tensors satisfying the symmetry properties under the permutation group S_n. A YT can correspond to an IR of $SU(n)$ provided the no. of rows does not exceed n. This is because in a YT the boxes in a column must be labelled by different indices and for $SU(n)$ the no. of available indices is n. Rules have been developed for giving the dimensionality of the IR of $SU(n)$ associated with a given YT and for deriving the Clebsch Gordon series obtained from the direct product of two such IR's.

To an IR $D(p_1 \cdots p_l)$ of a simple Lie group of rank l there corresponds a Young tableau $Y(\lambda_1, \ldots \lambda_l)$ where p_i's and λ_i's are integers with

$$p_i = \lambda_i - \lambda_{i+1},\ i < l$$
$$\text{and}\quad p_l = \lambda_l. \tag{6.56}$$

For the group $SU(n), l = n-1$. Conversely, a Young Tableau $Y(\lambda_1 \cdots \lambda_k)$ will correspond to an IR of $SU(n)$ provided $k \leq n$. A column with n boxes corresponds to a singlet representation of $SU(n)$ and therefore may be deleted. Thus the maximum no. of rows in a Young Tableau if it is to correspond to an IR of $SU(n)$ is $n - 1$.

$SU(2)$: Since $l = n - 1 = 2 - 1 = 1$, an IR of $SU(2)$ is characterized by a single integer p, $D = D(p)$ and the corresponding YT consists of a single row with p boxes.

$$D(p) : Y(p) = \square\square\square\square \;\ldots\ldots\; \square \tag{6.57}$$

For $SU(2)$ the available indices are 2 which may be denoted by a, b (corresponding to 2 possible states in the fundamental representation). The dimension of an IR of $SU(2)$ is then given by the no. ways. N in which one can fill p boxes with 2 indices a, b such that b does not preceed a.

$$N = N(p) = p + 1 = 2j + 1 \tag{6.58}$$
$$p = 2j = \text{integer.}$$

Direct Product of Representations. CG series

To obtain the CG series corresponding to the direct product of 2 IR's of a Lie group G one must first draw the tableaus Y_1 and Y_2 corresponding to the 2 IR's. Then add boxes from Y_2 to Y_1 in all possible ways such that one obtains only allowed YT's. For allowed Tableaus, (1) boxes are aligned on the left, (2) $\lambda_{i+1} \leq \lambda_i$ where λ_i = no. of boxes in the ith row, (3) boxes that were in the same row in Y_2 must not be put in the same column of the newly constructed tableaus. Similarly boxes in the same column must not be put in the same row of the newly constructed tableaus. The rules give for $SU(2)$,

(a) $\square \times \square = \square\square + \begin{array}{c}\square\\[-4pt]\square\end{array}$ (with a slash through the column) (6.59)

$$\text{D}(1) \times \text{D}(1) = \text{D}(2) \times \text{D}(0)$$
$$\text{or}\quad 2 \times 2 = 3 + 1$$

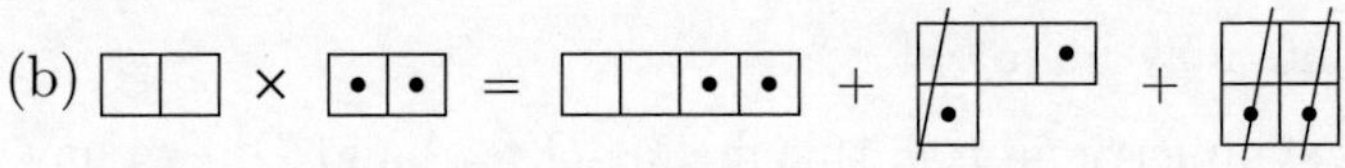

(6.60)

$$D(2) \times D(2) = D(4) + D(2) + D(0)$$

$$3 \times 3 = 5 + 3 + 1$$

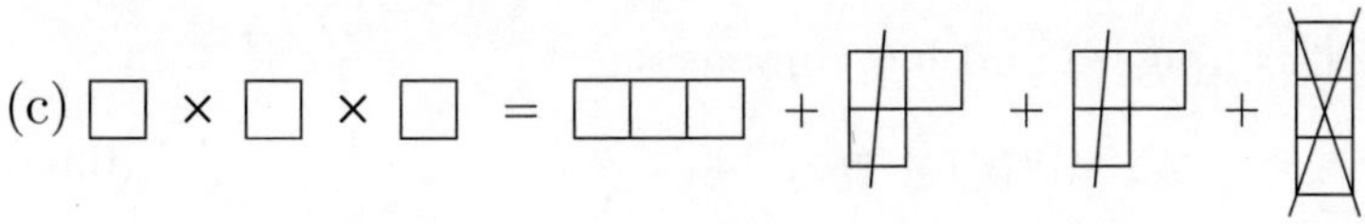

(6.61)

$$2 \times 2 \times 2 = 4 + 2 + 2$$

In general,

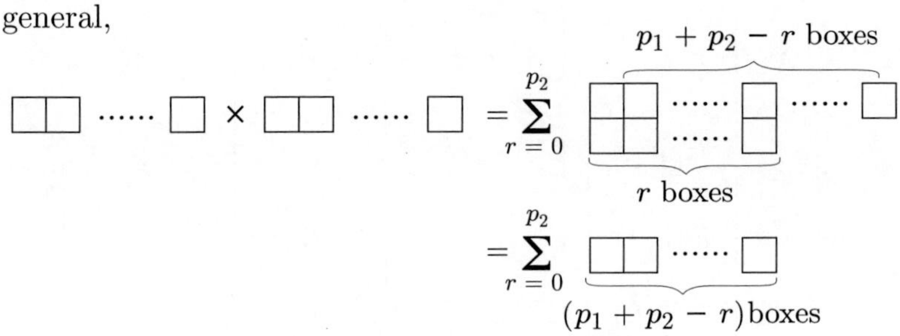

$$\text{Thus } D(p_1) \times D(p_2) = \sum_{r=0}^{p_2} D(p_1 + p_2 - 2r) = \sum_{p=p_1-p_2}^{p_1+p_2} D(p) \tag{6.62}$$

Writing, $p_1 = 2j_1$, $p_2 = 2j_2$, $p = 2j$ this is just the coupling rule

$\vec{J}_1 + \vec{J}_2 = \vec{J}, |j_1 - j_2| \leq j \leq j_1 + j_2$

***SU*(3):** Since $l = 2$, each IR of $SU(3)$ is labelled by 2 integers $D = D(p, q)$

$$D(p, q) \leftrightarrow Y(\lambda_1, \lambda_2)$$

$$= Y(p + q, q)$$

$$q = \lambda_2$$

$$p = \lambda_1 - \lambda_2$$

$$\text{or,} \quad \lambda_1 = p_1 + q \tag{6.63}$$

For $SU(3)$ the no of available indices is $n = 3$, say a, b, c. The dimensionality of the irreducible representation $D(p, q)$ of $SU(3)$ is given by the no. of ways in which the $p + q$ boxes can be filled with a, b, c such that in a row b must not proceed a and c must not proceed b and in a column the letters can not be same and also b cannot be above a and c cannot be above b. This gives,

$$N(p, q) = \frac{1}{2}(p + 1)(q + 1)(p + q + 2) \tag{6.64}$$

The following points may be noted:

(1) Since $N(p,q)$ is symmetric in p, q, this means $D(p,q)$ and $D(q,p)$ have the same dimension. One is said to be conjugate to the other. For $p > q$, $D(p,q)$ is denoted by N and $D(q,p)$ by N^*.

(2) $D(p,q)$ is self-conjugate and has dimension

$$N(p) = (p+1)^3 \tag{6.65}$$

$D(p,q)$	$Y(p+q,q)$	$N(p,q)$	Notation
$D(0,0)$		1	1
$D(1,0)$		3	3
$D(0,1)$	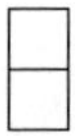	3	3^*
$D(2,0)$	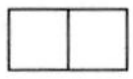	6	6
$D(0,2)$	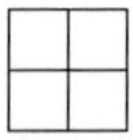	6	6^*
$D(1,1)$	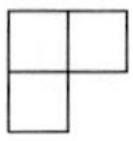	8	8
$D(2,1)$	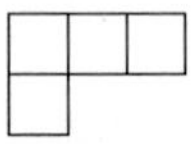	15	15
$D(1,2)$	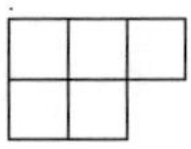	15	15^*
$D(3,0)$	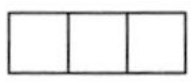	10	10
$D(0,3)$	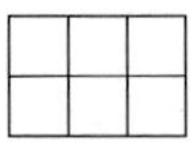	10	10^*
$D(2,2)$	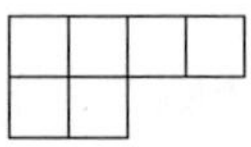	27	27

Direct product of $SU(3)$ representations, CG series

(a) 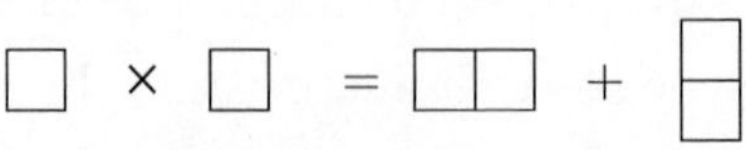

D(1,0) × D(1,0) = D(2,0) + D(0,2)

$$3 \times 3 = 6 + 3^* \tag{6.66}$$

(b) 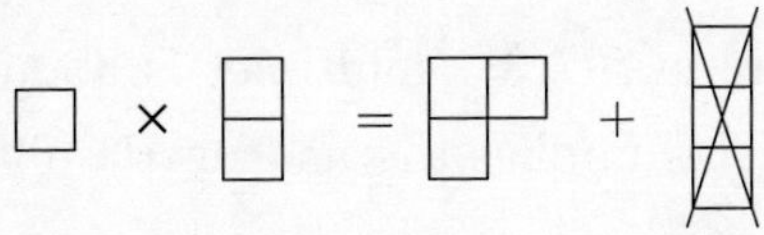

$$D(1,0) \times D(0,1) = D(1,1) + D(0,0)$$

$$3 \times 3^* = 8 + 1 \tag{6.67}$$

(c) 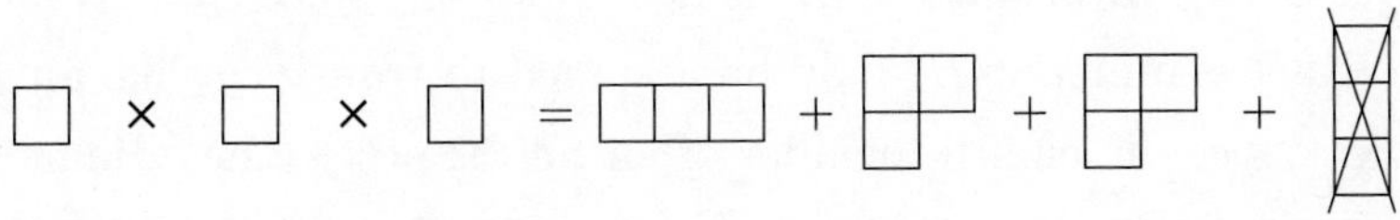

$$D(1,0) \times D(1,0) \times D(1,0) = D(3,0) + D(1,1) + D(1,1) + D(0,0)$$

$$3 \times 3 \times 3 = 10 + 8 + 8 + 1 \tag{6.68}$$

(d) 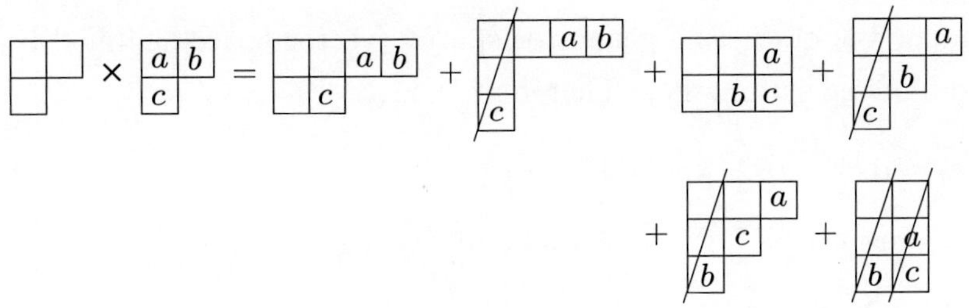

$$D(1,1)\times D(1,1) = D(2,2)+D(3,0)+D(0,3)+D(1,1)+D(1,1)+D(0,0)$$

$$8 \times 8 = 27 + 10 + 10^* + 8 + 8 + 1 \tag{6.69}$$

(e) $3 \times 3 \times 3^* =$ 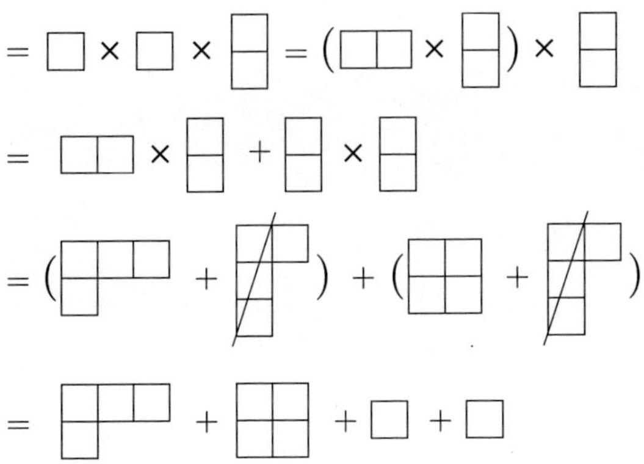

$$= 15 + 6^* + 3 + 3 \tag{6.70}$$

6.9 Hadron Mass Splitting. Broken $SU(3)_f$

The flavor $SU(3)$ symmetry, $SU(3)_f$, although a good symmetry for light hadrons, is not an exact symmetry. This is because hadrons belonging to an $SU(3)$ multiplet are not degenerate in mass, the mass splitting between neighbouring $SU(2)_I$ multiplets within an $SU(3)_f$ multiplet is about 10% of the average mass of the multiplet. This implies the hadron Hamiltonian is of the form

$$H = H_o + H' \tag{6.71}$$

where H_o is $SU(3)$ invariant but H' is not. However, since $SU(2)_{\overline{I}} \times U(1)_Y$ is a more exact symmetry, H' may be assumed to transform like an $I = 0$, $Y = 0$ object, e.g. the eighth member of an $SU(3)$ octet (the adjoint representation). This is similar to the case for an atom placed in a magnetic field when full rotational symmetry is broken but axial symmetry is preserved. Thus H' in (6.71) transforms like V_8. Now for $SU(3)$ there are 2 linearly independent vectors in the generator space. (vector space spanned by the generators of the symmetry). Therefore

$$V_a = \alpha F_a + \beta D_a \quad (a = 1, \ldots 8) \tag{6.72}$$

where F_a's are the $SU(3)$ generators and D_a is given by

$$D_a = \sum_{b,c} d_{abc} F_b F_c \tag{6.73}$$

Note that one can write

$$F_a = -\frac{2i}{3} \sum_{b,c} f_{abc} F_b F_c$$

Thus one can write

$$H' = \alpha F_8 + \beta D_8 \tag{6.74}$$

$$D_8 = \frac{1}{\sqrt{3}}(F_1^2 +_2^2 + F_3^2 - F_8^2) - \frac{1}{2\sqrt{3}}(F_4^2 + F_5^2 + F_6^2 + F_7^2)$$

$$(d_{8bc} = 0, \ b \neq c)$$

$$= \frac{1}{2\sqrt{3}}[3(F_1^2 + F_2^2 + F_3^2) - F_8^2] - \frac{1}{2\sqrt{3}} \sum_{a=1}^{8} F_a^2$$

$$= \frac{\sqrt{3}}{2}\left[I^2 - \frac{1}{4}Y^2\right] - \frac{1}{2\sqrt{3}} C_2 \tag{6.75}$$

where $I^2 = F_1^2 + F_2^2 + F_3^2$, $Y = \frac{2}{\sqrt{3}} F_8$

and $\quad C_2 = \sum_{a=1}^{8} F_a^2 =$ Quadratic Casimir operator for $SU(3)$

From (6.71) and (6.74)

$$m = \langle H_o \rangle + \alpha \langle F_8 \rangle + \beta \langle D_8 \rangle \tag{6.76}$$

we get from (6.75) using (6.76)

$$m = m_o + m_1 Y + m_2 \left[I(I+1) - \frac{1}{4} Y^2 \right] \tag{6.77}$$

Eq. (6.77) is known as the Gell-Mann-Okubo mass formula. Eq. (6.77) simplifies for baryon decuplet for which $I = \frac{Y}{2} + 1$. Eq. (6.77) then becomes

$$m = m_0' + m_1' Y \tag{6.78}$$

$\boldsymbol{J^P = \frac{1^+}{2}}$ **octet baryons:** using (6.77) one gets the sum rule

$$\frac{1}{2}(m_N + m_\Xi) = \frac{1}{4}(m_\Sigma + 3m_\Lambda) \tag{6.79}$$

$\boldsymbol{J^P = \frac{3^+}{2}}$ **decuplet baryons:** (6.78) gives the equal mass spacing rule

$$m(\Sigma^*) - m(\Delta) = m(\Xi) - m(\Sigma^*) = m(\Omega^-) - m(\Xi^*) \tag{6.80}$$

$\boldsymbol{J^P = 0^-}$ **octet mesons:** one gets from (6.77),with m replaced by m^2,

$$m_K^2 = \frac{1}{4}(m_\pi^2 + 3m_\eta^2) \tag{6.81}$$

Problems

1. Suppose F_a's are $N \times N$ matrices satisfying

$$[F_a, F_b] = \mathrm{i}f_{abc} F_c$$

and a_i^+, a_i are creation and annihilation operators, respectively, satisfying $(i = 1, \ldots, N)$

$$[a_j, a_j^+] = \delta_{ij}$$

$$[a_i, a_j] = [a_i^+, a_j^+] = 0$$

Show that the operators $X_a = \sum_{ij} a_i^+ (F_a)_{ij} a_j$ satisfy $[X_a, X_b] = \mathrm{i}f_{abc} X_c$

Write down the explicit forms for X_a for $F_a = \dfrac{\tau_a}{2}$, 2×2 Pauli matrices, and also for $F_a = \dfrac{\lambda_a}{2}$, 3×3 Gell-Mann matrices.

2. Consider the simple Lie algebra generated by the ten 4×4 matrices $\sigma_a, \sigma_a\tau_1, \sigma_a\tau_3$ and $\tau_2 (a = 1, 2, 3)$, where σ_a and τ_a are 2×2 Pauli matrices in orthogonal spaces. Take $H_1 = \sigma_3$ and $H_2 = \sigma_3\tau_3$ as the Cartan subalgebra. Find

   (a) the roots of the Lie algebra, and

   (b) the weights of the 4-dimensional representation generated by these matrices.

3. Find the simple roots, fundamental weights and the Dynkin diagram for the Lie algebra discussed in problem 6.2 above.

4. Find $10 \times 8, 10 \times 10$ and $10 \times \overline{10}$ for $SU(3)$ using the Young Tableau method.

5. Use the tensor method to prove that in $SU(3)$

   $8 \times 8 = 27 + 10 + 10^* + 8 + 8 + 1$

6. Calculate the dimensionality of the following Young Tableaus

   (a) 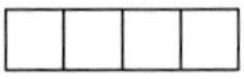 and 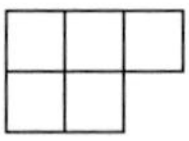 in SU(3)

   (b) 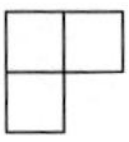 and 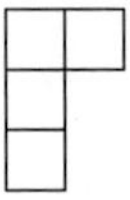 in SU(4)

   (c) 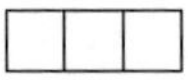 and 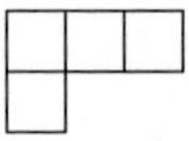 in SU(5)

7. Find the Clebsch-Gordon series for

   (a) $5 \times 5^*$ in $SU(5)$ (b) $6 \times 6 \times 6$ in $SU(6)$

   using Young Tableau

Chapter 7

The Quark Model

7.1 Motivation

The Quark model for hadrons was independently proposed by Gell-Mann and Zweig in 1964 in order to explain (1) the multiplet structure of light hadrons and (2) compositeness nature of hadrons. It turned out to be one of the three postulates on which the presently accepted standard model of particle physics is based.

Multiplet structure of hadrons: Although $SU(3)_f$ is a good symmetry for light hadrons, only a few of the irreducible representations of $SU(3)$ are realised as hadron multiplets. In particular, baryons appear in multiplets of 1, 8 and 10 of $SU(3)$ and mesons appear in multiplets of 1 and 8 only. Moreover, there are no hadrons that belong to the fundamental representations 3 and 3^* of $SU(3)$.

Compositeness of hadrons: Experimental data on magnetic moments and rms charge radii of hadrons seem to indicate that hadrons are extended and hence composite objects like atoms and not elementary and point-like objects like the electron. For example,

$$\mu_p = 2.79\ \mu_N (\neq 1\ \mu_N) \quad \text{and} \quad \mu_n = -1.91\ \mu_N (\neq 0)$$

and $\langle r_p^2 \rangle^{1/2} = 0.8 \times 10^{-13}\ \text{cm} (\neq 0)$

Furthermore, with so many hadrons one would like to think that they are made of more fundamental objects which are few in number. These constituents of hadrons were named "quarks" by Gell Mann. At the time they were first proposed one was not sure whether they were real or mere mathematical (hypothetical) objects.

7.2 Postulates

The quark model is based on the following postulates:

(1) There exists a triplet of objects $q(u, d, s)$ called quarks which belong to the fundamental representation 3 of $SU(3)_f$. The antiparticles of

quarks, the antiquarks $\bar{q}(\bar{u}, \bar{d}, \bar{s})$ then belong to the conjugate representation 3^* of $SU(3)_f$.

(2) Baryons are bound states of 3 quarks, while the mesons are bound states of one quark and one antiquark.

$B = qqq$, $M = q, \bar{q}$

The multiplet structure of hadrons then follow immediately from the rules for CG series decomposition of direct products of IR's of $SU(3)$.

(3) $SU(3)_f$ is a broken symmetry. However, $SU(2)_I$ symmetry is exact. This is implemented at the quark level by the mass relation, $m_u = m_d \neq m_s$.

(4) Quarks are spin $-\frac{1}{2}$ particles. This ensures that baryons have half-integral spin, whereas mesons have integral spins.

(5) Quarks have Baryon number $B = \frac{1}{3}$ (so that baryons have $B = {}_{+}1$). This implies that quarks have fractional electric charges.

(6) In addition to spin $SU(3)_f$ and quantum numbers (flavor quantum numbers) quarks carry another quantum number called "color". With respect to color, the quarks belong to the fundamental representation $\underline{3}$ of another $SU(3)$ called $SU(3)_c$. The 3 color states of quarks are named R, B and G. The antiquark color states are then named $\overline{R}$, $\overline{B}$ and $\overline{G}$.

7.3 Flavor Quantum Numbers of Quarks and Antiquarks

The flavor quantum numbers of quarks can be read off the Gell Mann matrices $\left(I_3 = \frac{1}{2}\lambda_3,\ Y = \frac{1}{\sqrt{3}}\lambda_8\right)$. Flavor quantum numbers of the antiquarks are negative of those of quarks.

Quark/antiquark	I_3	S	$Y(= B + S)$	$Q\left(= I_3 + \frac{Y}{2}\right)$
u	$+\dfrac{1}{2}$	0	$+\dfrac{1}{3}$	$+\dfrac{2}{3}$
d	$-\dfrac{1}{2}$	0	$+\dfrac{1}{3}$	$-\dfrac{1}{3}$

Quark/antiquark	I_3	S	$Y(=B+S)$	$Q\left(=I_3+\frac{Y}{2}\right)$
s	0	-1	$-\frac{2}{3}$	$-\frac{1}{3}$
$\overline{u}$	$-\frac{1}{2}$	0	$-\frac{1}{3}$	$-\frac{2}{3}$
$\overline{d}$	$+\frac{1}{2}$	0	$-\frac{1}{3}$	$+\frac{1}{3}$
$\overline{s}$	0	$+1$	$+\frac{2}{3}$	$+\frac{1}{3}$

7.4 Quark Structure (Flavor wave function) of Baryons

$$B = qqq \tag{7.1}$$

Rules for CG series decomposition of direct products of IR's give

$3 \times 3 \times 3 =$ 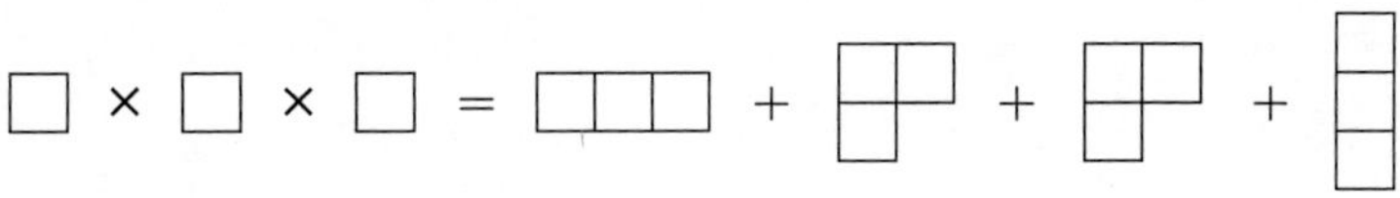

$$= 10_s + 8_\alpha + 8_\beta + 1_A \tag{7.2}$$

Decuplet Baryons, 10_S:

$$I = \frac{3}{2},\ Y = +1,\ \Delta^+ = uuu, = (uud)_S,\ \Delta^0 = (ddu)_S,\ \Delta^- = ddd$$

$$I = 1,\ Y = +0, \Sigma^{*+} = (uud)_S,\ \Sigma^{*0} = (uds)_S,\ \Sigma^{*-} = (dds)_S$$

$$I = \frac{1}{2},\ Y = -1, \Xi^{*0} = (ssu)_S, \Xi^{*-} = (ssd)_S,$$

$$I = 0,\ Y = -2, \Omega^- = sss \tag{7.3}$$

where, $(aab)_S = \dfrac{1}{\sqrt{3}}(aab + aba + baa)$

$$(abc)_S = \frac{1}{\sqrt{6}}\{(ab+ba)c + (bc+cb)a + (ca+ac)b\} \tag{7.4}$$

Octet Baryons, $8_\alpha, 8_\beta$:

$$I = \frac{1}{2},\ Y = 1 : p = (uud)_{\alpha,\beta}, n = (ddu)_{\alpha,\beta}$$

$$I = 1,\ Y = 0 : \Sigma^+ = (uus)_{\alpha,\beta}, \Sigma^0 = (\{ud\}s)_{\alpha,\beta}\Sigma^- = (dds)_{\alpha,\beta}$$

$$I = 1, Y = 0 : \wedge = ([ud]s)_{\alpha\beta}$$

$$I = \frac{1}{2}, Y = -1 : \Xi^0 = (ssu)_{\alpha,\beta}, \Xi^0 = (ssd)_{\alpha,\beta} \tag{7.5}$$

where $(aab)_\alpha = \frac{1}{\sqrt{6}}\{(ab+ba)a - 2aab\}$

$$(aab)_\beta = \frac{1}{\sqrt{2}}(ab - ba)a$$

$$\{ab\} = \frac{1}{\sqrt{2}}(ab + ba), [ab] = \frac{1}{\sqrt{2}}(ab - ba) \tag{7.6}$$

α denotes symmetric in first 2 particles, β denotes antisymmetric in first 2 particles.

Singlet Baryon, $\mathbf{1}_A$:

$$I = 0, \ Y = 0 : \wedge'(uds)_A \tag{7.7}$$

where $(abc)_A = \frac{1}{\sqrt{6}}\{(ab - ba)c + (bc - cb)a + (ca - ac)b\}$ (7.8)

7.5 Quark Structure (Flavor wave function) of Mesons

$$M = q\,\overline{q} \tag{7.9}$$

Rules for direct product reduction give for the $SU(3)$ case

$$3 \times 3^* = 8 + 1 \tag{7.10}$$

Let $\phi_a = \overline{q}\frac{\lambda_a}{\sqrt{2}}q, \ (a = 1, \ldots 8)$ (7.11)

$$\phi_0 = q^+\frac{\lambda_0}{\sqrt{2}}q$$

where $q = (uds)$, $\overline{q}(\overline{u}, \overline{d}, \overline{s})$, $\lambda_a's$ $(a = 1, \ldots 8)$ are 3×3 Gell-Mann matrices

and $\lambda_o = \sqrt{\frac{2}{3}} \cdot 1$ (7.12)

It is seen that $\phi_a \sim 8$ and $\phi_0 \sim 1$ of $SU(3)$

Octet Mesons, 8:

$$I = 1, \ Y = 0 : \pi^+ = u\overline{d} = \frac{1}{\sqrt{2}}(\phi_1 - i\phi_2)$$

$$\pi^0 = \frac{1}{\sqrt{2}}(u\overline{u} - d\overline{d}) = \phi_3$$

$$\pi^- = d\overline{u} = \frac{1}{\sqrt{2}}(\phi_1 + i\phi_2)$$

$$I = \frac{1}{2},\ Y = 1, K^+ = u\overline{s} = \frac{1}{\sqrt{2}}(\phi_4 - i\phi_5)$$

$$K^0 = d\overline{s} = \frac{1}{\sqrt{2}}(\phi_6 - i\phi_7)$$

$$I = \frac{1}{2},\ Y = -1,\ \overline{K}^0 = s\overline{d} = \frac{1}{\sqrt{2}}(\phi_6 + i\phi_7)$$

$$I = \frac{1}{2},\ Y = -1,\ K^- = s\overline{u} = \frac{1}{\sqrt{2}}(\phi_4 + i\phi_5)$$

$I = 0, Y = 0.$

$$\eta(550) = \frac{1}{\sqrt{6}}(u\overline{u} + d\overline{d} - 2s\overline{s}) = \phi_8$$

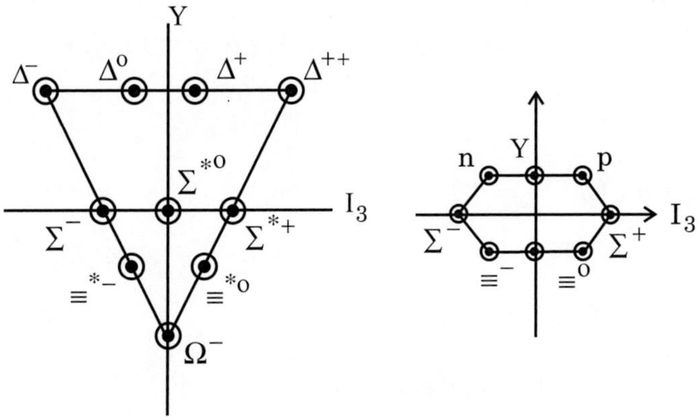

Fig. 7.1

Singlet meson, *I*:

$$\eta'(958) = \frac{1}{\sqrt{3}}(u\overline{u}) + d\overline{d} + s\overline{s}) = \phi_0 \tag{7.13}$$

7.6 Spin-parity of Mesons

Since, $M = q\overline{q}$

$$\overrightarrow{S} = \overrightarrow{S}_1 + \overrightarrow{S}_2, S_1 = S_2 = \frac{1}{2} \Rightarrow S = 0, 1 \tag{7.14}$$

$$S = 1 : |1,1\rangle = \uparrow\uparrow, |1,0\rangle = \frac{1}{\sqrt{2}}(\uparrow\downarrow + \downarrow\uparrow), |1,-1\rangle = \downarrow\downarrow$$

$$S = 0 : |0,0\rangle = \frac{1}{\sqrt{2}}(\uparrow\downarrow - \downarrow\uparrow)$$

$$\vec{J} = \vec{L} + \vec{S}, |L - S| \le J \le L + S \tag{7.15}$$

$\vec{L}$ = orbital angular momentum for relative motion of q and $\overline{q}$

$$L = 0,\ 1, 2 \ldots$$

$$P = (-)^{L+1}, C = (-)^{L+S} \tag{7.16}$$

$L = 0, J^{PC} = 0^{-+}, 1^{--}$ correspond to ground state mesons

$L = 1,\ J^{PC} = 0^{++},\ 1^{++},\ 2^{++};\ 1^{+-}$ correspond to first orbitally excited mesons.

The quark model predicts that there cannot be mesons with

$$J^{PC} = 0^{--}, (2n)^{+-}, (2n+1)^{-+}$$

Such mesons (exotic mesons) have so far not been seen.

Pseudoscalar mesons ($J^{PC} = 0^{-+}$): There are 9 lightest pseudoscalar mesons whose $SU(3)_f$ contents and flavor quantum numbers are given by (7.13)

Vector Mesons ($J^{PC} = 1^{--}$): There are 9 lightest vector mesons

$\rho(773)$;	$K^*(892)$,	$\overline{K^*}(892)$,	$\omega(783)$,	$\phi(1020)$
$I = 1, Y = 0$	$I = \frac{1}{2}, Y = 1,$	$I = \frac{1}{2}, Y = -1,$	$I = Y = 0,$	$I = Y = 0$

Unlike in the pseudoscalar case, the two $I = Y = 0$ vector mesons are not well separated in mass and therefore they cannot be assigned, to definite $SU(3)_f$ multiplets. Moreover, near equality of mass indicates that the ω meson has the same quark content as the ρ^0 meson. These facts can be understood if it is assumed that there is mixing between the ϕ_8 and ϕ_0 due to the fact that $SU(3)_f$ symmetry is not exact but "broken". Thus ω and ϕ are both mixtures of ϕ_8 and ϕ_0.

$\omega - \phi$ mixing:

$$\begin{pmatrix} \omega \\ \phi \end{pmatrix} = \begin{pmatrix} \cos\theta & \sin\theta \\ -\sin\theta & \cos\theta \end{pmatrix} \begin{pmatrix} \phi_0 \\ \phi_8 \end{pmatrix} \tag{7.17}$$

where θ is the mixing angle. Furthermore, if we assume there is "ideal" mixing, $\tan\theta = 1\sqrt{2}$ then the ω-meson is a mixture of $u\overline{u}$ and $d\overline{d}$, while the ϕ-meson is a pure $s\overline{s}$ state.

$$\omega = \frac{1}{\sqrt{2}}(u\bar{u} + d\bar{d}), \phi = -s\bar{s} \tag{7.18}$$

For ideal mixing, Gell-Mann-Okubo mass formula gives for vector mesons

$$m_{K^*}^2 = \frac{1}{4}(m_\rho^2 + 3m_8^2)$$

$$= \frac{1}{4}(m_\rho^2 + m_\omega^2 + 2m_\phi^2), \ (m_8^2 = m_\omega^2 \sin^2\theta + m_\phi^2 \cos^2\theta)$$

$$\text{or} \quad m_{K^*}^2 = \frac{1}{2}(m_\phi^2 + m_\rho^2), m_\omega \approx m_\rho \tag{7.19}$$

7.7 Spin-Parity of Baryons

$B = qqq$

$$\vec{S} = \vec{S_1} + \vec{S_2} + \vec{S_3}, S_1 = S_2 = S_3 = \frac{1}{2} \ \Rightarrow \ S = \frac{3}{2}, \frac{1}{2}, \frac{1}{2} \tag{7.20}$$

$$S = \frac{3}{2}: \ \uparrow\uparrow\uparrow, \frac{1}{\sqrt{3}}(\uparrow\uparrow\downarrow + \uparrow\downarrow\uparrow + \downarrow\uparrow\uparrow), \frac{1}{\sqrt{3}}(\downarrow\downarrow\uparrow + \downarrow\uparrow\downarrow + \uparrow\downarrow\downarrow), \downarrow\downarrow\downarrow$$

$$S = \left(\frac{1}{2}\right)_\alpha : \frac{1}{\sqrt{6}}\{(\uparrow\downarrow + \downarrow\uparrow)\uparrow -2\uparrow\uparrow\downarrow,\}\frac{1}{\sqrt{6}}\{(\uparrow\downarrow + \uparrow\downarrow)\downarrow -2\downarrow\downarrow\uparrow\}$$

$$S = \left(\frac{1}{2}\right)_\beta : \frac{1}{\sqrt{2}}(\uparrow\downarrow - \downarrow\uparrow)\uparrow, \frac{1}{\sqrt{2}}(\uparrow\downarrow - \downarrow\uparrow)\downarrow. \tag{7.21}$$

$\vec{L} = \vec{l_1} + \vec{l_2}$

$\vec{l_1}$ = orbital angular momentum for relative motion between quarks 1, 2; $\vec{l}_2$ = orbital angular momentum for relative motion between the CM of quarks 1 and 2 and quark 3.

$$\vec{J} = \vec{L} + \vec{S}, P = (-)^{l_1 + l_2} \tag{7.22}$$

The lightest (g.s.) baryons correspond to

$l_1 = l_2 = L = 0, \ J^P = \frac{1}{2}^+, \frac{3}{2}^+$

the first orbitally excited baryons correspond to

$l_1 = 1, l_2 = 0$ or $l_1 = 0, l_2 = 1$

and have $J^P = \frac{1}{2}^-, \frac{3}{2}^-, \frac{5}{2}^-$. Such negative parity baryons have been seen.

7.8 Spin-flavor Wave Functions of Baryons

In the original formulation of the quark model, quarks were assumed to have 3 kinds of degrees of freedom-space, spin and flavor. Thus the total wave function of a system of quarks will be a product of these 3 kinds of wave functions.

$$\psi = \psi_{space} \cdot \psi_{spin} \cdot \psi_{flavor} \tag{7.23}$$

For ground state (g.s.) baryons $L = 0$ and ψ_{space} is symmetric. Also for the lowest-lying decuplet baryons $\underline{10}$, $J^P = \frac{3}{2}^+$, both ψ_{space} and ψ_{flavor} are symmetric and so is their product. For g-s. octet baryons, 8, $J^P = \frac{1}{2}^+$ we have 2 choices for each of ψ_{spin} and ψ_{flavor} and hence $\psi_{spin-flavor}$ will, in general, be a mixture of the 4 possibilities.

$$\psi_{spin-flavor} = \sum_{i,j=\alpha,\beta} C_{ij}|8\rangle_i \left|\frac{1}{2}\right\rangle_j \tag{7.24}$$

However, if we assume that $\psi_{spin-flavor}$ is symmetric under exchange of quarks, then

$$C_{\alpha\alpha} = C_{\beta\beta} = \frac{1}{\sqrt{2}} \quad \text{and} \quad c_{\alpha\beta} = C_{\beta\alpha} = 0$$

This gives

$$\psi_{spin-flavor} = \frac{1}{\sqrt{2}}\left[|8\rangle_\alpha \left|\frac{1}{2}\right\rangle_\alpha + |8\rangle_\beta \left|\frac{1}{2}\right\rangle_\beta\right] \tag{7.25}$$

This for a spin-up proton

$$\left|p; S_z = \frac{1}{2}\right\rangle = \frac{1}{\sqrt{2}}\left[|uud\rangle_\alpha| \uparrow\uparrow\downarrow>_\alpha + |uud\rangle_\beta| \uparrow\uparrow\downarrow\rangle_\beta\right]$$

$$= \frac{1}{\sqrt{2}}\frac{1}{\sqrt{6}}\{|udu\rangle + |duu\rangle - 2|uud\rangle\} \times \frac{1}{\sqrt{6}}\{\uparrow\downarrow\uparrow> + | \downarrow\uparrow\uparrow\rangle - 2| \uparrow\uparrow\downarrow\rangle\}\Big]$$

$$+\frac{1}{\sqrt{2}}\{|udu\rangle - |duu\rangle\} \cdot \frac{1}{\sqrt{2}}\{| \uparrow\downarrow\uparrow\rangle - | \downarrow\uparrow\uparrow\rangle\}\Big]$$

$$= \frac{1}{3\sqrt{2}}[2|u \uparrow u \uparrow |\rangle - |u \uparrow u \downarrow d \uparrow\rangle - |u \downarrow u \uparrow d \uparrow\rangle + (2 \leftrightarrow 3) + (1 \leftrightarrow 3)] \tag{7.26}$$

7.9 Baryon Magnetic Moments

The (total) magnetic moment operator is a sum of an orbital part and a spin part

$$\vec{\mu}_{op} = \vec{\mu}^{\,l}_{op} + \vec{\mu}^{\,s}_{op} \tag{7.27}$$

For g.s. baryons $(L = 0)\vec{\mu}^{\,l}_{op}$ does not contribute and therefore

$$\vec{\mu}_{op} = \vec{\mu}^{\,s}_{op} = \sum_i \mu_i \vec{\sigma}_i, \;\; \mu_i = \mu_q = \frac{ee_q}{2m_q} \tag{7.28}$$

The magnetic moment of a baryon B is then given by

$$\mu_B = \langle B;\; S_Z = S|(\mu_{op})_z|B;\; S_z = S\rangle$$

Thus,
$$\mu_p = \left\langle p;\; S_z = +\frac{1}{2}|(\mu_{op})_z|p;\; S_z = +\frac{1}{2}\right\rangle$$

$$= \frac{1}{18} \times 3[4(2\mu_u - \mu_d) + 2(\mu_u - \mu_u + \mu_d)]$$

$$= \frac{1}{18} \cdot 3.2(4\mu_u - \mu_d) = \frac{1}{3}(4\mu_u - \mu_d) \tag{7.29}$$

The magnetic moment for the neutron is obtained from (7.29) by interchanging u and d quarks. Therefore

$$\mu_n = \frac{1}{3}(4\mu_d - \mu_u) \tag{7.30}$$

In the quark model with $m_u = m_d$, one has from (7.28)

$$\mu_u = -2\mu_d \tag{7.31}$$

This gives
$$\frac{\mu_p}{\mu_n} = \frac{4\mu_u - \mu_d}{4\mu_d - \mu_u} = -\frac{3}{2} = -1.50 \tag{7.32}$$

The experimental value is

$$\left(\frac{\mu_p}{\mu_n}\right)_{\text{expt}} = \frac{2.79\mu_N}{-1.91\mu_N} \approx -1.46 \tag{7.33}$$

The agreement is remarkable, considering the simplicity of the model.

Magnetic Moment of the $\wedge$-baryon: The spin-flavor wave function of the $\wedge$-baryon is given by

$$|\wedge;\; S_z = +\frac{1}{2}\rangle = \frac{1}{2}\,[|([ud]s\rangle_\alpha|\uparrow\uparrow\downarrow\rangle_\alpha + |[ud]s\rangle_\beta|\uparrow\uparrow\downarrow\rangle_\beta]$$

$$= \frac{1}{\sqrt{12}}[|u\uparrow d\downarrow s\uparrow\rangle - |d\uparrow u\downarrow s\uparrow\rangle - |u\downarrow d\uparrow s\uparrow\rangle$$

$$+ |d\downarrow u\uparrow s\uparrow\rangle + (2 \leftrightarrow 3) + (1 \leftrightarrow 3)] \tag{7.34}$$

This gives

$$\mu_\wedge = \left\langle \wedge;\ S_Z = +\frac{1}{2}|\mu^z_{op}|\wedge;\ S_Z = +\frac{1}{2}\right\rangle = \frac{1}{12}\cdot 3.4\ \mu_s = \mu_s \tag{7.35}$$

From (7.30) and (7.35) we get using (7.31) and (7.28)

$$\frac{2\mu_\wedge}{\mu_n} = \frac{2\mu_s}{2\mu_d} = \frac{\mu_s}{\mu_d} = \frac{m_d}{m_s} \tag{7.36}$$

The experimental value of the lhs of (7.36) is

$$\left(\frac{2\mu_\wedge}{\mu_n}\right)_{\text{expt}} = 0.64 \tag{7.37}$$

From (7.36) and (7.37) one gets

$$\frac{m_d}{m_s} \approx \frac{2}{3} \tag{7.38}$$

From (7.29) and (7.31) one gets

$$\mu_p = \frac{1}{3}(4\mu_u - \mu_d) = -3\mu_d \tag{7.39}$$

This gives, using the experimental value of μ_p

$$m_d = \frac{M_p}{2.79} \approx \frac{1}{3}M_p \tag{7.39}$$

Thus the quark model predicts

$$m_u \approx m_d \approx 335 \text{ MeV} \quad \text{and} \quad m_s \approx 500 \text{ MeV} \tag{7.40}$$

7.10 Color and $SU(3)_c$

In the initial formulation, the quark model faced two serious difficulties—(1) non-observability of free quarks and (2) spin-statistics problem of baryons regarded as bound states of quarks. The first difficulty was an experimental one and could be got around by assuming that the quarks were heavy and the available accelerator energies were not high enough to produce them. But the second difficulty was a theoretical one and one had to do something about it in order to save the quark model.

In the quark model, for the ground state baryons all orbital angular momenta are zero and ψ_{space} is symmetric and also ψ_{spin} and ψ_{flavor} are both symmetric for the 10, $J^P = \frac{3}{2}^+$ baryons, This means that $\psi = \psi_{space}\cdot\psi_{spin}\cdot\psi_{flavor}$ for these baryons, is symmetric. Furthermore for the

8, $J^P = \frac{1}{2}^+$ baryons one gets the correct value for the magnetic moments if $\psi_{spin-flavor}$ is taken to be symmetric. This violates the spin statistics-connection. Because quarks are spin-$\frac{1}{2}$ particles, one expects the total wave function ψ_{tot} to be antisymmetric under particle exchange for a system of quarks. Therefore either ψ defined above is not the total wave function or quarks do not obey normal spin-statistics relation. One chooses the first option and assumes quarks to have an additional degree of freedom, called "Color" so that

$$\psi_{tot} = \psi_{space} \cdot \psi_{spin} \cdot \psi_{flavor} \cdot \psi_{color} \tag{7.41}$$

Now if one further assumes ψ_{color} to be antisymmetric for a system of 3 quarks, normal spin-statistics connection is restored. For this, the number of color states (N_c) for a quark must be at least 3 ($N_c \geq 3$). Moreover, quarks and antiquarks must belong to different (non-equivalent) representations of the color symmetry group. Thus the minimal choice for the color symmetry group is $SU(3)$. In order to distinguish it from the flavor symmetry group which is also an $SU(3)$, the 2 groups are, respectively, called $SU(3)_c$ and $SU(3)_f$. Since for $SU(3)$.

$$3 \times 3 \times 3 = 10_s + 8_\alpha + 8_\beta + 1_A \tag{7.42}$$

This means that the antisymmetric color state of a system of 3 quarks is an $SU(3)$ singlet. Next one invokes the "color hypothesis" which states that only the color singlet state is physical (observable) and all color nonsinglet states are unphysical and therefore cannot be seen experimentally. The only possible ways to get the color singlet state out of quarks and antiquarks are

$$\left.\begin{aligned} 3 \times 3^* &= 8 + 1 \\ 3 \times 3 \times 3 &= 10 + 8 + 8 + 1 \\ 3^* \times 3^* \times 3^* &= 10^* + 8 + 8 + 1 \end{aligned}\right\} \tag{7.43}$$

They correspond to mesons ($q\bar{q}$), baryons (qqq) and antibaryons ($\overline{qqq}$). According to the color hypothesis free quarks and antiquarks are not observable since they belong to nonsinglet representations 3 and 3^*, respectively of $SU(3)$. Also since

$$3 \times 3 = 6 + 3^* \tag{7.44}$$

one sees $q\bar{q}$ hadrons but no qq (diquark) hadrons. The color wave functions of mesons and baryons are

$$\left.\begin{aligned}\psi^M_{color}(q^i\overline{q}^j) &= \frac{1}{\sqrt{3}}\sum_{\alpha,\beta}\delta_{\alpha\beta}q^i_\alpha\overline{q}^j_\beta \\ \psi^B_{color}(q^iq^jq^k) &= \frac{1}{\sqrt{6}}\sum_{\alpha,\beta\gamma}\epsilon_{\alpha\beta\gamma}\, q^i_\alpha q^j_\beta q^k_\gamma\end{aligned}\right\} \tag{7.45}$$

Here (i, j, k) refer to flavor indices and (α, β, γ) refer to colour indices.

Gluons: Quarks must interact strongly in order to form hadrons which are bound states of quarks (or quarks and antiquarks). In a field theoretic description, the strong interaction between quarks is thought to be mediated by some quanta called "gluons" (which glue the quarks together) just like the electromagnetic interactions between charged particles (e.g. electrons) are thought to be mediated by the photon which is the quantum of the electromagnetic field.

Color quantum number of gluons: Since $3 \times 3^* = 8 + 1$

color conservation requires gluons to be either in 8 or 1 of $SU(3)_c$. But like quarks, gluons have not been seen in free states. Thus according to the color hypothesis, gluons must belong to the adjoint representation 8 of $SU(3)_c$. One can formally write down the eight color states of the gluon in terms of those of $q, \overline{q}$. These are

$$R\overline{B}, B\overline{R}, R\overline{G}, G\overline{R}, B\overline{G}, G\overline{B}, \frac{1}{\sqrt{2}}(R\overline{R} - G\overline{G}), \frac{1}{\sqrt{6}}(R\overline{R} + G\overline{G} - 2B\overline{B})$$

Spin of gluons: Since quarks are spin $-\frac{1}{2}$ particles, angular momentum conservation at quark-quark-gluon vertex implies that gluons must have spin 0 or 1. Experimental evidence ($e^+e^- \rightarrow 3jets$) indicates that gluons have spin 1. Thus they are vector bosons ($J^{PC} = 1^{--}$) like the photon.

Flavor quantum numbers of gluons: Strong interactions do not change quark flavor just as electromagnetic interactions do not change electric charge of particles. Thus gluons are flavour-neutral, that is, they belong to the singlet representation of $SU(3)_f$.

Other evidence of color: The Color degree of freedom for quarks was invented to save the spin-statistics connection of baryons regarded as bound states of quarks which are taken to be spin $-\frac{1}{2}$ fermions. However, it was realised later that there appeared three other experimental facts which can be easily explained if one invokes a new (colour) degree of freedom for quarks with the number of quark colors $N_c = 3$. These are

(i) $\pi^0 \to 2\gamma$: In the quark model, $\pi^0 \to 2y$ decay is given by the diagram

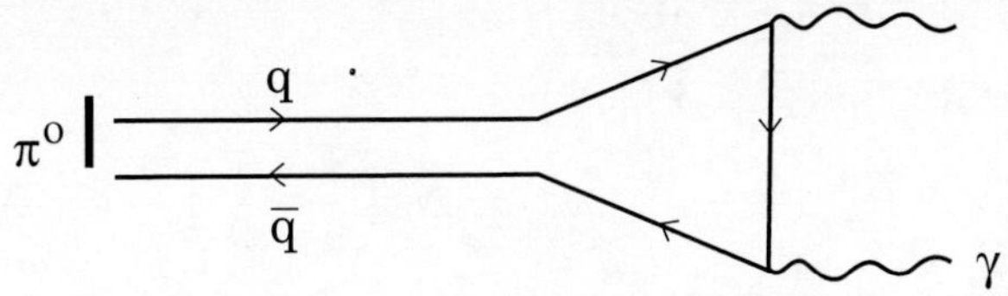

Fig. 7.2. The $\pi^0 \to 2\gamma$ decay in the quark model

and one gets for the decay width

$$\Gamma(\pi^0 \to 2\gamma) = \frac{\alpha^2}{8\pi} \cdot \frac{m_\pi^3}{f_\pi^2} \cdot f^2 \tag{7.46}$$

Here $\alpha = e^2/4\pi = 1/137$ is the electromagnetic fine structure constant, f_π = Pion decay constant.

and $f = \sum_q \langle I_3 Q^2 \rangle_q$

I_3 = Third component of i-spin operator, Q = electric charge operator

The flavor wave function of π^0 is

$$\left|\pi^0 >= \frac{1}{\sqrt{2}}(|u\bar{u} > -|d\bar{d} >)\right.$$

and $\qquad Q|q >= e_q|q >, I_3\left|u >= \frac{1}{2}\right|u >, I_3\left|d >= -\frac{1}{2}\right|d >$

one gets $f = N_c \cdot \frac{1}{2}(e_u^2 - e_d^2) = N_c \cdot \frac{1}{2}\left\{\left(\frac{2}{3}\right)^2 - \left(\frac{-1}{3}\right)^2\right\} = N_c\left(\frac{1}{6}\right)$

$$f_{\text{expt}} = \frac{1}{2} \quad \text{gives } N_c = 3.$$

(ii) $e^+e^- \to$ hadrons: In the quark model, the process is described by the diagram.

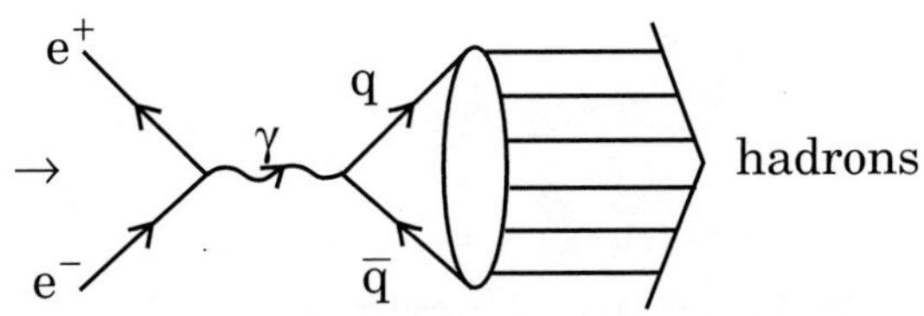

Fig. 7.3. $e^+e^- \to$ hadrons in the quark model

One measures

$$R(s) = \frac{\sigma(e^+e^- \to \text{ hadrons})}{\sigma(e^+e^- \to \mu^+\mu^-)} = \sum_q e_q^2 \theta(s - 4m_q^2) \tag{7.47}$$

For 2 GeV $< \sqrt{s} < 3$ GeV,

$$R(s) = N_c(e_u^2 + e - d^2 + e_s^2) = N_c\left(\frac{4}{9} + \frac{1}{9} + \frac{1}{9}\right) = N_c\left(\frac{2}{3}\right)$$

The experimental value is $R(s) = 2$

This implies $N_c = 3$.

(iii) τ-decay: The third evidence for quark color comes from the measured branching ratio of leptonic decay of the heavy lepton $\tau(1.8 \text{ GeV})$. In the quark model,

$$B_e = \frac{\Gamma(\tau^- \to \nu_\tau e^- \overline{\nu}_e)}{\Gamma(\tau^- \to \text{all})} = \frac{G_F^2}{G_F^2 + G_F^2 + N_c(G_F^2 \cos^2\theta_c + G_F^2 \sin^2\theta_c)}$$

$$= \frac{1}{2 + N_c} = \frac{1}{5}, \quad \text{for } N_C = 3 \tag{7.48}$$

The experimental value of B_c is $(B_c)_{\text{expt}} = 17.5\%$

This is consistant with $N_c = 3$.

7.11 Quark-Quark Interaction

The strong interaction between quarks is mediated by gluons. This interaction is assumed to be invariant under local gauge transformations with $SU(3)_c$ as the underlying gauge group. The corresponding theory, in analogy with QED, is called quantum chromodynamics (QCD). While the gauge group $U(1)$ of QED is Abelian, the gauge group $SU(3)_c$ of QCD is non-abelian. This causes major differences between the two theories in spite of many similarities. The gauge-invariant, Lorentz-invariant QCD interaction Lagrangian is

$$\mathcal{L}_{\text{int}}^{QCD} = g_s \overline{\psi}(x) \gamma^\mu \frac{\lambda^a}{2} \psi(x) A_\mu^a(x) \tag{7.49}$$

The quark field $\psi(x)$ transforms like 3 of $SU(3)_c$ and as a 4-component Dirac spinor under Lorentz transformation,

$$\psi(x) = \begin{pmatrix} \psi_1(x) \\ \psi_2(x) \\ \psi_3(x) \end{pmatrix} \tag{7.50}$$

where $\psi_i(x)(i = 1, 2, 3)$ is a 4-component Dirac spinor and $\overline{\psi}(x)$ is the adjoint spinor, $\overline{\psi} = \psi^+\gamma^0$. The gluon field $A_\mu^a(x)(a = 1, \ldots 8)$ transforms like the adjoint representation 8 of $SU(3)_c$ and is a 4-vector under Lorentz transformation, λ^a's $(a = 1, \ldots, 8)$ are the 3×3 Gell Mann matrices, g_s is the strong coupling constant.

7.12 One-gluon Exchange Potential and Hyperfine Splitting of Hadron Mass Spectrum

The 1-gluon exchange potential in the non-relativistic limit has the same form as the 1-photon exchange potential in QED except for the color factor.

$$V^{ij}(r) = V^{ij}_{\text{central}}(r) + V^{ij}_{\text{non-central}}(r), \quad (i, j = q, \overline{q}) \tag{7.51}$$

$$V_{\text{central}}(r) = \frac{C\alpha_s}{r}, \quad \alpha_s = \frac{g_s^2}{4\pi}, \quad C = \text{color factor}, \; r = |\overline{r}^i - \overline{r}^j| \tag{7.52}$$

$$V_{\text{non-central}}(r) = V_{ss}(r) + V_{LS}(r) + V_T(r) \tag{7.53}$$

For the ground state hadrons $V_{LS}(r)$ and $V_T(r)$ do not contribute and one is left with only the $V_{ss}(r)$.

$$V^{ij}_{ss}(r) = \frac{\vec{S}^{(i)} \cdot \vec{S}^{(j)}}{m_i m_j} \cdot V_s(r) = -C\alpha_s \frac{\vec{S}^{(i)} \cdot \vec{S}^{(j)}}{m_i m_j} \cdot \frac{8\pi}{3} \delta_3(\vec{r}) \tag{7.54}$$

In writing (7.54) we have assumed

$$V_s(\vec{r}) = \frac{2}{3} \nabla^2 V_c(\vec{r}) = -C\alpha_s \cdot \frac{8\pi}{3} \delta_3(\vec{r})$$

$C = \frac{4}{3}$ for $q\overline{q}$ in 1 of $SU(3)_c$, i.e. for mesons.

$C = -\frac{2}{3}$ for qq in 3^* of $SU(3)_c$, i.e. for baryons.

In baryons the diquark (qq) must be in 3^* of $SU(3)_c$ so that the baryon (qqq) can be in 1 of $SU(3)_c$. Since $C < 0$ in both cases, higher spin states are pushed up in energy for both mesons and baryons.

Mesons: Since $M = q\bar{q}$, $\vec{S} = \vec{S}^{(1)} + \vec{S}^{(2)}$, $S^{(1)} = S^{(2)} = \frac{1}{2}$

$S = 0, 1$

$$\Delta m = \langle V_{ss} \rangle = A \left\langle \frac{\vec{S}^{(1)} \cdot \vec{S}^{(2)}}{m_1 m_2} \right\rangle, A = \langle V_s(\vec{r}) \rangle \tag{7.56}$$

$$\left\langle \vec{S}^{(1)} \cdot \vec{S}^{(2)} \right\rangle = \frac{1}{2}\left[S(S+1) - 2 \cdot \frac{1}{2} \cdot \frac{3}{2}\right]$$

$$= \begin{cases} -\frac{3}{4}, \ S = 0 \\ +\frac{1}{4}, \ S = 1 \end{cases} \tag{7.57}$$

$$m(\rho) - m(\pi) = \left[\frac{1}{4} - \left(-\frac{3}{4}\right)\right] \frac{A}{m_d^2} = A/m_d^2$$

$$m(K^*) - m(K) = \left[\frac{1}{4} - \left(-\frac{3}{4}\right)\right] \frac{A}{m_d m_s} = A/m_d m_s \tag{7.58}$$

This gives $\dfrac{m(K^*) - m(K)}{m(\rho) - m(\pi)} = \dfrac{m_d}{m_s}$ (7.59)

Baryons: Since $B = qqq$, $\vec{S} = \vec{S}^{(1)} + \vec{S}^{(2)} + \vec{S}^{(3)}, \vec{S}^{(1)} = \vec{S}^{(2)} = \vec{S}^{(3)} = \frac{1}{2}$

$$S = \frac{3}{2}, \frac{1}{2}, \frac{1}{2}$$

$$\Delta m = A' \left\langle \frac{\vec{S}^{(1)} \cdot \vec{S}^{(2)}}{m_1 m_2} + \frac{\vec{S}^{(2)} \cdot \vec{S}^{(3)}}{m_2 m_3} + \frac{\vec{S}^{(1)} \cdot \vec{S}^{(3)}}{m_1 m_3} \right\rangle, A' = \langle V_s(r) \rangle$$

For Δ and N,

$$m_1 = m_2 = m_3 = m_d$$

and $\Delta m = \dfrac{A'}{m_d^2} \langle \vec{S}^{(1)} \cdot \vec{S}^{(2)} + \vec{S}^{(2)} \cdot \vec{S}^{(3)} + \vec{S}^{(3)} \cdot \vec{S}^{(1)} \rangle$

$$= \frac{A'}{m_d^2} \cdot \frac{1}{2}\left[S(S+1) - 3 \cdot \frac{1}{2} \cdot \frac{3}{2}\right] \tag{7.60}$$

This gives $m(\Delta) - m(N) = \dfrac{3}{2} \cdot \dfrac{A'}{m_d^2}$ (7.61)

For $\wedge, \sum^*, \sum$,

$$m_1 = m_2 = m_d, m_3 = m_s,$$

$$\Delta m = A'\left\langle \frac{\vec{S}^{(1)}\cdot\vec{S}^{(2)}}{m_d^2} + \frac{\vec{S}'\cdot\vec{S}^{(3)}}{m_d m_s}\right\rangle,\ \vec{S}' = \vec{S}^{(1)} + \vec{S}^{(2)},\ \vec{S} = \vec{S}' + \vec{S}^{(3)} \tag{7.62}$$

Therefore, $\Delta m = A'\left[\frac{1}{m_d^2}\cdot\frac{1}{2}\left\{S'(S'+1) - 2\cdot\frac{1}{2}\cdot\frac{3}{2}\right\}\right.$

$$\left. + \frac{1}{m_d m_s}\frac{1}{2}\left\{S(S+1) - S'(S'+1) - \frac{1}{2}\cdot\frac{3}{2}\right\}\right] \tag{7.63}$$

Also, since in $\sum^*, \sum$ u and d quarks are in symmetric flavor state, they must be in symmetric spin state so that $\psi_{spin-flavor}$ is symmetric, and therefore $S' = 1$, but in $\wedge, u$ and d quarks are in an antisymmetric flavor state and therefore they must also be in an antisymmetric spin state so that $\psi_{spin-flavor}$ is again symmetric and thus $S' = 0$.

$$\text{Thus, } m(\Sigma^*) - m(\Sigma) = \frac{A'}{m_d m_s}\cdot\frac{1}{2}\left[\frac{3}{2}\cdot\frac{5}{2} - \frac{1}{2}\cdot\frac{3}{2}\right] = \frac{A'}{m_d m_s}\cdot\frac{3}{2} \tag{7.64}$$

$$m(\textstyle\sum) - m(\Delta) = \frac{A'}{m_d^2}\cdot\frac{1}{2}[1.2 - 0] + \frac{A'}{m_d m_s}\cdot\frac{1}{2}[-1.2 + 0]$$

$$= \frac{A'}{m_d^2} - \frac{A'}{m_d m_s} = \frac{A'}{m_d^2}\left(1 - \frac{m_d}{m_s}\right) \tag{7.65}$$

So, we get

$$\frac{m(\sum^*) - m(\sum)}{m(\Delta) - m(N)} = \frac{m_d}{m_s} \tag{7.66}$$

and
$$\frac{m(\sum) - m(\Delta)}{m(\Delta) - m(N)} = \frac{2}{3}\left(1 - \frac{m_d}{m_s}\right) \tag{7.67}$$

From (7.59) and (7.66)

$$\frac{m(K^*) - m(K)}{m(\rho) - m(\pi)} = \frac{m(\sum^*) - m(\sum)}{m(\Delta) - m(N)} = \frac{m_d}{m_s} \tag{7.68}$$

The experimental data agree with (7.67) and (7.68) for $m_d/m_s = 2/3$, the same value one obtains from fitting baryon magnetic moments data with quark model predictions.

Furthermore,

$$\frac{m(\Delta) - m(N)}{m(\rho) - m(\pi)} = \frac{m(\sum^*) - m(\sum)}{m(K^*) - m(K)} = \frac{3A'}{2A} = \frac{3}{2}\cdot\frac{C_B}{C_M}\left(\frac{a_M}{a_B}\right)^3$$

$$\cong \frac{C_B}{C_M} = \frac{1}{2}. \tag{7.69}$$

In obtaining (7.69) we have assumed,

$$\langle V_s(r)\rangle \propto C\alpha_s\langle\delta_3(\vec{r})\rangle \propto C\alpha_s|\psi(\vec{0})|^3 \propto C\alpha_s\cdot\frac{1}{a^3} \tag{7.70}$$

where $a = \langle r^2\rangle^{1/2}$ = rms radius of the hadron

From experimental data

$$\left(\frac{a_M}{a_B}\right)^3 = \left(\frac{0.7}{0.8}\right)^3 = \frac{2}{3}$$

Thus the ratio of hyperfine splittings of baryons and mesons with the same quark flavors is given by the ratio of color factors. The prediction (7.69) agrees with the observed mass splittings.

7.13 Zweig Rule

The Zweig rule was originally invented to explain the decay characteristics of the ϕ meson considered as an $s\bar{s}$ bound state in the quark model. The branching fraction for decay $\phi \to \pi^+\pi^-\pi^0$ is small (15%) compared to the decay $\phi \to K\overline{K}$(84%) although the former is kinematically much more favored (larger available phase space). According to the Zweig rule hadron decays which are described by quark flow diagrams with connected quark lines are favored while those decays with disconnected quark lines are highly suppressed.

In the framework of QCD the Zweig rule is attributed to the number of gluons involved. Since the initial and final hadron states are both color singlets, the number of gluons emitted must be at least two in the $\phi \to 3\pi$ decay. Furthermore, C-invariance requires an odd number of gluons. Thus the minimum number of gluons is three in the $\phi \to 3\pi$ decay. The decay $\phi \to K\overline{K}$ goes via emission of one gluon.

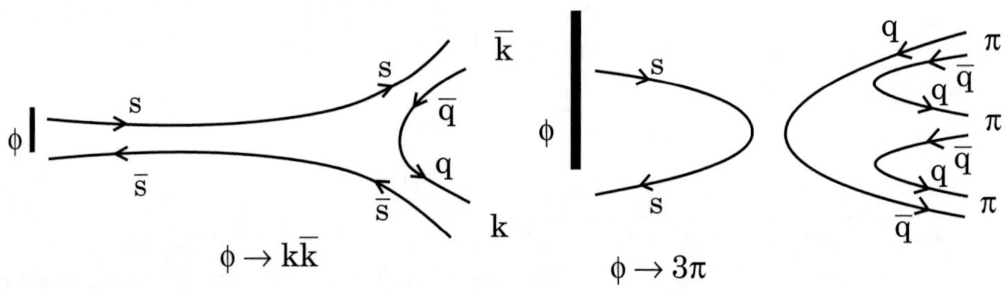

Fig. 7.4

Problems

1. Consider the hypothetical "antisymmetric quark model" in which color is ignored and the spin-flavor wave function of the baryon considered as a bound state of 3 quarks ($B = qqq$) is taken to be antisymmetric. Compute μ_p and μ_n with this wave function and show that in this model.

$$\mu_p/\mu_n = -1/2$$

contrary to the experimental data. (so this option is ruled out and the color hypothesis is justified).

2. Show that the quark model predicts

$$\frac{\Gamma(\rho \to \pi\gamma)}{\Gamma(\omega \to \pi\gamma)} = \frac{(\mu_\mu + \mu_d)^2}{(\mu_u - \mu_d)^2} = \frac{1}{9}$$

assuming that the radiative decay involves a quark spin flip (magnetic dipole) transition.

3. Show that the quark model predicts (neglecting a possible dependence on the vector meson mass).

$$\Gamma(\rho^0 \to e^-e^+) : \Gamma(\omega \to e^-e^+) : \Gamma(\phi \to e^-e^+) = 9 : 1 : 2$$

assuming that the decay proceeds via a vitnal photon,

$$V(q\bar{q}) \to \gamma^* \to e^-e^+$$

4. The spin-flavor symmetry group of hadrons is often taken to be $SU(6)$ particularly in the analysis of the baryon spectra. Obtain the multiplet structure of mesons and baryons in the $SU(6)$ quark model, assuming as usual $M = q\bar{q}$ and $B = qqq$. Also find the flavor and spin contents of each of these $SU(6)$ multiplets and check that the lightest baryons are correctly reproduced.

5. The mass difference Δm between 2 hadrons in the same $SU(3)_f$ multiplet is considered to be the sum of a strong interaction part Δm_s and an electromagnetic part Δm_e, that is

$$\Delta m = \Delta m_s + \Delta m_e$$

If is further assumed that Δm_s is due to partial braking of the $SU(3)_f$ symmetry such that $SU(2)_I$ symmetry is preserved. (This means that for 2 hadrons belonging to the same i-spin multiplet $\Delta m_s = 0$) and Δm_e is due to the i-spin symmetry breaking which in the quark model amounts to taking $m_d \neq m_u$. Hence show that the Coleman-Glashow formula for the lightest baryon octet

$$[m(n) - m(p)] + [m(\Xi^-) - m(\Xi^0)] = 2[m(\Sigma^-) - m(\Sigma^+)]$$

naturally follows in the quark model description.

6. Calculate the color factor for $V(q\bar{q})$ in mesons ($M = q\bar{q}$) and for $V(qq)$ in baryons ($B = qqq$).

Chapter 8

Heavy Flavors

8.1 The Need for a Fourth Quark

The quark model proposed by Gell-Mann and Zweig in 1964 was motivated by observed $SU(3)_f$ symmetry for hadrons and hence involved 3 quarks (u, d, s) that belonged to the fundamental representation 3 of $SU(3)$. The model was very successful in understanding the spectroscopy of light hadrons (mass $\leq$ 2 GeV). However, soon it was felt that there should exist a fourth quark with charge +(2/3)e. There were three motivations for this. These were (1) lepton-quark symmetry (2) absence of strangeness-changing neutral currents and (3) discovery of heavy vector resonances (mass > 3 GeV) and their interpretation.

(1) Lepton-quark symmetry: In the 1960's two lepton pairs were known (ν_e, e) and (ν_μ, μ). However, there were only three quarks. Soon after the Gell-Mann-Zweig quark model was proposed, Bjorken and Glashow (1964) suggested that lepton-quark symmetry implied the existence of a fourth quark with charge $+(2/3)e$ and carrying a new non-zero quantum number called charm C similar to the strangeness quantum number. C is $+1$ for charmed quark, -1 for anticharm quark and 0 for all other quarks.

(2) Absence of strangeness-changing neutral currents: The early measurements showed that β-decays of strange baryons, e.g. $\wedge \to pe^-\overline{\nu}_e$ where $\Delta S = 1$ was suppressed by a factor of about 20 compared to neutron β-decay $n \to pe^-\overline{\nu}_e(\Delta S = 0)$. Moreover, the value of the weak coupling constant obtained from nuclear β-decays was slightly less than that obtained from muon decay $\mu^- \to e^-\overline{\nu}_e\nu_\mu$. In the quark model description, the quark level processes in neutron and $\wedge$ decays are, respectively $d \to ue^-\overline{\nu}_e$ and $s \to ue^-\overline{\nu}_e$. If one calls the coupling constants in the $\wedge, n$ and μ decays $G_\wedge$, G_β and G_μ respectively then one finds that $G_\beta \approx G_\mu$ and $G_\wedge/G_\beta \ll 1$. Moreover, $G_\beta^2 + G_\wedge^2 \approx G_\mu^2$. Thus one can parameterize them, as first done by Cabibbo, as $G_\beta = G_\mu \cos\theta_c$, $G_\wedge = G_\mu \sin\theta_c$ and $G_\mu = G_F/\sqrt{2}$ with $\theta_c \approx$

12°. Thus in V-A theory one can write the weak interaction Hamiltonian as

$$H^w_{int} = \frac{G_F}{\sqrt{2}}(J^w_\alpha J^{w,\alpha\dagger} + h \cdot c) \tag{8.1}$$

with $J^w_\alpha = (J^w_\alpha)_l + (J^w_\alpha)_h$ (8.2)

$$(J^w_\alpha)_\ell = \sum_\ell \overline{\nu}_\ell \Gamma_\alpha \ell,\ \Gamma_\alpha = \gamma_\alpha(1-\gamma_5) \tag{8.3}$$

and $(J^w_\alpha)_h = \overline{u}\Gamma_\alpha d_c,\ d_c = d\cos\theta_c + s\sin\theta_c$ (8.4)

$\theta_c =$ Cabibbo angle $\approx 12^\circ$

However, one found that strangeness-changing decays like $K^\circ \to \mu^+\mu^-$, $K^+ \to \pi^+\nu\overline{\nu}$, $\wedge \to ne^+e^-$, $\sum^+ \to ne^+\nu_e$ were highly suppressed. In fact, it was found

$$\frac{\Gamma(K^+ \to \pi^+\nu\overline{\nu})}{\Gamma(K^+ \to \pi^\circ e^+\nu_e)} < 10^{-5} \tag{8.5}$$

The hadronic part of all these processes is either $d\overline{s}$ annihilation or $d \to s$ transition in which strangeness-changing-neutral current is involved. However, strangeness-nonchanging neutral current processes like $\nu_\mu e^- \to \nu_\mu e^-$ or $\nu_\mu N \to \nu_\mu N$ have been seen.

It was first suggested by Glashow, Iliopoulos and Maiani in 1970 that $\Delta S \neq 0$ weak neutral current processes will be suppressed if there existed a fourth quark with charge $+\frac{2}{3}e$ called c-quark carrying a new quantum number called charm which couples with $s_c = -d\sin\theta_c + s\cos\theta_c$ (orthogonal to d_c) to give an additional contribution to $J^{w,hadron}_\alpha$ in (8.4) so that the modified $J^{w,hadron}_\alpha$ becomes

$$(J^w_\alpha)_h = \overline{u}\Gamma_\alpha d_c + \overline{c}\Gamma_\alpha s_c \tag{8.6}$$

where $$\begin{pmatrix} d_c \\ s_c \end{pmatrix} = \begin{pmatrix} \cos\theta_c & \sin\theta_c \\ -\sin\theta_c & \cos\theta_c \end{pmatrix} \begin{pmatrix} d \\ s \end{pmatrix} \tag{8.7}$$

In Fig. 8.1 contribution of the diagram (a) will be cancelled by the contribution of the diagram (b) provided the mass of the c-quark is not very different from the mass of the u-quark. This cancellatron is known as the GIM mechanism. As a consequence, the hadronic weak neutral current is given by

$$\begin{aligned}(J^{w,N}_\alpha)_h &= \overline{u}\Gamma_\alpha u + \overline{c}\Gamma_\alpha c + \overline{d}_c\Gamma_\alpha d_c + \overline{s}_c\Gamma_\alpha s_c \\ &= \overline{u}\Gamma_\alpha u + \overline{c}\Gamma_\alpha c + \overline{d}\Gamma_\alpha d + \overline{s}\Gamma_\alpha s\end{aligned} \tag{8.8}$$

Thus there is no flavour-changing neutral current (FCNC).

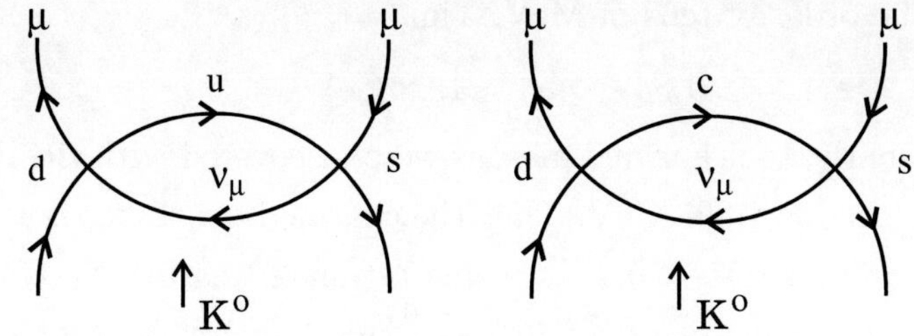

Fig. 8.1 The decay $K^o \rightarrow \mu^+ \mu^-$

(3) Discovery of heavy vector resonances and their interpretation: In 1974, two $J^{PC} = 1^{--}$ resonances $\psi(3.1)$ and $\psi'(3.7)$ with masses around 3.1 and 3.7 GeV, respectively, were discovered in e^+e^- colliding beam experiment at SLAC. (The former was also seen in p-nucleus collision at BNL around the same time). Their widths were extremely small ($\Gamma_\psi \approx 70$ KeV and $\Gamma_{\psi'} \approx 220$ KeV) compared to their masses. For example ρ (773) with mass 773 Mev has a width $\Gamma \approx 150$ MeV. Their properties are best explained if they are regarded as $Q\overline{Q}$ bound states where Q is a hitherto unseen heavy quark and one invokes the Zweig rule. It was tempting to identify the heavy quark with the charmed quark c introduced earlier for different reasons. Just before the discovery of the ψ resonnaces, it was shown by Applquist and Politzer that the $c\bar{c}$ spectrum is very similar to the positronium (e^+e^-) spectrum and in analogy to the positronium system was called the "charmonium". Now the Zweig rule says that the favored decay of charmonium is $c\bar{c} \rightarrow c\bar{q} + q\bar{c}(q = u, d, s)$ and the decay $c\bar{c} \rightarrow q\bar{q}$ states is highly suppressed.

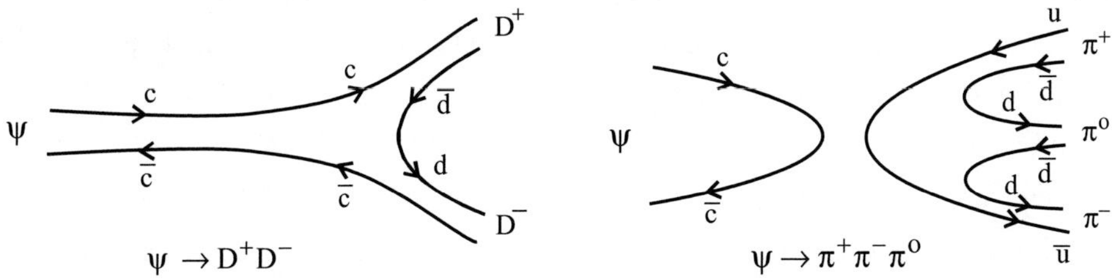

$\psi \rightarrow D^+D^-$
Zweig rule favored

$\psi \rightarrow \pi^+\pi^-\pi^o$
Zweig rule suppressed

Fig. 8.2

Thus ψ will be a narrow (small width) resonance provided $m_\psi < 2m_D$ where m_D is the mass of the lightest charmed meson. Later on, higher charmonium states $\psi(3.77)$, $\psi(4.04)$, $\psi(4.16)$ and $\psi(4.41)$ were seen with widths of the order of tens of MeV. Thus

$$1.84 < m_D < 1.88 \text{ GeV}. \tag{8.9}$$

Subsequently, the charmed mesons were discovered with the lightest one having mass $m_D = 1.865$ GeV. Thus the charmed quark has mass $m_c \approx 1.5$ GeV. This value agrees with the value estimated earlier by Gaillard, Lee and Rosner from calcutation of $K^0 - \overline{K}^0$ mass difference. Moreover, in the Cabibbo theory as extended by GIM $c \to s$ transition is favored and thus the decay of charmed mesons $(c\overline{q})$ will always contain a $\overline{K}$-meson $(s\overline{q})(q = u, d)$. For example,

$$D^\circ \to K^-\pi^+, D^+ \to \overline{K}^0\pi^+$$

In general, $D \to \overline{K}\pi, \overline{D} \to K\pi$ (8.10)

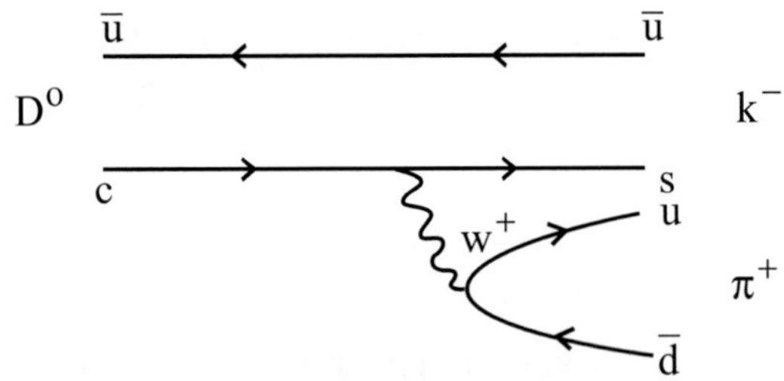

Fig. 8.3

Thus the identification of the heavy quark with the c-quark is complete.

8.2 Charm and $SU(4)_f$ Symmetry

With four quarks $(u, d, s, c,)$ one can assume that the hadron flavor symmetry is higher that $SU(3)$. Probably the flavor symmetry group is $SU(4)$ with the quark multiplet forming the fundamental representation 4 of $SU(4)_f$ and the antiquarks belonging to the conjugate representation 4^* of $SU(4)_f$. One can very easily extend the results of $SU(3)$ to $SU(4)_f$. Thus a Young tableau corresponding to a nontrivial representation of $SU(4)$ will atmost have 3 rows. A column with 4 rows will correspond to the singlet of $SU(4)$ and may be deleted from the tableau. Therefore, a representation of $SU(4)$ will be characterized by 3 integers

$D = D(p, q, r) \leftrightarrow Y(\lambda_1, \lambda_2, \lambda_3)$

$p = \lambda_1 - \lambda_2,\ q = \lambda_2 - \lambda_3,\ r = \lambda_3$

The dimension of a $SU(4)$ representation $D(p,q,r)$ is given by

$$N(p,q,r)=\frac{1}{12}(p+1)(q+1)(r+1)(p+q+2)(q+r+2)(p+q+r+3) \quad (8.11)$$

$D(r,q,p)$ is conjugate to $D(p,q,r)$ and the two have the same dimension. If one is denoted by N, the other will be denoted by N^*. Since mesons are $q\bar{q}$ bound states and baryons are qqq bound states one can obtain the possible $SU(4)$ representations by using Young tableau techniques.

$q\bar{q}$: 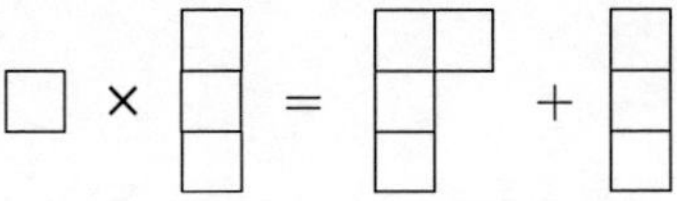

$$4 \times 4^* = 15 + 1 \quad (8.12)$$

qqq : 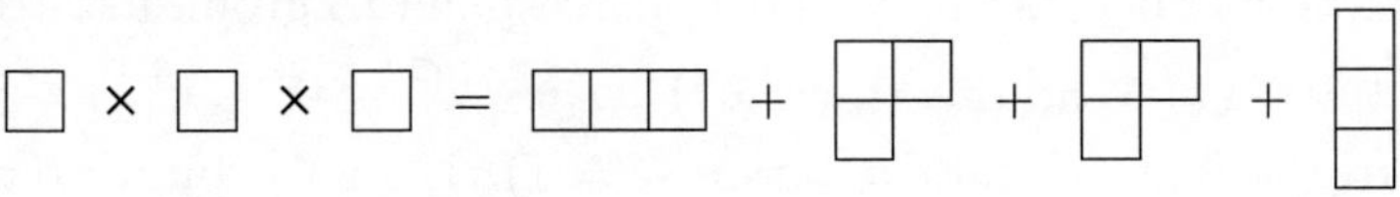

$$4 \times 4 \times 4 = 20_s + 20_\alpha + 20_\beta + 4^* \quad (8.13)$$

The Gell-Mann-Nishijima relation now becomes

$$Q = I_3 + \frac{1}{2}(B + S + C) \quad (8.14)$$

There will be 7 mesons containing c or $\bar{c}$ quark in each $SU(4)$ meson multiplet. They are

$$\psi, \eta_c = c\bar{c},\ C = 0$$

$$\left.\begin{array}{r} D^{*+}, D^{+} = c\bar{d} \\ D^{*0}, D^{0} = c\bar{u} \\ D_s^{*+}, D_s^{+} = c\bar{s} \end{array}\right\} C = +1 \quad \left.\begin{array}{r} \overline{D}^{*0}, \overline{D}^{0} = u\bar{c} \\ D^{*-}, D^{-} = d\bar{c} \\ D_s^{*-}, D_s^{-} = s\bar{c} \end{array}\right\} C = -1 \quad (8.15)$$

There will be 9 baryons with $J^P = \frac{1}{2}^+$ containing one charm quark i.e with quark content $(cqq), q = u, d, s$. Since for $SU(3)$, $3 \times 3 = 6 + 3^*$ the quark structure of the baryons are given by

$\boldsymbol{C = +1}$, $J^P = \frac{1}{2}^+$ **baryons**

$\Sigma_c^{++} = uuc,\ \equiv_c^{+} = \{us\}c,\ \Omega_c^{o} = ssc$

$\Sigma_c^{+} = \{ud\}c,\ \equiv_c^{0} = \{ds\}c$

$\Sigma_c^{0} = ddc$

$\equiv_c^{\prime +} = [us]c,\ \wedge_c^{+} = [ud]c$

$$\Xi_c^{\prime 0} = [ds]c$$

$(\Sigma_c, \Xi_c,\ \Omega_c)$ constitute 6 of $SU(3)_f$

and $(\Xi_c^{\prime}, \Lambda_c)$ constitute 3^* of $SU(3)_f$

Here $\{ab\} = \frac{1}{\sqrt{2}}(ab + ba)$ and $[ab] = \frac{1}{\sqrt{2}}(ab - ba)$ (8.16)

All the above states have been seen.

8.3 Charmonium States

Apart from $\psi(J^{PC} = 1^{--})$ and $\eta_c(J^{PC} = 0^{-+})$ charmonium ($c\bar{c}$) states with other J^{PC} have also been seen. ψ and η_c correspond to ground states ($\ell = 0$). The χ states $\chi(3.41), \chi(3.45), \chi(3.55)$ have $J^{PC} = 0^{++}, 1^{++}$, 2^{++} and correspond to the first orbital excitation ($\ell = 1$). Higher orbital excitation states and radial excitation states have also been seen. $\psi(3.77)$ correspond to first $\ell = 2$ excitation (1^3D_1) and the $\psi'(3.68)\psi(4.04)$, $\psi(4.16)$ and $\psi(4.41)$ correspond to radial excitation states of charmonium. $\psi'(3.68)$ is regarded as the 2^3S_1 state of charmonium.

The energy levels of the charmonium system have been obtained by solving the nonrelativisitic Schrodinger equation with a QCD-motivated potential consisting of a short-distance part and a long-distance part. Because of the asymptotic freedom of QCD the short-distance part is given by the 1-gluon exchange contribution which in the NR limit is coulombic, that is, $(1/r)$ similar to the 1-photon-exchange potential in QED except for a color factor. The long-distance part represents the quark confinement and can not be computed because of multigluon exchanges. Many phenomenological forms have been tried and the best choice seems to be the linear potential. Thus the potential is assumed to be

$$V(r) = \frac{a}{r} + br,\ a = C\alpha_s,\ C = \text{colour factor} = -4/3,\ \alpha_s = 0.3,$$

and $b = 0.2\ (\text{Gev})^2$.

8.4 The Third Generation of Fermions

The possibility of the existence of a third family of quarks was suggested by Kobayashi and Masakawa in 1972 in order to explain CP violation in the neutral kaon system. The lepton-quark symmetry then required the presence of a third family of leptons with properties similar to the properties of the two families of leptons already known at that time. Closely following the discovery of the c-quark, these fermions were discovered in high energy accelerator experiments.

τ-lepton: In 1975 Perl et al detected oppositely charged $e\mu$ pairs ($e^{\pm}\mu^{\mp}$) in the final states of e^+e^- collisions. Their occurrence can be best explained by the process.

$$e^+e^- \to \tau^+\tau^-$$

with the τ's undergoing weak decays to lighter leptons,

$$\tau^+ \to \ell^+\nu_e\overline{\nu}_t, \; \tau^- \to \ell^-\overline{\nu}_e\nu_t, \; \ell = e, \mu \tag{8.17}$$

The mass of the τ-lepton turned out to be $m_\tau = 1.784 \pm 0.004$. GeV and its lifetime was found to be $\tau_\tau = (2.910 + 0.015) \times 10^{-13}$ sec. Moreover, the branching fraction of its decay to e or μ was measured to be

$$B_e = \frac{\Gamma(\tau^- \to e^-\overline{\nu}_e \cdot \nu_\tau)}{\Gamma(\tau^- \to all)} \approx 17.5\% \tag{8.18}$$

which agrees with the quark model prediction

$$B_e = \frac{1}{2 + N_c} = \frac{1}{5}, \tag{8.19}$$

b-quark: In 1977 Herb et al observed a number of resonances in the energy range 9.5 GeV–10.5 GeV similar to the charomonium ($c\overline{c}$) states in the invariant mass spectrum of $\mu^+\mu^-$ pairs produced through the Drell-Yan process in 400 GeV proton-nucleons coclisons at FNAL

$$p + \text{nucleus} \to \mu^+\mu^- X \tag{8.20}$$

They were also seen one year later in e^+e^- collider experiments at DESY. There were 3 narrow resonances at 9.46, 10.02 and 10.35 GeV and a broad resonance at 10.57 GeV with quantum numbers of the photon ($J^{PC} = 1^{--}$). They were best interpreted as the ground state and the radially excited states of the bound system of a still heavier b-quark with charge $-\frac{1}{3}e$ and its antiparticle $\overline{b}$ called "bottominium" ($\gamma = b\overline{b}$). The narrow widths of the first 3 resonances and the much larger width of the fourth resonance can again be explained by application of the Zweig rule provided the mass of the lightest pseudoscalar B-meson ($b\overline{q}, q\overline{b}$ with $q = u, d$) lies in the range given by

$$10.35 \text{ GeV} < 2m_B < 10.57 \text{ GeV} \tag{8.21}$$

The B-mesons were subsequently discovered and the mass of the lightest B-meson was found to be $m_{B_d} = 5.27$ GeV.

The t-quark: The partner of the b-quark, a charge $+\frac{2}{3}e$ quark called the t-quark (top quark) was finally found after a long search at FNAL in 1995 in $\bar{p}p$ collider experiments with $\sqrt{s} = 1.8$ TeV. The $t\bar{t}$ pair is first produced in $q\bar{q}$ or gluon-gluon collision subprocess

$$q\bar{q}, gg \to t\bar{t}$$

followed by dominant decays

$$t \to W^+ b, \ \bar{t} \to W^- \bar{b}.$$

The process with the lowest background is the one in which both W's undergo leptonic decay. One looked for events jets + 2 charged leptons + missing p_T (corresponding to ν's). The mass of the t-quark was found to be $m_t = (175.5 \pm 5.1)$ GeV$/c^2$.

At the present level of knowledge there are thus three pairs of leptons and three pairs of quarks arranged in three generations. These are

$$\text{First generation}: \begin{pmatrix} \nu_e \\ e^- \end{pmatrix} \begin{pmatrix} u \\ d \end{pmatrix}$$

$$\text{Second generation}: \begin{pmatrix} \nu_\mu \\ \mu^- \end{pmatrix} \begin{pmatrix} c \\ s \end{pmatrix}$$

$$\text{Third generation}: \begin{pmatrix} \nu_\tau \\ \tau^- \end{pmatrix} \begin{pmatrix} t \\ b \end{pmatrix} \tag{8.22}$$

They constitute the fundamental fermions of what is known as the "standard model" of particle physics. Ordinary matter is made of only the first generation fermions. The second and third generation fermions are heavier and unstable (except for the neutrinos). They are produced in high energy accelerators and have been seen in cosmic rays. However, they were all present copiously in the early universe. An interesting fact to note is that the sum of electric charge of the fermions in any generation equals zero.

$$3\left(\frac{2}{3} - \frac{1}{3}\right) - 1 = 0 \tag{8.23}$$

This is necessary for cancellation of "triangle anomalies" and thus provides an explanation of the charge quantization (fractional electric charges of quarks).

Addendum to chapter 8

$SU(3)_f$ decomposition of $SU(4)_f$ multiplets:

Mesons ($q\overline{q}$): $4 \times 4^* = 15 + 1 = (8, 0) + (3, -1) + (3, 1) + (1, 0) + (1, 0)$

The first number in each set indicates the $SU(3)$ multiplet, while the second number indicates the value of the charm quantum number.

Baryons (qqq): $4 \times 4 \times 4 = 20_s + 20_\alpha + 20_\beta + 4_A$

$20_s = (10, 0) + (6, 1) + (3, 2) + (1, 3)$

$20_{\alpha,\beta} = (8, 0) + (6, 1) + (3^*, 1) + (3, 2)$

$4_A = (1, 0) + (3^*, 1)$

All mesons containing a, c or $\overline{c}$ quark or both have been seen. However, charmed baryons containing only one c-quark have so far been seen.

Problems

1. According to weak interaction theory, the dominant hadronic weak decay proceeds via the quark transmutations $c \to s$ and/or $u \to d$. Assuming that these, and only these, transmutations can occur, show that the $D^+ \to K^-\pi^+\pi^+$ is an allowed weak decay, but that $D^+ \to K^+\pi^+\pi^-$ is forbidden.

2. The expected radiative transitions between the charmonium ($c\overline{c}$) states are $\psi \to \eta_c\gamma$, $\psi' \to \chi_c\gamma$, $\psi' \to \eta_c\gamma$ and $\chi_c \to \psi\gamma$. Indicate which are electric and which are magnetic dipole transitions.

3. The hadronic decay widths of $\eta_c(2.98)$ and $\psi(3.1)$ are estimated assuming

   $\eta_c(c\overline{c}) \to\ n$ gluons $\to$ hadrons

   $\psi(c\overline{c}) \to n'$ gluons $\to$ hadrons

   show that the minimum values of n and n' are 2 and 3, respectively.

4. The state $J/\psi(3100)$ has a full width $\Gamma = 87$ KeV and 88% of its decays are to a hadronic final state. Assuming that the hadronic

decay proceeds via 3 gluons, $\psi \to$ 3 gluons $\to$ hadrons with a rate given by the same formula as that for the triplet positronium decay, namely,

$$\Gamma[e^- e^+(3_{S1}) \to 3\gamma] = \frac{2(\pi^2 - 9)}{9\pi}\alpha^6 \text{ m}$$

but with the em fine structure constant α replaced by $C\alpha_s$, where α_s is the QCD fine structure constant ($\alpha_s = g_s^2/4\pi$) and C is the color factor, ($C = 4/3$) and the electron mass m replaced by the c-quark mass m_c. Estimate the value of α_s from the observed data and the modified decay rate formula.

5. Use the meson and baryon mass formula derived in chapter 7, namely,

$$m(q_1\bar{q}_2) = m_1 + m_2 + \frac{a}{m_1 m_2}(\vec{\sigma}_1 \cdot \vec{\sigma}_2)$$

and $$m(q_1 q_2 q_3) = m_1 + m_2 + m_3 + a' \sum_{i<j} (\vec{\sigma}_i \cdot \vec{\sigma}_j) \cdot \frac{1}{m_i m_j}$$

where m_1, m_2, m_3 are quark masses and a an a' are positive constants to show that

a) $$\frac{m(D^*) - m(D)}{m(K^*) - m(K)} = \frac{m_s}{m_c}$$ and

b) $$\frac{m(\Sigma_c) - m(\Lambda_c)}{m(\Sigma) - m(\Lambda)} = \frac{m_s}{m_c} \cdot \frac{m_c - m_u}{m_s - m_u}$$

Check the agreement with observed values of the masses.

Chapter 9

Deep-Inelastic Lepton-Hadron Scattering and the Quark-Parton Model

The importance of scattering experiments in unfolding the internal structure of composite systems was recognised ever since Rutherford's famous α-scattering experiments in 1911 led to the discovery of the nuclear structure of the atom. The job is facilitated if the projectile in the scattering process is an elementary, that is, a point-like object without any internal structure. The electron satisfies the requirements best. Moreover, being a stable charged particle it can be accelerated to very high energies in an accelerator. Thus electron scattering experiments have been done to probe the structure of atoms and nuclei and more recently, the structure of the nucleon itself. Heisenberg's uncertainty principle tells that the central distance that can be probed is of order $1/Q$, where Q is the magnitude of the momentum transfer. Thus to probe distances $\leq$ 1fm, one needs electrons of energies $\geq$ 1 GeV and at such high energies the scattering is highly inelastic. This became possible after the 21 GeV electron beam was produced at SLAC. The SLAC-MIT electron-proton scattering experiments in 1967–68 started a new era in particle physics. Subsequently, the ep scattering results were confirmed by μN and $\nu_\mu N$ scattering experiments. Studies of lepton-hadron scattering have played a major role in establishing the standard model of particle physics.

9.1 Elastic Electron-Proton Scattering

9.1.1 Rutherford Scattering

Before we study deep-inelastic scattering, we consider the scattering of a nonrelativistic electron from a proton regarded as a static charge source when the scattering is purely elastic. If the kinetic energy is large compared to the potential energy one can use Born approximation. In the first Born approximation, the scattering amplitude is given by

$$f(\theta) = -\frac{1}{4\pi} \cdot \frac{2m}{\hbar} \int e^{i\vec{q}\cdot\vec{r}} V(\vec{r}) d^3r \tag{9.1}$$

$$\vec{q} = (\vec{p} - \vec{p}')/\hbar = \vec{k} - \vec{k}', |\vec{k}| = |\vec{k}'|$$

$$q^2 = (k - \vec{k}')^2 = 2|\vec{k}|^2(1 - \cos\theta) = 4|\vec{k}|^2 \sin^2\frac{\theta}{2}$$

$$E = \frac{\hbar^2|\vec{k}|^2}{2m} \tag{9.2}$$

$$V(\vec{r}) = -e\phi(\vec{r}) \tag{9.3}$$

$\phi(\vec{r})$ is the solution of the Poisson equation.

$$\nabla^2\phi(\vec{r}) = -4\pi(Ze)\rho(\vec{r}) \tag{9.4}$$

$\rho(\vec{r})$ is the charge density with normalization.

$$\int \rho(\vec{r}) d^3r = 1 \tag{9.5}$$

and Ze = total electric charge of the static source, we get

$$f(\theta) = -\frac{1}{4\pi} \cdot \frac{2m}{\hbar^2}(-4\pi)\left(\frac{Ze^2}{-q^2}\right) \int e^{i\vec{q}\cdot\vec{r}} \rho(\vec{r}) d^3r = \frac{-2mZe^2}{\hbar^2 q^2} F(q^2)$$

$$= -\left(\frac{Ze^2}{4E\sin^2\frac{\theta}{2}}\right) F(q^2) \tag{9.6}$$

where $F(q^2) = \int e^{i\vec{q}\cdot\vec{r}} \rho(r) d^3r$ (9.7)

is known as the form factor of the charge distribution. For a point charge

$$\rho(\vec{r}) = \delta_3(\vec{r}) \quad \text{and} \quad F(q^2) = 1 \tag{9.8}$$

The differential scattering cross section is given by

$$\frac{d\sigma}{d\Omega} = |f(\theta)|^2 = \left(\frac{Ze^2}{4E\sin^2\frac{\theta}{2}}\right)^2 |F(q^2)|^2 \tag{9.9}$$

For a point charge (9.9) reduces to the Rutherford expression for the differential cross section of scattering of a nonrelativistic electron from a static point charge.

For $|qa| \ll 1$ (a is the extent of the charge distribution) one gets for a spherical charge distribution.

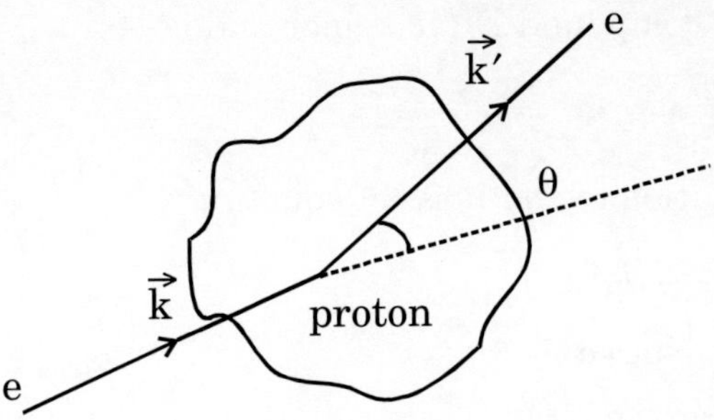

Fig. 9.1

$$F(q^2) = \int e^{i\vec{q}\cdot\vec{r}} \rho(\vec{r}) a^3 r = \int \left[1 + i\vec{q}\cdot\vec{r} - \frac{(\vec{q}\cdot\vec{r})^2}{2!} + \cdots\right] \rho(\vec{r}) d^3 r$$
$$= 1 - \frac{q^2}{6}\langle r^2 \rangle + \cdots \tag{9.10}$$

where $\langle r^2 \rangle = \int r^2 \rho(\vec{r}) d^3 r$

$\langle r^2 \rangle^{\frac{1}{2}}$ is known as rms charge radius. For an exponential charge distribution.

$$\rho(r) = \frac{1}{8\pi a^3} e^{-r/a} \tag{9.11}$$
$$F(q^2) = \frac{1}{(1+q^2a^2)^2} \tag{9.12}$$

For $qa \gg 1$.

$$F(q^2) = \frac{1}{(q^2a^2)^2} \tag{9.13}$$

This is known as the "dipole" fit.

9.1.2 Scattering of a Relativistic Electron from a Static Point Charge (Mott Scattering)

We next consider the scattering of a relativistic electron from a proton still regarded as a static charge source. For simplicity, we take it to be a point charge.

The S-matrix element to the lowest order in electromagnetism is given by

$$S_{fi} = (-ie) \int d^4x \overline{\psi}_f(x) \gamma^\mu \psi_i(x) A_\mu^{\text{ext}}(x) \tag{9.14}$$
$$(\overline{\psi} = \psi^+ \gamma^0)$$

where $\psi_a(x) = \sqrt{\frac{m}{(2\pi)^3 E_a}} u(p_a, \lambda_a) e^{-ip_a\cdot x}$ (9.15)

$(a = i, f)$

$u(p_a, \lambda_a)$ is the 4-compound Dirac spinor and

$$E_a = \sqrt{\vec{p}_a^{\,2} + m^2}$$

$A_\mu^{\text{ext}}(x)$ is the solution of the Poisson equation

$$\Box\, A_\mu^{\text{ext}}(x) = J_\mu^{\text{ext}}(x) \tag{9.16}$$

For a static point charge

$$J_\mu^{\text{ext}}(x) = (Ze)\delta_3(\vec{x})\delta_{\mu 0}$$

This gives

$$A_\mu^{\text{ext}}(x) = \frac{Ze}{4\pi|\vec{x}|}\delta_{\mu 0} = \delta_{\mu 0}(Ze)\int \frac{e^{i\vec{q}\cdot\vec{x}}}{|\vec{q}|^2} \cdot \frac{d^3q}{(2\pi)^3} \tag{9.17}$$

Substituting (9.17) in (9.14) and carrying out the integration we get

$$S_{fi} = -2\pi i\delta(E_i - E_f)N \cdot A \tag{9.18}$$

where $N = \sqrt{\dfrac{m^2}{(2\pi)^6 E_i E_f}}$

$$A = i(Ze^2)\frac{\overline{u}(p_f, \lambda_f)\gamma^0 u(p_i, \lambda_i)}{|\vec{q}|^2 + i\epsilon}$$

$$\vec{q} = \vec{p}_f - \vec{p}_i, |\vec{q}|^2 = (\vec{p}_f - \vec{p}_i)^2 = 4|\vec{p}|^2 \sin^2\frac{\theta}{2}, |\vec{p}_f| = |\vec{p}_i| = |\vec{p}|$$

Now the transition probability is

$$\propto |S_{fi}|^2 \propto [2\pi\delta(E_i - E_f)]^2$$

and $\quad |2\pi\delta(E_i - E_f)|^2 = 2\pi\delta(E_i - E_f)2\pi\delta(0) = \lim_{T\to\infty} 2\pi\delta(E_i - E_f)T$

$$\left[\delta(w) = \frac{1}{2\pi}\int_{-\infty}^{\infty} e^{iwt}dt = \frac{1}{2\pi}\cdot\lim_{T\to\infty}\int_{-T/2}^{T/2} e^{iwt}dt \text{ gives } 2\pi\delta(0) = \lim_{T\to\infty} T\right]$$

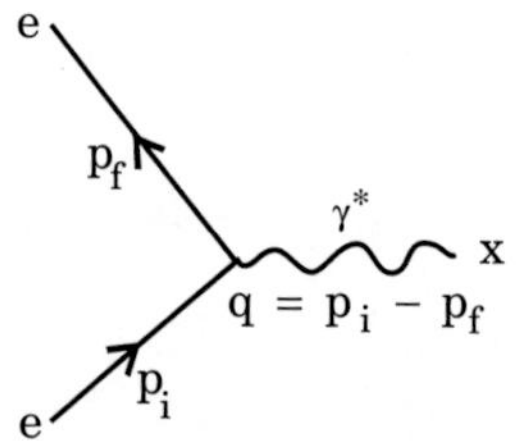

Fig. 9.2

Thus transition probability per unit time, given by

$$\omega = \lim_{T\to\infty}\frac{1}{T}|S_{fi}|^2 \text{ is finite.} \tag{9.19}$$

Experimentally, one measures "Rate", defined as transition probability per unit time per particle integrated over final state phase space.

Hence Rate $R = \lim_{T\to\infty} \int \frac{1}{T}|S_{fi}|^2 \frac{V d^3p_f}{(2\pi)^3}$

$$= \lim_{T\to\infty} \int \frac{1}{T}|S_{fi}|^2 \cdot d^3p_f \quad [V = (2\pi)^3]$$

$$= \int |N|^2|A|^2 \cdot 2\pi\delta(E_i - E_f) d^3p_f \tag{9.20}$$

Also Rate = Cross section × Flux of incident particle

Flux = no. of particles incident per unit area per unit time

$$= n_i|\vec{v}_i| = \frac{1}{V}|\vec{v}_i| = \frac{1}{(2\pi)^3}|\vec{v}_i|, \quad |\vec{v}_i| = \frac{|\vec{p}_i|}{E_i}$$

$$\text{cross section } \sigma = \frac{\text{Rate}}{\text{Flux}} = \frac{(2\pi)^3}{|\vec{v}_i|} \int |N|^2 \cdot |A|^2 \cdot 2\pi\delta(E_i - E_f) d^3p_f$$

$$= \frac{(2\pi)^3}{|\vec{v}_i|} \cdot \frac{m^2}{(2\pi)^6 E_i} \int |A|^2 \cdot 2\pi\delta(E_i - E_f) \frac{d^3p_f}{E_f}$$

$$= \frac{(2m)^2}{(2E_i)|\vec{v}_i} \int |A|^2 \cdot 2\pi\delta(E_i - E_f) \frac{d^3p_f}{(2\pi)^3(2E_f)}$$

$$= \frac{(2m)^2}{16\pi^2} \cdot \frac{|\vec{p}_f|}{|p_i|} \int |A|^2 d\Omega \tag{9.21}$$

elastic scattering, $E_i = E_f = E, \quad |\vec{p}_i| = |\vec{p}_f| = |\vec{p}|$

$$E^2 = |\vec{p}|^2 + m^2$$

We finally get

$$\frac{d\sigma}{d\Omega} = \frac{(2m)^2}{16\pi^2}|A|^2 \tag{9.22}$$

For spin-averaged cross section we have to replace $|A|^2$ by

$$|\overline{A}|^2 = \frac{1}{2}\sum_{\text{spin}} |A|^2$$

Therefore the spin-averaged cross section is given by

$$\overline{\left(\frac{d\sigma}{d\Omega}\right)} = \frac{(2m)^2}{16\pi^2}|\overline{A}|^2 \tag{9.23}$$

$$= \frac{(2m)^2}{16\pi^2} \cdot \left(\frac{Ze^2}{|\overline{q}|^2}\right)^2 \frac{1}{2}\sum_{\text{spins}} |\overline{u}(p_f, \lambda_f)\gamma^0 u(p_i, \lambda_i)|^2 \tag{9.24}$$

$$\text{Now} \quad = \frac{1}{2}\sum_{\text{spins}} |\overline{u}(p_f,\lambda_f)\gamma^0 u(p_i',\lambda_i)|^2$$

$$= \frac{1}{2}\sum_{\text{spins}} \overline{u}(p_f,\lambda_f)\gamma^0 u(p_i'\lambda_i)u^+(p_i\lambda_i)(\gamma^0)^+\gamma^0 u(p_f,\lambda_f)$$

$$= \frac{1}{2}\sum_{\text{spins}} \overline{u}(p_f,\lambda_f)\gamma^0 u(p_i\lambda_0)\overline{u}(p_i,\lambda_i)\gamma^0 u(p_f,\lambda_f)$$

$$= \frac{1}{(2m)^2}\cdot\frac{1}{2}\text{Trace }[\gamma^0(p_i+m)\gamma^0(p_f+m)]$$

$$= \frac{1}{2}\cdot\frac{1}{(2m)^2}\text{Trace }[\gamma^0 p_i\gamma^0 p_f + m^2\gamma^0\gamma^0]$$

$$= \frac{1}{2}\cdot\frac{1}{(2m)^2}4[E_iE_f - p_i\cdot p_f + E_fE_i + m^2]$$

$$= \frac{1}{2}\cdot\frac{1}{(2m)^2}4[2E_iE_f - (E_iE_f - |\vec{p}_i||\vec{p}_f|\cos\theta) + m^2]$$

$$= \frac{1}{2}\cdot\frac{1}{(2m)^2}4[2E^2 - (E^2 - |\vec{p}|^2\cos\theta) + m^2]$$

$$= \frac{1}{2}\cdot\frac{1}{(2m)^2}4[2E^2 - |\vec{p}|^2(1-\cos\theta)]$$

$$= \frac{1}{2}\cdot\frac{1}{(2m)^2}8E^2\left[1-\beta^2\sin^2\frac{\theta}{2}\right]\quad\left[\beta=\frac{|\vec{p}|}{E}\right] \tag{9.25}$$

Substituting (9.25) in (9.24) we get

$$\overline{\left(\frac{d\sigma}{d\Omega}\right)} = \frac{(Z\alpha)^2}{|\vec{q}|^4}\cdot 4E^2\left(1-\beta^2\sin^2\frac{\theta}{2}\right),\quad\left(\alpha=\frac{e^2}{4\pi}\right)$$

$$= \frac{(Z\alpha)^2}{4\beta^4E^2\sin^4\frac{\theta}{2}}\left(1-\beta^2\sin^2\frac{\theta}{2}\right) \tag{9.26}$$

In the *ER* limit, $\beta\to 1$ (9.26) becomes

$$\overline{\left(\frac{d\sigma}{d\Omega}\right)} = \left(\frac{d\sigma}{d\Omega}\right)_{\text{Mott}} = \frac{(Z\alpha)^2}{4E^2\sin^4\frac{\theta}{2}}\left(1-\sin^2\frac{\theta}{2}\right)$$

In the *NR* limit, $\beta \ll 1$ (9.26) becomes

$$\left(\frac{d\sigma}{d\Omega}\right) = \left(\frac{d\sigma}{d\Omega}\right)_{\text{Rutherford}} = \frac{(Z\alpha)^2}{16\in^2\sin^4\frac{\theta}{2}},$$

Here $\in=\frac{1}{2}m\beta^2$ is the *NR* kinetic energy and the spin effect disappears.

9.1.3 Elastic Scattering of Two Point-like Charged Dirac Particles (Pure QED Case)

We now consider the elastic *e-p* scattering with the following conditions:

(i) Both the electron and the proton are point-like charged Dirac particles and together form a dynamical system.

(ii) The electron is ultrarelativistic so that we can neglect its mass compared to its momentum ($|\vec{p}| \gg m$).

(iii) The *e-p* interaction is dominantly electromagnetic i.e. we neglect weak interaction effects.

(iv) The electromagnetic coupling is weak enough so that we consider only the lowest nonvanishing contribution in the perturbative expansion of the S-matrix

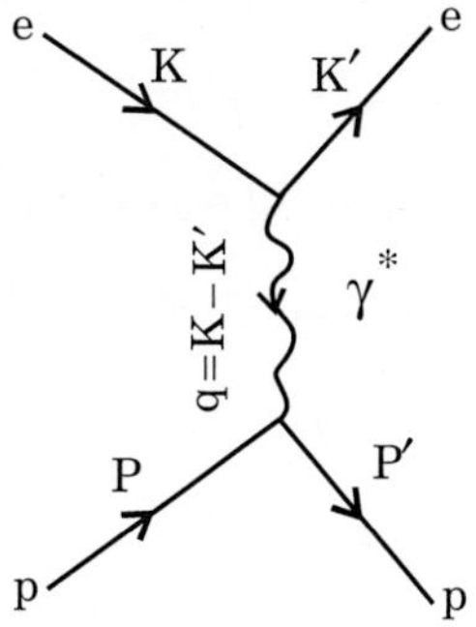

Fig. 9.3

$$L_I(x) = -eJ_\mu^{em}(x)A^\mu(x) \tag{9.27}$$

$$J_\mu^{em} = J_\mu^{em,e} + J_\mu^{em,p}, \; J_\mu^{em,e} = \overline{\psi}_e(x)\gamma_\mu\psi_e(x)$$

$$A^\mu(x) = \text{Photon field}$$

$$K + P = K' + P' \tag{9.28}$$

$$K^2 = K'^2 = m^2, \; P^2 = P'^2 = M^2$$

$$\text{Lab frame : } P = (M, \vec{0}), \; K = (E, \vec{K}), \; K' = (E', \vec{K}')$$

$$\text{electron ultra-relativistic} \Rightarrow \; E \approx |\vec{K}|, \; E' \approx |\vec{K}'|, |\vec{K}|, |\vec{K}| \gg m$$

$$s = (K + P)^2 \approx M^2 + 2ME$$

$$t = q^2 = (K - K')^2 \approx -4EE' \sin^2 \frac{\theta}{2}$$

$$u = (K' - P)^2 \approx M^2 - 2ME' \tag{9.29}$$

For transition $i \to f$, the S-matrix element is given by

$$S_{fi} = \delta_{fi} - i(2\pi)^4(P_i - P_f)A \tag{9.30}$$

A = invariant matrix element

In perturbation theory,

$$A = \sum_n A^{(n)}, |A^{(n)}|^2 \propto \alpha^n, \ \alpha = \frac{e^2}{4\pi} = \frac{1}{137}$$

Retaining only the lowest nonvanishing term is the perturbative expansion

$$\begin{aligned} A \approx A^{(2)} &= -i(-ie)^2 < K'|J_\mu^{em}(0)|K>_e \frac{-ig^{\mu\nu}}{q^2 + i\in} < P'|J_\mu^{em}(0)|P>_{\text{proton}} \\ &= \frac{4\pi\alpha}{q^2}\ell^\mu h_\mu \end{aligned} \tag{9.31}$$

$$\begin{aligned} \text{Here } \ell^u &=< K'|J^\mu_{em}(0)|K>_e = \overline{u}_e(K')\gamma^\mu u_e(K) \\ h_\mu &=< P'|J_\mu^{em}(0)|P>_p = \overline{u}(P')\gamma_u u(P) \end{aligned} \tag{9.32}$$

assuming the proton to be a point-like Dirac particle.

The spin-averaged scattering cross section is given by

$$\sigma = \frac{1}{L}\int |\overline{A}|^2(2\pi)^4\delta_4(P_i - P_f)d\phi_f \tag{9.33}$$

$$\begin{aligned} \text{where } L = \text{luminosity} &= n_e n_p|\vec{v}_e - \vec{v}_p| = (2E_e)(2E_p)|\vec{v}_e - \vec{v}_p| \\ &= 4\sqrt{(K.P)^2 - m^2M^2} \end{aligned} \tag{9.34}$$

Neglecting the electron mass, the luminosity in the lab frame is given by

$$L \approx 4(K.P.) = 4ME \tag{9.35}$$

$$P_i - P_f = K + P - K' - P' = P + q - P', \quad q = K - K'$$

$$d\phi_f = \text{Final state phase space element} = \frac{d^3K'}{(2\pi)^3(2E_e')} \cdot \frac{d^3P'}{(2\pi)^3(2E_p')}$$

$$|\overline{A}|^2 = \frac{1}{4}\sum_{\text{spin}}|A|^2 \tag{9.36}$$

$$\text{Now} \quad \int \frac{d^3P'}{2E'p} = \int d^4P'\delta(P'^2 - M^2)\theta(P_0)$$

Carrying out the $\int d^3P'$ integration in (9.33) we get

$$\sigma = \frac{1}{4ME}\int |\overline{A}|^2 \cdot \frac{1}{2(2\pi)^2}\delta((P+q)^2 - M^2)\frac{d^3K'}{E'} \tag{9.37}$$

Now $\dfrac{d^3K'}{E'} = \dfrac{|\vec{K}'|^2 d|\vec{K}'| d\Omega}{E'} = \dfrac{|\vec{K}'|^2 E' dE' d\Omega}{E'} = E' dE' d\Omega$

and $(P+q)^2 - M^2 = P^2 + q^2 + 2P\cdot q - M^2 = q^2 + 2P\cdot q = q^2 + 2M(E-E')$

$$= -2EE'(1-\cos\theta) + 2M(E-E') = f(E')$$

using $\delta(f(x)) = \sum_i \dfrac{1}{|f'(x_i)|}\delta(x - x_i)$

where $f(x_i) = 0$, $f'(x_i) = \dfrac{d}{dx}f(x)$, at $x = x_i$, $f(E') = 0$ gives

$$-2EE'(1-\cos\theta) + 2M(E-E') = 0$$

or
$$E' = \frac{E}{1 + \frac{E}{M}(1-\cos\theta)} \tag{9.38}$$

Also
$$f'(E') = \frac{d}{dE'}f(E') = -2M\left[1 + \frac{E}{M}(1-\cos\theta)\right] = -2M\left(\frac{E}{E'}\right) \tag{9.39}$$

for E' given by (9.38)

Carrying out the E' integration in (9.37) we get

$$\left(\frac{d\sigma}{d\Omega}\right)^{el}_{\text{Lab}} = \frac{1}{64\pi^2 M^2}\cdot\left(\frac{E'}{E}\right)^2\cdot|\overline{A}|^2 \tag{9.40}$$

It remains to calculate $|A|^2$ which depends on the assumed dynamics.

From the definition of $|\overline{A}|^2$ and (9.31) we get

$$|\overline{A}|^2 = \frac{1}{4}\sum_{\text{spin}}|A|^2 = \left(\frac{4\pi\alpha}{q^2}\right)^2 L^{\mu\nu}H_{\mu\nu} \tag{9.41}$$

$$L^{\mu\nu} = \frac{1}{2}\sum_{\text{spins}}\ell^\mu\ell^{\nu+} = \frac{1}{2}\sum_{\text{spins}}\left[\overline{u}_e(K')\gamma'^\mu u_e(K)\right]\left[\overline{u}_e(K')\gamma^\nu u_e(K)\right]^+$$

$$= \frac{1}{2}\sum_{\text{spins}}\overline{u}_e(K')\gamma^\mu u_e * K)\overline{u}_e(K)\gamma^\nu u_e(K')$$

$$= \frac{1}{2}\text{Tr}\left[\gamma^\mu(K+m)\gamma^\nu(K+m)\right]$$

$$= \frac{1}{2}\text{Tr}\left[\gamma^\mu K\gamma^\nu K' + m^2\gamma^\mu\gamma^\nu\right]$$

$$= 2[K^\mu K'^\nu + K'^\mu K^\nu + g^{\mu\nu}(m^2 - K\cdot K')]$$

and $H_{\mu\nu} = \frac{1}{2}\sum_{\text{spins}} h_\mu h_\nu^+ = 2[P_\mu P'_\nu + P'_\mu P_\nu + g_{\mu\nu}(M^2 - P\cdot P')]$

This gives

$$L^{\mu\nu}H_{\mu\nu} = 8[(K.P)(K'.P') + (K.P')(K'.P) - M^2(K.K') \\ -m^2(P.P') + m^2M^2] \quad (9.42)$$

with $m^2 \approx 0$ and using Lab frame kinematics

$$K\cdot P = K'P' = ME, \quad K'P = K\cdot P' = ME'$$

$$q^2 + 2M(E - E') = 0, \quad q^2 \approx -2K.K' \approx -4EE'\sin^2\frac{\theta}{2}$$

(9.42) becomes

$$\begin{aligned} L^{\mu\nu}H_{\mu\nu} &= 8M^2\left[E^2 + E'^2 + \frac{q^2}{2}\right] \\ &= 8M^2\left[(E-E')^2 + 2EE' + \frac{q^2}{2}\right] \\ &= 16M^2EE'\left[1 + \frac{q^2}{4EE'} + \left(\frac{q^2}{2M}\right)^2\cdot\frac{1}{2EE'}\right] \\ &= 16M^2EE'\left[1 - \sin^2\frac{\theta}{2}\left(1 + \frac{q^2}{2M^2}\right)\right] \\ &= 16M^2EE'\left[\cos^2\frac{\theta}{2} - \frac{q^2}{2M^2}\sin^2\frac{\theta}{2}\right] \end{aligned} \quad (9.43)$$

Substituting (9.43) in (9.41) and the result in (9.40) we finally get

$$\begin{aligned} \left(\frac{d\sigma}{d\Omega}\right)^{ep\to ep}_{\text{Lab}} &= \frac{\alpha^2}{4E^2\sin^4\frac{\theta}{2}}\left(\frac{E'}{E}\right)\left[\cos^2\frac{\theta}{2} - \frac{q^2}{2M^2}\sin^2\frac{\theta}{2}\right] \\ &= \frac{\alpha^2}{4E^2\sin^4\frac{\theta}{2}}\cos^2\frac{\theta}{2}\left(\frac{E'}{E}\right)\left[1 - \frac{q^2}{2M^2}\tan^2\frac{\theta}{2}\right] \\ &= \left(\frac{d\sigma}{d\Omega}\right)_{\text{Mott}}\left(\frac{E'}{E}\right)\left[1 - \frac{q^2}{2M^2}\tan^2\frac{\theta}{2}\right] \end{aligned} \quad (9.44)$$

where $$\left(\frac{d\sigma}{d\Omega}\right)_{\text{Mott}} = \frac{\alpha^2}{4E^2\sin^4\frac{\theta}{2}}\cos^2\frac{\theta}{2} \quad (9.45)$$

is the differential cross section for elastic scattering of an ultrarelativistic electron from a static point charge.

However, it is found that the experimental data on high energy ep elastic scattering in the Lab frame is best fitted by a formula of the type

$$\left(\frac{d\sigma}{d\pi}\right)_{\text{Lab}}^{ep\to ep} = \left(\frac{d\sigma}{4E^2\sin^4\frac{\theta}{2}}\right)\left(\frac{E'}{E}\right)\left[A(q^2)\cos^2\frac{\theta}{2} + 2\tau B(q^2)\sin^2\frac{\theta}{2}\right] \quad (9.46)$$

$\tau = -q^2/4M^2$

The discrepancy is attributed to the fact that in deriving (9.41) we have assumed that the proton is also a point-like Dirac particle just like the electron. This assumption is clearly not correct since the proton is known to be an extended object with a finite charge radius and "anomalous" magnetic moment. So we have to take into account proton structure and strong interaction effects in order to derive the correct cross section formula.

However, the above pure QED result is applicable to elastic scattering of two highly relativistic point-like Dirac particles, e.g. the $e\mu$ scattering (or electron-quark scattering). It can be shown that when the masses of both particles can be neglected compared to their 3-momenta (9.41) becomes

$$|\overline{A}|^2 \approx (4\pi\alpha)^2\frac{2(s^2+u^2)}{t^2} \quad (9.47)$$

where $s = (K+P)^2 \approx 2K\cdot P = (K'+P')^2 \approx 2K'\cdot P'$

and $u = (K'-P)^2 \approx -2K'\cdot P = (K-P')^2 \approx -2K\cdot P'$

Carrying out the phase space integration in the CM frame, it can be shown that

$$\left(\frac{d\sigma}{d\Omega}\right)_{CM}^{el} = \frac{1}{64\pi^2 s}|\overline{A}|^2 \quad (9.48)$$

Substituting (9.47) in (9.48) one easily sees that for highly relativistic elastic scattering the differential cross-section formula becomes

$$\left(\frac{d\alpha}{d\pi}\right)_{CM}^{e\mu\to e\mu} = \frac{\alpha^2}{8E^2}\cdot\frac{1+\cos^4\frac{\theta}{2}}{\sin^4\frac{\theta}{2}} \quad (9.49)$$

where E is the beam energy and θ is the scattering angle in the CM frame.

9.1.4 Proton Structure Effect and Electromagnetic Form Factors

Due to strong interaction effects and compositeness, the matrix element of the electromagnetic current between proton states is not same as that for an elementary Dirac particle like the electron that is,

$$h_\mu = \langle P'|J_\mu^{em}(0)|P\rangle \neq \overline{u}(P')\gamma_\mu u(P)$$

However, because of Lorentz transformation properties, the most general form of h_μ is

$$h_\mu = \overline{u}(P')[f_1\gamma_\mu = f_2 P_\mu + f_3 P'_\mu]u(P) \tag{9.50}$$

where $f_i's(i = 1, 2, 3)$ are function's of Lorentz invariant variables

$$f_i = f_i(q^2),\ q = P' - P$$

(There is in this case only one Lorentz-invariant variable.) Because of gauge invariance, the *em* current is conserved.

$$\partial_\mu J^{\mu,em}(x) = 0$$

This implies

$$q^\mu h_\mu = 0 \tag{9.51}$$

From (9.50) and (9.51) we get

$$\begin{aligned} 0 &= \overline{u}(P')[f_1 q + f_2(q.P) + f_3(q.P')]u(P) \\ &= \overline{u}(P')[f_1(P' - P) + (f_2 - f_3)(P' \cdot P - M^2)]u(P) \\ &= \overline{u}(P')[f_1(M - M) + (f_2 - f_3)(P' \cdot P - M^2)]u(P) \end{aligned} \tag{9.52}$$

because $(\not{P} - m)u(P) = 0,\ \overline{u}(P')(\not{P}' - M) = 0,\ P^2 = P'^2 = M^2$

$$\text{Therefore, } f_2 = f_3 \tag{9.53}$$

Substituting in (9.50), we get

$$h_\mu(P, P') = \overline{u}(P')[f_1(q^2)\gamma_\mu + f_2(q^2)(P' + P)_\mu]u(P) \tag{9.54}$$

$f_1(q^2)$ and $f_2(q^2)$ are called form factors. Thus the proton has two electromagnetic form factors. In general, a particle of spin S has $(2S+1)$ independent from factors.

Gordon decomposition of the electromagnetic current

We can write

$$\begin{aligned} \overline{u}(P')\gamma^\mu u(P) &= \frac{1}{2M}\overline{u}(P')[M\gamma^\mu + \gamma^\mu M]u(P) \\ &= \frac{1}{2M}\overline{u}(P')[\not{P}'\gamma^\mu + \gamma^\mu \not{P}]u(P) \end{aligned}$$

$$= \frac{1}{2M}\overline{u}(P')[P'_\nu\gamma^\nu\gamma^\mu + \gamma^\mu\gamma^\nu P_\nu]u(P)$$

$$= \frac{1}{2M}\cdot\frac{1}{2}\overline{u}(P')[(P+P')_\nu(\gamma^\mu\gamma^\nu + \gamma^\nu\gamma^\nu)$$

$$+(P-P')_\nu(\gamma^\mu\gamma^\nu - \gamma^\nu\gamma^\mu)]u(P)$$

$$= \frac{1}{2M}\cdot\frac{1}{2}\overline{u}(P')[(P+P')_\nu\cdot 2g^{\mu\nu} - 2i\sigma^{\mu\nu}q_\nu]u(P)$$

$$= \frac{1}{2M}\overline{u}(P')[(P+P')^\mu - i\sigma^{\mu\nu}q_\nu]u(P) \qquad (9.55)$$

(9.55) can be written in the form

$$\overline{u}(P')(P+P')_\mu u(P) = \overline{u}(P')[2M\gamma_\mu + i\sigma_{\mu\nu}q^\nu]u(P) \qquad (9.56)$$

Eq. (9.56) is known as the Gordon decomposition of the electromagnetic current. Substituting (9.56) in (9.54) we get

$$\hbar_\mu = \left\langle P' \left| J_\mu^{em}(0) \right| P \right\rangle_{\text{proton}} = \overline{u}(P)\left[F_1(q^2)\gamma_\mu + \frac{i\sigma^{\mu\nu}q_\nu}{2M}F_2(q^2)\cdot u(P)\right] \quad (9.57)$$

where $F_1 = f_1 + 2Mf_2,\ F_2 = 2Mf_2$

$F_1(q^2)$ and $F_2(q^2)$ are known as Dirac form factors for the proton. By making a nonrelativistic reduction of the rhs. of (9.57) one gets the 'normalization' of the form factors, namely,

$$F_1(0) = Z_p = +1,\ F_2(0) = \Re_p = 1.79 \qquad (9.58)$$

where Z_p is the electric charge of the proton (in unit of e) and $\Re_p$ is the anomalous part of the magnetic moment of the proton (in unit of $e/2M$).

Proceeding in the same way as for the pure QED case with h_μ now given by eq. (9.57) one gets for the differential cross section of electron-proton scattering in the Lab frame the expression

$$\left(\frac{d\alpha}{d\pi}\right)^{el}_{\text{Lab}} = \left(\frac{\alpha^2}{4E^2\sin^4\frac{\theta}{2}}\right)\left(\frac{E'}{E}\right)\left[A(q^2)\cos^2\frac{\theta}{2} + 2\tau B(q^2)\sin^2\frac{\theta}{2}\right] \qquad (9.59)$$

where $A(q^2) = F_1^2 + \tau F_2^2,\ B(q^2) = (F_1+F_2)^2,\ \tau = -q^2/4M^2,\ F_i = F_i(q^2)$ $(i = 1, 2)$

Eq. (9.59) is known as the Rosenbluth formula.

Often one defines electric and magnetic form factors.

$$G_E(q^2) = F_1(q^2) - \tau F_2(q^2),$$

$$G_M(q^2) = F_1(q^2) + F_2(q^2), \qquad (9.60)$$

with $G_E(0) = F_1(0) = Zp = +1$,

$$G_M(0) = F_1(0) + F_2(0) = 1 + \Re_p = 2.79 = \mu_p$$

$G_E(0)$ is the electric charge (is unit of e) and $G_M(0)$ is the magnetic moment (in unit of $e/2M$) of the proton. In terms of $G_E(q^2)$ and $G_M(q^2)$, $A((q^2)$ and $B(q^2)$ become

$$A(q^2) = \frac{G_E^2 + \tau G_M^2}{1+\tau}, \quad B(q^2) = G_M^2 \tag{9.61}$$

Experimentally, one gets a "dipole fit" for these form factors, viz.

$$\frac{G_M(q^2)}{\mu_p} = G_E(q^2) \approx (1 - q^2/0.71)^{-2} \text{ for large } q^2 \text{ where } q^2 \text{ is in } (\text{GeV/C}^2)^2.$$

The "dipole fit" corresponds to an exponential charge distribution in a non-relativistic description.

$$G_E(q^2) = \int e^{i\vec{q}\cdot\vec{r}} \rho_E(r) d^3r$$

$$G_M(q^2) = \int e^{i\vec{q}\cdot\vec{r}} \rho_M(r) d^3r$$

$$\rho_E(r) \approx \rho_M(r) \alpha e^{-r/a} \tag{9.62}$$

One can similarly define electromagnetic form factors of the neutron with the normalization

$$F_1^n(0) = Z_n = 0, \; F_2^n(0) = \Re_n = -1.91$$

$$G_E^n(0) = Z_n = 0, \; G_M^n(0) = \mu_n = \Re_n = -1.91 \tag{9.63}$$

Experimentally, one cannot measure the electromagnetic form factors of the neutron directly. One studies electron-deuteron scattering at high energies and neglecting nuclear binding energies compared to incident electron energies one writes $\sigma_{eD} \approx \sigma_{ep} + \sigma_{en}$. From the eD and ep scattering data one can compute the electromagnetic form factors of the neutron.

9.2 Deep Inelastic (DI) Electron-proton Scattering

As the energy of the incident electron increases, particle production starts and the number of possible final states increases enormously. It becomes increasingly difficult, almost impossible, to study the final states individually. To extract information from experimental data one detects only the scattered electron and measures its angular and energy distribution and lets all the other final state particles go undetected. Thus one studies what is known as a one-particle inclusive process,

$$e^- + p \to e^- + X \tag{9.64}$$

where X is anything subject to the conservation laws. ($X = p$, $p\pi^0$, $n\pi^+$, $p\pi^+\pi^-$, pK^+K^-). The total 1-particle inclusive cross section in given by

$$\sigma^I = \sum_X \int d\sigma(e^- + p \to e^- + X) \tag{9.65}$$

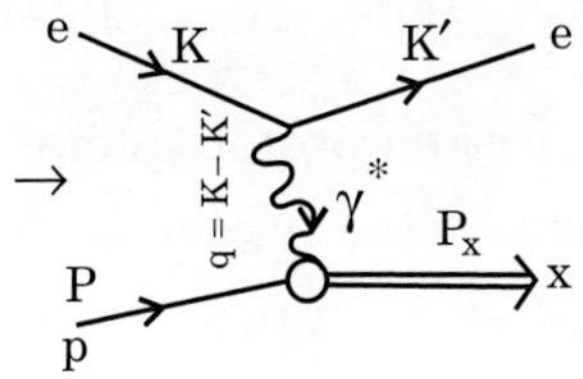

Fig. 9.4

Energy momentum conservation gives

$$K + P = K' + P_X$$

$$\text{or} \quad P_X = P + K - K' = P + q, \ q = K - K' \tag{9.66}$$

mass shell condition gives

$$K^2 = K^2 = m^2, \ P^2 = M^2, \ P_X^2 = W^2 \geq M^2$$

We shall use Lab frame: $P = (M, \vec{0}), K = (E, \vec{K}), K' = (E, \vec{E}')$

The electron is assumed to be ultrarelativistic $|\vec{K}|, |\vec{K}| \gg m$ so that $E \approx |\vec{K}|$, $E' \approx |\vec{K}|$. Thus we shall neglect the electron mass and wherever possible the proton mass also.

The Mandelstam variables for the process are

$$s = (K + P)^2 \approx 2K \cdot P = 2ME$$

$$t = (K - K')^2 \approx -2K \cdot K' = -2EE'(1 - \cos\theta) = -4EE' \sin^2 \frac{\theta}{2}$$

$$u = (K' - P)^2 \approx -2K' \cdot P = -2ME' \tag{9.67}$$

Instead of (t, u) one often works with what are known as DI variables

$$Q^2 = -q^2 \approx +4EE' \sin^2 \frac{\theta}{2}$$

$$\text{and}\quad \nu = \frac{P\cdot q}{M} = E - E' \tag{9.68}$$

or the "scaling variables"

$$x = \frac{Q^2}{2M\nu} = \frac{-t}{s+u} = \frac{2EE'\sin^2\frac{\theta}{2}}{M(E-E')}$$

$$\text{and}\quad y = \frac{\nu}{E} = \frac{E-E'}{E} = \frac{s+u}{s} \tag{9.68a}$$

$$S_{fi} = \delta_{fi} - i(2\pi)^4\delta_4(P_i - P_f)A$$

$$A = \sum_n A^{(n)},\ \left|A^{(n)}\right|^2 \propto \alpha^n,\ \alpha = \frac{e^2}{4\pi} = \frac{1}{137} \tag{9.69}$$

One assumes that the electron-proton interaction is dominantly electromagnetic. Moreover, because of the smallness of the fine structure constant one can retain only the lowest non-vanishing term in the perturbation expansion. Thus

$$A \approx A^{(2)} = \frac{4\pi\alpha}{q^2}\ell^\mu h_\mu$$

$$\text{where}\ \ell^\mu = \langle K'|J^{\mu\ell m}(0)|K\rangle_e = \overline{u}_e(K')\gamma^\mu u_e(K)$$

$$h_\mu = \langle X; P_x|J_\mu^{em}(0)|p; P\rangle \tag{9.70}$$

The total inclusive (spin-averaged) cross section is given by

$$\begin{aligned}\sigma^I &= \sum_x \int d\sigma(e+p\to e+X)\\ &= \frac{1}{L}\sum_x \int |\overline{A}|^2\cdot(2\pi)^4\delta_4(P+q-P_x)d\phi_f\\ &= \frac{1}{L}\sum_x \int |\overline{A}|^2\cdot(2\pi)^4\delta_4(P+q-P_x)d\phi_X \times \frac{d^3K'}{(2\pi)^3(2E')}\end{aligned} \tag{9.71}$$

$$\begin{aligned}\text{where}\ L &= (2E_e)(2E_p)|\vec{v}_e - \vec{v}_p| = 4\sqrt{(K.P)^2 - m^2M^2}\\ &\approx 4(K.P)\\ &= 2s = 4ME\end{aligned} \tag{9.72}$$

We can write (9.71) as

$$E'\frac{d\sigma^I}{d^3K'} = \frac{1}{2s}\sum_x \int |\overline{A}|^2\cdot(2\pi)^4(P+q-P_X)d\phi_X \tag{9.73}$$

Now $$\frac{d^3K'}{E'} = \frac{|\vec{K}'|^2 d|\vec{K}'|d\Omega}{E'} = E'dE'd\Omega \tag{9.74}$$

From (9.70) we can write

$$|\overline{A}|^2 = \frac{1}{4}\sum_{\text{spins}} |A|^2 = \left(\frac{4\pi\alpha}{q^2}\right)^2 L^{\mu\nu}H_{\mu\nu} \tag{9.75}$$

where $$L^{\mu\nu} = \frac{1}{2}\sum_{\text{spin}} \ell^\mu \ell^{\nu+} = 2[K^\mu K'^\nu + K'^\mu K^\nu + g^{\mu\nu}(m^2 - K.K')] \tag{9.76}$$

and $$H_{\mu\nu} = \frac{1}{2}\sum_{\text{spin}} h_\mu h_\nu^+$$

Substituting (9.75) and (9.74) in (9.73) we get

$$\left(\frac{d^2\sigma^I}{dE'd\Omega}\right)_{\text{Lab}} = \frac{\alpha^2}{Q^4}\cdot\frac{E'}{E}L^{\mu\nu}W_{\mu\nu} \tag{9.77}$$

where $L^{\mu\nu}$ is given by (9.76) and

$$W_{\mu\nu} = \frac{1}{8\pi M}\sum_{\text{spins}}\sum_{x}\int h_\mu h_\nu^+ (2\pi)^4 \delta_4(P+q-P_X)d\phi_X \tag{9.78}$$

using the relations

$$(2\pi)^4\delta_4(P+q-P_X) = \int d^4x e^{i(P+q-P_x)x}$$

$$J_\mu(x) = e^{-i\hat{p}\cdot x}J_\mu(0)e^{i\hat{p}\cdot x} \quad [\hat{p} = \text{momentum operator}]$$

and $$\sum_X \int d\phi_x |X;P_x><X;P_x| = 1 \tag{9.79}$$

We can write (9.78) as

$$W_{\mu\nu} = \frac{1}{8\pi M}\sum_{\text{spins}}\int d^4x e^{iq\cdot x}\langle P|J_\mu(0)J_\nu(x)|P\rangle \tag{9.80}$$

clearly $W_{\mu\nu} = W_{\mu\nu}(P,q)$

General form of $W_{\mu\nu}(P,q)$

One can write down a general form of $W_{\mu\nu}(P,q)$ from symmetry considerations alone. First, $W_{\mu\nu}$ transforms like a second-rank tensor under Lorentz transformations. Secondly, the fact that J_μ^{em} is hermitean implies that $W_{\mu\nu} = W_{\nu\mu}^*$. Thirdly, since $L_{\mu\nu}$ is symmetric, only the symmetric part of $W_{\mu\nu}$ contributes (This also follows from time-reversal invariance of electromagnetic interactions). Thus $W_{\mu\nu} = W^*_{\nu\mu} = W^*_{\mu\nu}$. This means $W_{\mu\nu}$ is a real symmetric tensor of rank-2 to be constructed out of the metric tensor $g_{\mu\nu}$ and the available 4-vectors, namely, P_μ and q_ν. ($W_{\mu\nu}$ can not contain $\gamma_\mu\, \gamma_\nu$ since proton spins have been summed over in the definition of $W_{\mu\nu}$).

Therefore we can write

$$W_{\mu\nu}(P,q) = w_1 g_{\mu\nu} + w_2 P_\mu P_\nu + w_3 q_\mu q_\nu + w_4(P_\mu q_\nu + q_\mu P_\nu) \tag{9.81}$$

where w_i's are real Lorentz-invariant quantities, being functions of Lorentz invariant variables $P \cdot q$ and q^2. Finally gauge invariance implies that the electromagnetic current is conserved, $\partial_\mu J^{\mu,em}(x) = 0$. Then from the definition of $W_{\mu\nu}$ it follows that

$$q^\mu W_{\mu\nu} = 0 = W_{\mu\nu} q^\nu \tag{9.82}$$

From (9.81) and (9.82) it follows that

$$0 = q^\mu W_{\mu\nu} = w_1 q_\nu + w_2 (q \cdot P) P_\nu + w_3 q^2 q_\nu + w_4\{(P \cdot q) q_\nu + q^2 P_\nu\}$$

Equating to zero coefficients of q_ν and P_ν, independently we get

$$w_1 + w_3 q^2 + w_4 (P \cdot q) = 0 \quad \text{and} \quad w_2 (q \cdot P) + w_4 q^2 = 0$$

Thus we can write

$$w_4 = -w_2 \left(\frac{q \cdot P}{q^2}\right) \quad \text{and} \quad w_3 = -\frac{1}{q^2} w_1 + \left(\frac{q \cdot P}{q^2}\right)^2 w_2 \tag{9.83}$$

Substituting (9.83) in (9.81) we finally get

$$W_{\mu\nu}(q,P) = W_1 \left(-g_{\mu\nu} + \frac{q_\mu q_\nu}{q^2}\right) + \frac{W_2}{M^2} \left(P_\mu - \frac{q \cdot P}{q^2} q_\mu\right) \left(P_\nu - \frac{q \cdot P}{q^2} q_\nu\right)$$

where $W_1 = -w_1$ and $W_2 = w_2 M^2$ (9.84)

$W_i = W_i(Q^2, \nu), (i = 1, 2)$ are called the "structure functions" of the proton.

Substituting (9.76) and (9.84) in (9.77) we get

$$\left(\frac{d^2\sigma^I}{dE' d\Omega}\right)_{\text{Lab}} = \left(\frac{\alpha^2}{4E^2 \sin^4 \frac{\theta}{2}}\right) \left[W_2(Q^2,\nu) \cos^2 \frac{\theta}{2} + 2W_1 \sin^2 \frac{\theta}{2}\right] \tag{9.85}$$

Eq. (9.85) was first derived by Drell and Walecka.

Bjorken Scaling

Analysis of experimental data showed that for $Q^2 \to \infty, \nu \to \infty$ with $x = \frac{1}{w} = Q^2/2M\nu$ held fixed and finite νW_2 and W_1 become functions of x only,

$$\nu W_2(Q^2,\nu) \to F_2(x), \quad 2MW_1(Q^2\nu) \to F_1(x)$$

Moreover, $\frac{1}{x}F_2(x) \approx F_1(x)$ (9.86)

which is known as the Callan-Gross relation.

This was first predicted by Bjorken from current algebra analysis and is known as "Bjorken Scaling". Initially, the variable ω was used but later on, the variable x was found to be physically more meaningful. (ω is known as the Bjorken variable and x as Feynman variable)
with $0 \leq x \leq 1, \ 1 \leq w \leq \infty$

9.3 Inclusive ep Scattering as (Virtual) γP Total Inclusive Scattering

The hadron vertex in DI electron-proton scattering may be regarded as the scattering of the proton by a virtual photon.

$$\gamma^* + p \to X$$

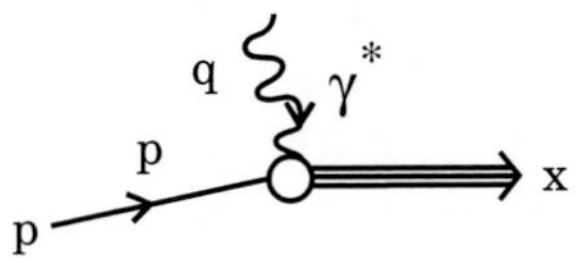

Fig. 9.5

energy-momentum conservation gives

$$q + P = P_X$$

$$q^2 < 0, \ P^2 = M^2, \ P_X^2 \geq M^2$$

$$W^2 = P_X^2 = (q+p)^2 = M^2 + 2M\nu + q^2$$

$$\nu = \frac{p \cdot q}{M}, \ Q^2 = -q^2 > 0$$

Define $K = \dfrac{W^2 - M^2}{2M} = \nu + \dfrac{q^2}{2M}$

For a real photon, $q^2 = 0$ and $K = \nu = q_0 =$ photon energy in the rest frame of the proton.

In analogy with scattering by a real photon, we write (for a virtual photon, energy is not a well defined quality)

$$\sigma_{\text{tot}}^{\lambda}(\gamma^* + p \to X) = \frac{4\pi^2\alpha}{K} \in_{\lambda}^{\mu} \in_{\lambda}^{\nu *} W_{\mu\nu}(P,q) \tag{9.87}$$

A virtual photon can be both transversely and longitudinally polarized. For transverse polarization, $\lambda = \pm 1, \in^{\mu} = \mp\frac{1}{\sqrt{2}}(0, 1, \pm i, 0)$ and for longitudinal polarization,

$$\lambda = 0,\ \in^{\mu} = \frac{1}{\sqrt{q^2}}((\sqrt{\nu^2 - q^2}, 0, 0, \nu))$$

$$q = (\nu, 0, 0, \sqrt{\nu^2 - q^2}) \quad (q \cdot \in = 0,\ \in^2 = -1)$$

$$\sigma_T = \frac{1}{2}(\sigma^+_{tot} + \sigma^-_{tot}) = \frac{4\pi^2\alpha}{K} W_1(Q^2, \nu) \tag{9.88}$$

$$\sigma_L = \sigma^0_{tot} = \frac{4\pi^2\alpha}{K}\left[\left(1 - \frac{\nu^2}{q^2}\right) W_2(Q^2, \nu) - W_1(Q^2, \nu)\right] \tag{9.89}$$

One can write

$$\left(\frac{d^2\sigma^I}{dE'd\Omega}\right)_{\text{Lab}} = \Gamma(\sigma_T + a\sigma_L) \tag{9.90}$$

where σ_T and σ_L are given by (9.88) and (9.89) and

$$a = \left[1 - 2\left(\frac{\nu^2 - q^2}{q^2}\right)\tan^2\frac{\theta}{2}\right]^{-1}$$

and $$\Gamma = \frac{\alpha}{2\pi^2} \cdot \frac{K}{Q^2} \cdot \frac{E'}{E} \cdot \frac{1}{1 - a} \tag{9.91}$$

From (9.88) and (9.89) one gets

$$R = \frac{\sigma_L}{\sigma_T} = \frac{\left(1 - \frac{\nu^2}{q^2}\right) W_2(Q^2\nu) - W_1(Q^2\nu)}{W_1(Q^2, \nu)} \tag{9.92}$$

In the scaling limit $\left(Q^2 \to \infty, \nu \to \infty, x = \frac{Q^2}{2M\nu} = \text{fixed}\right)$

$$\nu W_2(Q^2\nu) \to F_2(x),\ 2MW_1(Q^2, \nu) \to F_1(x)$$

and (9.92) becomes

$$R \approx \frac{\frac{1}{x}F_2(x) - F_1(x)}{F_1(x)} \approx 0, \quad \text{since } F_2(x) \approx xF_1(x) \tag{9.93}$$

Finally, for comparison with the "parton model" results, to be discussed later it is convenient to write the double differential cross section for inclusive ep scattering in terms of variables x, y defined in (9.68a). One finds

$$\frac{d^2\sigma_1}{dxdy} = \frac{2\pi\alpha^2}{Q^4} s\left[\left(1 - y - \frac{Mxy}{2E}\right)\hat{F}_2(x, q^2) + xy^2\hat{F}_1(x, Q^2)\right] \tag{9.94}$$

where $\hat{F}_2(x, Q^2) = \nu W_2(Q^2, \nu)$ and $\hat{F}_1(x, Q^2) = 2MW_1(Q^2, \nu)$

We can state "scaling" in the following way.

For $Q^2 \to \infty$

$$\hat{F}_2(x_1, Q^2) \to F_2(x)$$
$$\hat{F}_1(x_1, Q^2) \to F_1(x) \tag{9.95}$$

i.e. the structure functions become independent of Q^2 in the scaling limit.

9.4 The Parton Model

The parton model was suggested by Feynman and developed subsequently by Bjorken and Paschos (1968) in order to explain the observed "Scaling" behavior of the structure functions involved in the description of the DI scattering. The model is basically a subnuclear version of the impulse approximation used in NRQM to explain the observed scattering of high energy particles from a composite system with loosely bound constituents. In the parton model, the proton is assumed to be made of weakly interacting point-like constituents called "partons" (= parts of the proton) which in the infinite-momentum frame move almost in the same direction as the parent proton ($\langle p_T \rangle / |\vec{P}| \ll 1, p_L = z|\vec{P}|$, $0 \le z \le 1$. The partons can be both charged or neutral. Only the charged partons participate in the scattering process and are assumed to have spin $-\frac{1}{2}$ like the electron. The electron interacts with only one parton at a time. The rest of the partons go unaffected and recombine with the scattered parton in order to form the final state hadrons. The total inclusive cross section is thus taken to be an incoherent sum of individual electron-parton scattering cross sections convoluted over the parton momentum distribution function. This means that in computing the total cross section one neglects both the initial state and final state parton-parton interactions. Thus in the parton model, the process is given by the following diagram.

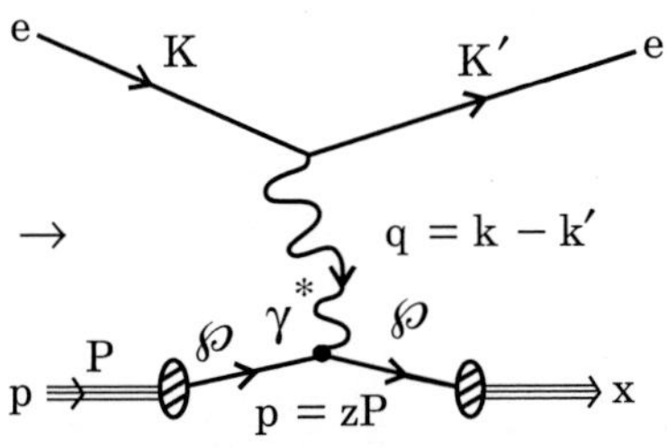

Fig. 9.6

Kinematics

Observed process

$$e^- + p \to e^- + X$$

$$K + P = K' + P'$$

$$q = K - K'$$

$$Q^2 = -q^2 = -(K - K')^2 \approx +4EE' \sin^2 \frac{\theta}{2}$$

$$\nu = \frac{2p \cdot q}{M}$$

$$K^2 = K'^2 = m^2 = 0$$

$$P^2 = M^2,\ W^2 = (P + q)^2 \geq M^2$$

Parton-level process

$$e^- + \wp \to e^- + \wp$$

$$K + p = K' + p'$$

$$p = zP,\ 0 \leq z \leq 1$$

$$q = K - K'\ p' = p + q$$

$$K^2 = K'^2 = m^2 \approx 0$$

$$p^2 = p'^2 = m_\wp^2 \approx 0 \tag{9.96}$$

In the parton model

$$\begin{aligned} \sigma^I &= \sum_X \int d\sigma(e^- + p \to e^- + X) \\ &\approx \sum_\wp \int \int d\sigma(e^- + \wp \to e^- + \wp) f_\wp(z) dz \end{aligned} \tag{9.97}$$

$f\wp(z)dz$ = probability of the parton $\wp$ having a fraction between z and $z + dz$ of the proton momentum.

Momentum conservation gives the normalization

$$\sum_\wp \int_0^1 z f_\wp(z) dz = 1 \tag{9.98}$$

Now $d\sigma(e^- + \wp \to e^- + \wp)$

$$= \frac{1}{L_{e\wp}} |\overline{A}|^2 \cdot (2\pi)^4 \delta_4(p + q - p') \frac{d^3K'}{(2\pi)^3(2E')} \frac{d^3p'}{(2\pi)^3(2E'_{\wp'})} \tag{9.99}$$

where $L_{e\wp} = 4\sqrt{(K\cdot p)^2 - m^2 m_\wp^2}$

$$\approx 4(K\cdot p) = 4(K\cdot P)z = (4ME)z \tag{9.100}$$

$$|\overline{A}|^2 = \frac{1}{4}\sum_{\text{spins}} |A|^2$$

$$A \approx A^{(2)} = \frac{4\pi\alpha}{q^2} e_\wp \ell^\mu h_\mu^{(\wp)}$$

$$\ell^\mu = \langle k'|J^{\mu,em}(0)|K\rangle_e = \overline{u}_e(K')\gamma^\mu u_e(K)$$

$$h_\mu^{(\wp)} = \langle p'|J_\mu^{em}(0)|p\rangle_\wp = \overline{u}_\wp(p')\gamma_\mu u_\wp(p) \tag{9.101}$$

Using (9.101) we get

$$|\overline{A}|^2 = \left(\frac{4\pi\alpha}{Q^2}\right)^2 e_\wp^2 L^{\mu\nu} H_{\mu\nu}^{(\wp)} \tag{9.102}$$

$$L^{\mu\nu} = \frac{1}{2}\sum_{\text{spins}} \ell^\mu \ell^{\nu+} = 2[K^\mu K'^\nu + K'^\mu K^\nu + g^{\mu\nu}(m^2 - K\cdot K')]$$

$$H_{p\mu}^{(\wp)} = \frac{1}{2}\sum_{\text{spins}} h_\mu h_\nu^+ = 2[p_\mu p'_\nu + p'_\mu p_\nu + g_{\mu\nu}(m\wp^2 - p\cdot p')] \tag{9.103}$$

$$\frac{d^3K'}{E'} \approx E' dE' d\Omega \tag{9.104}$$

Using (9.99)–(9.104) we get from (9.97)

$$\left(\frac{d\sigma^I}{dE'd\Omega}\right)_{\text{Lab}} = \frac{\alpha^2}{Q^4}\cdot\frac{E'}{E} L^{\mu\nu} W_{\mu\nu}^{(\wp)} \tag{9.105}$$

where $$W_{\mu\nu}^{(\wp)} = \frac{1}{(2\pi)(2M)}\sum_\wp \int\int \frac{1}{z} e_\wp^2 H_{\mu\nu}^{(\wp)}\cdot(2\pi)^4\delta_4(p+q-p')$$
$$\times\frac{d^3p'}{(2\pi)^3(2E_{p'})}\cdot f_\wp(z)dz \tag{9.106}$$

Here $W_{\mu\nu}^{(\wp)}$ is the hadronic tensor as calculated in the parton model.

Now $$\int \frac{d^3p'}{2Ep'} = \int d^4p'\delta(p'^2 - m^2)\theta(p'0) \tag{9.107}$$

Carrying out the parton momentum integration in (9.106) using (9.107) we get

$$W_{\mu\nu}^{(\wp)} = \frac{1}{2M}\sum_\wp \int \frac{1}{z} f_\wp e_\wp^2 H_{\mu\nu}^{(\wp)}\delta((p+q)^2 - m_\wp^2)dz \tag{9.108}$$

Now $\delta((p+q)^2 - m_\wp^2) = \delta(q^2 + 2p\cdot q) = \delta(-Q^2 + 2zP\cdot q) = \delta(-Q^2) + 2zM\nu)$

$$= \frac{1}{2M\nu}\delta(z - x), \; x = Q^2/2M\nu \tag{9.109}$$

Using (9.109) we can carry out the z-integration in (9.108) to get

$$W_{\mu\nu}^{(\wp)} = \frac{1}{2M}\sum_{\wp}\frac{1}{x}f_\wp(x)e_\wp^2\frac{1}{2M\nu}H_{\mu\nu}^{(\wp)} \; (p = xP, \; p' = p + q) \tag{9.110}$$

$$\begin{aligned} H_{\mu\nu}^{(\wp)} &= 2[p_\mu p'_\nu + p'_\mu p_\nu + g_{\mu\nu}(m_\wp^2 - p\cdot p')] \\ &= 2\left[p_\mu(p+q)_\nu + (p+q)_\mu p_\nu + g_{\mu\nu}\left\{m_\wp^2 - p\cdot(p+q)\right\}\right] \\ &= 2[2p_\mu p_\nu + (p_\mu q_\nu + q_\mu p_\nu) - g_{mu\nu}(p\cdot q)] \\ &= 2[2x^2 P_\mu P_\nu + x(P_\mu q_\nu + q_\mu P_\nu) - x g_{\mu\nu}(P\cdot q)] \\ &= 2x[2xP_\mu P_\nu + (P_\mu q_\nu + q_\mu P_\nu) - g_{\mu\nu}(M\nu)] \end{aligned} \tag{9.111}$$

Substituting (9.111) in (9.110) we finally get

$$W_{\mu\nu}^{(\wp)} = \frac{1}{2M^2\nu}\sum_p e_\wp^2 f_\wp(x)[-g_{\mu\nu}(M\nu) + 2xP_\mu P_\nu + (P_\mu q_\nu + q_\mu P_\nu)] \tag{9.112}$$

Comparing with the general expression for $W_{\mu\nu}$ obtained earlier from symmetry considerations, viz.

$$W_{\mu\nu} = W_1(Q^2,\nu)\left(-g_{\mu\nu} + \frac{q_\mu q_\nu}{q^2}\right) + \frac{W_2(Q^2,\nu)}{M^2}\left(P_\mu - \frac{P\cdot q}{q^2}q_\mu\right)\times\left(P_\nu - \frac{P\cdot q}{q^2}q_\nu\right) \tag{9.113}$$

(Because of gauge invariance, only the $g_{\mu\nu}$ and $P_\mu P_\nu$ term contribute to the product $L^{\mu\nu}W_{\mu\nu}$), we get in the parton model,

$$W_1(Q^2,\nu) = \frac{1}{2M}\sum_\wp e_\wp^2 f_\wp(x)$$

$$W_2(Q^2,\nu) = \frac{1}{\nu}\sum_\wp e_\wp^2 x f_\wp(x)$$

or $$2MW_1(Q^2,\nu) = \sum_\wp e_\wp^2 f_\wp(x) = F_1(x) \tag{9.114}$$

$$\nu W_2(Q^2,\nu) = \sum_\wp x e_\wp^2 f_\wp(x) = F_2(x) \tag{9.115}$$

Thus in the parton model, the structure functions scale exactly because they are functions of x only. Furthermore,

$$\frac{1}{x}F_2(x) = F_1(x) \tag{9.116}$$

The Callan-Gross relation (9.116) is a consequence of the assumed spin-$\frac{1}{2}$ nature of the charged partons. If the charged partons were spin-0 particles, then

$$h_\mu^{(\wp)} = \langle p'|J_\mu^{em}(0)|p\rangle = (p'+p)_\mu$$

and $$H_{\mu\nu}^{(\wp)} = h_\mu^{(\wp)} h_\nu^{(\wp)+} = (p'+p)_\mu (p'+p)\nu = (2p+q)_\mu (2p+q)_\nu$$

$$= 4p_\mu p_\nu + 2(P_\mu q_\nu + q_\mu p_\nu) + q_p q_\nu$$

$$= 4x^2 P_\mu P_\nu + 2x(P_\mu q_\nu + q_\mu P_\nu) + q_\mu q_\nu \tag{9.117}$$

As a consequence, there will be no $g_{\mu\nu}$ term in $W_{\mu\nu}^{(\wp)}$ and hence $W_1(Q^2,\nu) = 0$ which means

$$\sigma_T = \frac{4\pi^2\alpha}{K} W_1(Q^2,\nu) = 0$$

and $$R = \frac{\sigma_L}{\sigma_T} \to \infty \tag{9.118}$$

contrary to experimental observation ($R = 0$ in the scaling limit).

Experimentally, DI electron scattering from many isoscalar ($I = 0$) nuclear targets (e.g. H^2, C^{12}, Fe^{56} etc.) have been studied. To describe them, one defines average nucleon structure functions.

$$F_i^{eN}(x) = \frac{1}{2}[F_i^{ep}(x) + F_i^{en}(x)] \quad (i = 1,2) \tag{9.119}$$

e.g. Neglecting nuclear binding,

$$F_i^{eD}(x) \approx F_i^{ep}(x) + F_i^{en}(x) \approx 2F_2^{eN}(x) \tag{9.120}$$

It is tempting to identify the charged partons with the quarks and antiquarks $(q,\overline{q})$ of the quark model that was earlier successfully applied to explain low-energy hadron properties. Thus in the quark-parton model.

$$F_2^{ep}(x) = \sum_{\wp/P} e_p^2 x f_\wp^p(x) = x[e_u^2\{f_u^p(x) + f_{\overline{u}}^p(x)\} + e_d^2\{f_d^p(x) + f_{\overline{d}}^p(x)\} + e_s^2\{f_s^p(x) + f_{\overline{s}}^p(x)\}] \tag{9.121}$$

Similarly,

$$F_2^{en}(x) = \sum_{\wp/n} e_p^2 x f_\wp^n(x) = x[e_u^2\{f_u^n(x) + f_u^n(x)\}e_d^2\{f_d^n(x) + f_d^n(x)\}$$
$$+ e_s^2\{f_s^n(x) + f_x^n(x)\}] \quad (9.122)$$

Isospin symmetry gives

$$f_u^p(x) = f_d^n(x) = u(x)$$
$$f_d^p(x) = f_u^n(x) = d(x)$$
$$f_s^p(x) = f_s^n(x) = s(x) \quad (9.123)$$

One has similar, relations for antiquarks.

For simplicity, we restrict to 2-flavor case. Then

$$F_2^{ep}(x) = x[e_u^2\{u(x) + \overline{u}(x)\} + e_d^2\{d(x) + \overline{d}(x)\}]$$

and $$F_2^{en}(x) = x[e_u^2\{d(x) + \overline{d}(x)\} + e_d^2\{u(x) + \overline{u}(x)\}] \quad (9.124)$$

Then $$F_2^{eN}(x) = \frac{1}{2}[F_2^{ep}(x) + F_2^{en}(x)]$$
$$= \langle e_\wp^2\rangle x\{u(x) + d(x) + \overline{u}(x) + \overline{d}(x)\} \quad (9.125)$$

where $$\langle e_\wp^2\rangle = \frac{1}{2}(e_u^2 + e_d^2) = \frac{1}{2}\left(\frac{4}{9} + \frac{1}{9}\right) = \frac{5}{18} = 0.278 \quad (9.126)$$

Experimental data gives

$$\int_0^1 \cdot F_2^{eN}(x)dx = 0.15$$

Using (9.125) and (9.126) we get

$$\int_0^1 x[u(x) + d(x) + \overline{u}(x) + \overline{d}(x)]dx = 0.54 \quad (9.127)$$

However, momentum conservation gives

$$\sum_\wp \int_0^1 x f_\wp(x)dx = 1 \quad (9.128)$$

This implies that the charged partons carry about 50% of the total proton momentum. The neutral partons carry away the remaining 50% of the proton momentum. It is natural to identify the neutral partons with "gluons", the quanta that mediate the strong interaction between quarks.

9.4.1 Sum Rules

If the proton is made up of constituents, its quantum numbers should be calculable from those of its constituents. Of the three flavor quantum numbers (electric charge Q, third component of i-spin I_3 and baryon number B) only 2 are independent because of Gell Mann-Nishijima relation $Q = I_3 + B/2$.

Therefore, we have

$$\frac{1}{2}\int_0^1 \{[u(x) - \overline{u}(x)] - [d(x) - \overline{d}(x)]\}dx = I_3^{(p)} = +\frac{1}{2}$$

and $$\frac{1}{3}\int_0^1 \{[u(x) - \overline{u}(x)] - [d(x) - \overline{d}(x)]\}dx = B(p) = +1$$

Solving the two simultaneous equations we get

$$\int_0^1 [u(x) - \overline{u}(x)]dx = 2$$

$$\int_0^1 [d(x) - \overline{d}(x)]dx = 1 \tag{9.129}$$

The above sum rules agree with the low energy quark model description of the proton, namely, the proton is a bound state of 2 u-quarks and 1 d-quark.

9.4.2 Gottfried Sum Rule

Consider the following integral

$$I_G = \int_0^1 \frac{1}{x}[F_2^{ep}(x) - F_2^{en}(x)]dx \tag{9.130}$$

In the quark-parton model, the nucleon structure functions are related to quark and antiquark distribution functions for the proton. Thus we get

$$\begin{aligned}(I_G)_{\text{Parton model}} &= \int_0^1 [e_u^2\{u(x) + \overline{u}(x)\} + e_d^2\{d(x) + \overline{d}(x)\} \\ &\qquad - e_u^2\{d(x) + \overline{d}(x)\} - e_d^2\{u(x) + \overline{u}(x)\}] \\ &= (e_u^2 - e_d^2)\int_0^1 [\{u(x) + \overline{u}(x)\} - \{d(x) + \overline{d}(x)\}] \\ &= (e_u^2 - e_d^2)\int_0^1 [\{u(x) - \overline{u}(x)\} - \{d(x) - \overline{d}(x)\} \\ &\qquad + 2\{\overline{u}(x) - \overline{d}(x)\}]dx\end{aligned}$$

$$= \left(\frac{4}{9} - \frac{1}{9}\right)\left[(2-1) + 2\int_0^1 \{\overline{u}(x) - \overline{d}(x)\}dx\right]$$
$$= \frac{1}{3} - \frac{2}{3}\delta, \ \delta = \int_0^1 [\overline{d}(x) - \overline{u}(x)]dx$$
$$= \frac{1}{3} \tag{9.131}$$

assuming an i-spin symmetric sea for which $\delta = 0$.

The latest experimental value of I_G as measured by the NMC is

$$(I_G)_{\text{expt}} = 0.240 \pm 0.016 \tag{9.132}$$

This implies that the assumption of an i-spin symmetric quark-gluon sea is not correct (There may also be contributions from other sources.)

9.5 Deep-Inelastic (DI) Neutrino-Nucleon (νN) Scattering

In order to get more information, if any, of the nucleon structure and also for further tests of the parton model, one would like to use other high energy point-like particles as probes. Among the other options μN scattering is expected to have the same properties as the eN scattering. νN scattering might give new information because the neutrino has only weak interaction. However, neither the muon nor the neutrino can be accelerated in an accelerator since the muon is charged but unstable and the neutrino is stable but uncharged and one can accelerate only stable charged particles in an accelerator. High energy muon and neutrino beams are produced by secondary means. High energy π's and K's are first produced by hadron-hadron collisions. The decay products of πs and Ks contain μ's and ν_μ's and their antiparticles. μ's are separated out by applying magnetic fields, ν_μ's are not affected. The decay products are then made to pass through lead blocks. All particles other than the ν_μ's get absorbed and one obtains a pure ν_μ bcam.

At low energies, one has only elastic and quasi-elastic charge exchange scattering.

$$\nu_\mu + N \to \nu_\mu + N, \overline{\nu}_\mu + N \to \overline{\nu}_\mu + N (N = p, n)$$
$$\text{and} \quad \nu_\mu + n \to \mu^- + p, \ \overline{\nu}_\mu + p \to \mu^+ + n \tag{9.133}$$

As the energy of the neutrino increases, particle production takes place and the number of final states increases. As in the case of DI eN scattering, one detects only the final state lepton ($\mu^{\pm}$) and leaves all other particles undetected. That is, one studies one-particle inclusive processes.

$$\nu_\mu + N \to \mu^- + X,\ \overline{\nu}_\mu + N \to \mu^+ X' \tag{9.134}$$

To be specific, let us consider the process

$$\nu_\mu + p \to \mu^- + X \tag{9.135}$$

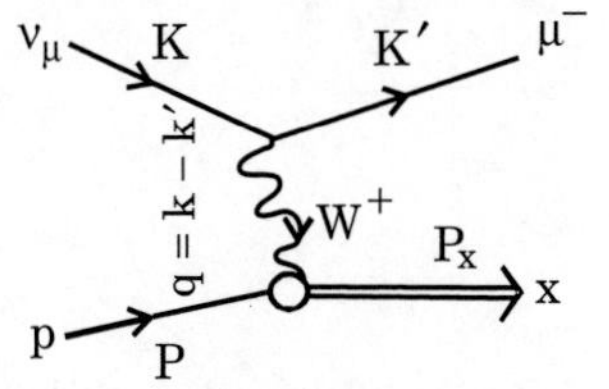

Fig. 9.7

The total inclusive cross section is given by

$$\sigma^{I,\nu} = \sum_X \int d\sigma(\nu_\mu + p \to \mu^- + X) \tag{9.136}$$

$$k + P = k' + P_X \ \Rightarrow\ P_X = p + q, q = k - k'$$

The kinematics for DI νN scattering is the same as in the ep case. Therefore, we have

$$s = (k+P)^2 \approx 2k \cdot P = 2ME$$

$$t = q^2 = (k-k')^2 \approx -2k \cdot k' \approx -4EE' \sin^2 \frac{\theta}{2}$$

$$u = (k'-P)^2 \approx -2k' \cdot P \approx -2ME'$$

$$Q^2 = -q^2 \approx +4EE' \sin^2 \frac{\theta}{2}$$

$$\nu = \frac{p \cdot q}{M} \approx E - E'$$

$$x = \frac{Q^2}{2M\nu}, y = \frac{\nu}{E} = \frac{E-E'}{E}, 0 \leq x, y \leq 1 \tag{9.137}$$

9.5.1 Dynamics

The dynamics is simplified by making the following valid assumptions

(a) leptons are pt. like Dirac particles.

(b) Neutrinos have only weak interactions.

(c) Perturbation theory is applicable and therefore only the non-vanishing lowest order term in the perturbative expansion of the S-matrix is retained.

For, $Q^2 \langle\langle M_W^2$, we can use the V-A theory of weak interactions.

$$L_{\text{int}} = \frac{G_F}{\sqrt{2}}(J_\mu^w J^{\mu,w^+} + \hbar \cdot c) \tag{9.138}$$

We have

$$d\sigma(\nu_\mu + p \to \mu^- + X) = \frac{1}{L_{\nu p}} \int |\overline{A}|^2 (2\pi)^4 \delta_4(P + q - P_X) d\phi_f \tag{9.139}$$

where $L_{\nu p} \approx 4(k \cdot P) = 4ME$

$$d\phi_f = \frac{d^3k'}{(2\pi)^3(2E_{k'})} d\phi_X$$

and $|A|^2 = \frac{1}{2} \sum_{spins} |A|^2$ (no averaging over initial neutrino spin)

$$A \approx A^{(2)} = \frac{G_F}{\sqrt{2}} \ell^\mu h_\mu \tag{9.140}$$

$$\frac{G_F}{\sqrt{2}} = \frac{10^{-5}}{M_p^2}$$

$$\ell^\mu = \langle \mu(k') | J^{\mu,W}(0) | \nu_\mu(k) \rangle = \overline{u}^{(\mu)}(k') \gamma^\mu (1 - \gamma_5) u^{(\nu)}(k)$$

and $h_\mu = \langle X, P_x | J^{w+} \mu(0) | p; P \rangle$ (9.141)

As before,

$$\frac{d^3k'}{E_{k'}} \approx E' dE' d\Omega \tag{9.142}$$

Using (9.140), (9.141) and (9.142) we get from (9.139)

$$\left(\frac{d\sigma^{I,\nu}}{dE' d\Omega}\right)_{\text{lab}} = \frac{G_F^2}{32\pi^2} \frac{E'}{E} L^{\mu\nu} W_{\mu\nu} \tag{9.143}$$

where $L^{\mu\nu} = \sum_{\text{spin}} \ell^\mu \ell^{\nu+}$

$$= 8[k^\mu k'^\nu + k'^\mu k^\nu + g^{\mu\nu}(m^2 - k \cdot k') - i \in^{\mu\nu\alpha\beta} k_\alpha k'_\beta] \tag{9.144}$$

and $W_{\mu\nu} = \frac{1}{4\pi M} \sum_X \int d\Phi_X [H_{\mu\nu}(P, P_X) \cdot (2\pi)^4 \delta_4(P + q - P_X)]$ (9.145)

with $H_{\mu\nu} = \frac{1}{2} \sum_{\text{spin}} h_\mu h_\nu^+$

As in the case of DI eN scattering we can write

$$W_{\mu\nu} = \frac{1}{8\pi M}\sum_{\text{spin}}\int d^4xe^{iq\cdot x}\langle P|J^w_\mu(x)J^w_\nu(0)^+|P\rangle \tag{9.146}$$

The Eq. (9.146) shows that

$$W_{\mu\nu} = W_{\mu\nu}(P,q)$$

From general considerations, namely, that under Lorentz transformations, $W_{\mu\nu}$ transforms like a rank-2 tensor to be constructed out of the two 4-vectors P_μ and $q\mu$ allowing for parity violation in weak interaction and the assumption $W_{\mu\nu} = W^*_{\nu\mu}$ (similar to the properties of $L_{\mu\nu}$) we can write

$$W_{\mu\nu}(P,q) = -W_1g_{\mu\nu} + \frac{W_2}{M^2}P_\mu P_\nu - \frac{iW_3}{2M^2}\in_{\mu\nu\alpha\beta}P^\alpha q^\beta + \frac{W_4}{M^2}q_\mu q_\nu$$
$$+\frac{W_5}{M^2}(P_\mu q_\nu + q_\mu P_\nu) - \frac{iW_6}{M^2}(P_\mu q_\nu - q_\mu P_\nu) \tag{9.147}$$

where $W_i = W_i(Q^2,\nu), (i = 1,\ldots 6)$

W_i's are called the "weak structure functions" of the proton and are real, Lorentz-invariant quantities

Using the expressions for $L^{\mu\nu}$ and $W_{\mu\nu}$ given by (9.144) and (9.147), respectively, using the relation $\in^{\mu\nu\alpha\beta}\in\mu\nu\alpha'\beta' = -2(g^\alpha_{\alpha'},g^\beta_{\beta'},-g^\alpha_{\beta'}g^\beta_{\alpha'})$ and neglecting the contribution of W_4, W_5 and W_6 (since their contributions are proportional to m^2) we get

$$L^{\mu\nu}W_{\mu\nu} = 16\left[W_1(k\cdot k') + \frac{W_2}{M^2}\left\{(p\cdot k)(p\cdot k') - \frac{P^2}{2}(k\cdot k')\right\}\right]$$
$$+\frac{W_3}{2M^2}\cdot\frac{1}{2}2\{(k\cdot P)(k'\cdot q) - (k\cdot q)(k'\cdot P)\} \tag{9.148}$$

In the DI limit and in terms of lab frame variables (9.148) becomes

$$L^{\mu\nu}W_{\mu\nu} = 16EE'\left[2W_1\sin^2\frac{\theta}{2} + W_2\cos^2\frac{\theta}{2} + W_3\left(\frac{E+E'}{M}\right)\sin^2\frac{\theta}{2}\right] \tag{9.149}$$

Substituting (9.145) in (9.143) we finally get

$$\left(\frac{d\sigma^{I,V}}{dE'd\Omega}\right)_{\text{lab}} = \frac{G_f^2}{2\pi^2}E'^2\left[W_2(Q^2,\nu)\cos^2\frac{\theta}{2} + 2W_1(Q^2,\nu)\sin^2\frac{\theta}{2}\right.$$
$$\left.+\left(\frac{E+E'}{M}\right)W_3(Q^2,\nu)\sin^2\frac{\theta}{2}\right] \tag{9.150}$$

9.5.2 Scaling Limit

Analysis of experimental data on DI νN scattering shows that in the scaling limit, namely for $Q^2 \to \infty$, $\nu \to \infty$ with $x = \frac{Q^2}{2M\nu}$ = fixed $(0 \leq x \leq 1)$ all three structure functions become function of x only.

$$2MW_1 \to F_1(x)$$

$$\nu W_2 \to F_2(x)$$

and $$\nu W_3 \to F_3(x) \tag{9.151}$$

Moreover, $\frac{1}{x}F_2 \approx F_1$

In terms of dimensionless scaling variables x, y defined in (9.137) one finds

$$\frac{d^3k}{E_{K'}} \approx E'dE'd\Omega = \frac{\pi}{E}dQ^2 d\nu = \pi(2M\nu)dxdy \tag{9.152}$$

and one can write the expression for double differential cross-section given by Eq. (9.150) in a manifestly Lorentz invariant manner with transparent scaling behavior.

$$\frac{d\sigma^{I,\nu}}{dxdy} = \frac{G_F^2}{2\pi}s\left[\tilde{F}_2(x,Q^2)\left(1 - y - \frac{Mxy}{2E}\right) + \tilde{F}_1(x,Q^2)\frac{xy^2}{2} + \tilde{F}_3(x,Q^2)\left(1 - \frac{y}{2}\right)xy\right] \tag{9.153}$$

where we have written

$$2MW_1(Q^2,\nu) = \tilde{F}_1(x,Q^2), \nu W_2(Q^2,\nu) = \tilde{F}_2(x,Q^2)$$

and $$\nu W_3(Q^2,\nu) = \tilde{F}_3(x,Q^2) \tag{9.154}$$

In terms of $\tilde{F}_i$'s the "scaling" means

$$Q^2 \to \infty, \tilde{F}_i(x,Q^2) \to F_i(x) \tag{9.155}$$

For $\overline{\nu}p$ scattering the sign before the W_3 term will be negative.

9.5.3 The Parton Model

As in the case of DI eN scattering,scaling in DI νN scattering can also be understood in the framework of the parton model. In the parton model, the nucleon is assumed to be made up of point-like constituents called partons. Partons can be charged or neutral. Like eN scattering in the DI νN scattering also only charged partons participate.

Moreover, charged partons are assumed to be spin-$\frac{1}{2}$ fermions. In the parton model one approximates the total inclusive DI νN scattering cross section by an incoherent sum of (charge-exchange) elastic neutrino-parton scattering cross sections convoluted with parton momentum distribution function. One retains the nonvanishing lowest order term in the S-matrix expansion.

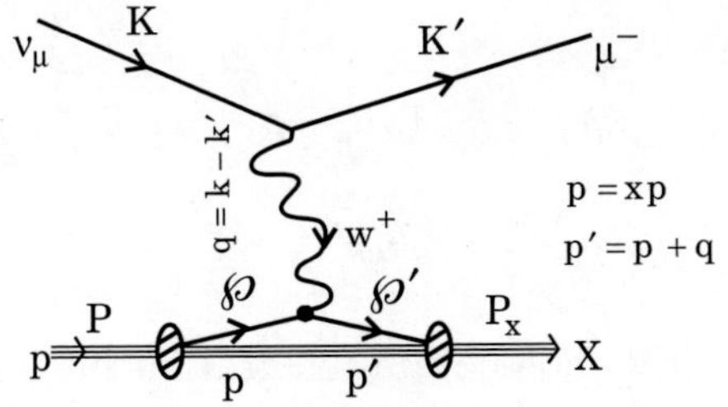

Fig. 9.8

$$\text{Hence} \quad \sigma^{I,\nu} \approx \sum_{\wp} \int\int d\sigma(\nu_\mu + \wp \to \mu^- + \wp')f_{\wp}(z)dz \tag{9.156}$$

$f_{\wp}(z)dz =$ probability that the parton carries a fraction of proton momentum between z and $z + dz\, p = zP, 0 \leq z \leq 1$

$$\sum_w p \int z f_{\wp}(z)dz = 1 \tag{9.157}$$

$$d\sigma(\nu_\mu + \wp \to \mu^- + \wp') = \frac{1}{L_{\nu\wp}}|\overline{A}|^2(2\pi)^4\delta_4(p + q - p')\frac{d^3k'}{(2\pi)^3(2E_{k'})} \frac{d^3p'}{(2\pi)^3(2E_{p'})} \tag{9.158}$$

where $L_{\nu\wp} \approx 4(k \cdot p) = 4(k \cdot P)z \approx z(4ME)$ and $|\overline{A}|^2 = \frac{1}{2}\sum_{\text{spin}}|A|^2$

$$\text{with} \quad A = \frac{G_F}{\sqrt{2}}\ell^\mu h_\mu(\wp) \tag{9.159}$$

$$\ell^\mu = \overline{u}^{(k')}_{(\mu)}\gamma^\mu(1 - \gamma_5)u^{(k)}_{(\nu)}$$

$$h^{(\wp)}_\mu = \overline{u}^{(\wp)}(p')\gamma_\mu(1 - \gamma_5)u^{(\wp)}(p), \text{ if } \wp \text{ is a particle}$$

$$\hbar^{(\wp)}_\mu = \overline{v}^{(\wp)}(p)\gamma_\mu(1 - \gamma_5)v^{(\wp)}(p'), \text{ if } \wp \text{ is an antiparticle.}$$

We can write Eq. (9.158) in the form

$$\left(\frac{d\sigma^{I,\nu}}{dE'd\Omega}\right)_{\text{lab}} = \frac{G_F^2}{32\pi^2} \cdot \frac{E'}{E} L^{\mu\nu}W^{(\wp)}_{\mu\nu} \tag{9.160}$$

where $L^{\mu\nu}$ is given by Eq. (9.144)

The hadron tensor $W_{\mu\nu}^{(\wp)}$ in the parton model is given by the following expression.

$$W_{\mu\nu}^{(\wp)} = \frac{1}{4\pi M}\sum_p \int\int \frac{1}{z} H_{\mu\nu}^{(\wp)}(p = zP, p')(2\pi)^4\delta_4(p+q-p'), \frac{d^3p'}{(2\pi)^3(2E_{p'})} f_\wp(z)dz \quad (9.161)$$

where $H_{\mu\nu}^{(\wp)} = \frac{1}{2}\sum_{\text{spin}} h_\mu^{(\wp)} h_\nu^{(\wp)+}$

$$= 4[p_\mu p'_\nu + p'_\mu p_\nu + g_{\mu\nu}(m_p^2 - p\cdot p') \mp i \in_{\mu\nu\alpha\beta} p^\alpha p'^\beta] \quad (9.162)$$

Using $\int \frac{d^3p'}{2E_{p'}} = \int d^4p'\delta(p'^2 - m_\wp^2)\theta(p'^0)$

We can carry out the momentum integration in (9.161) and get

$$W_{\mu\nu}^{(\wp)} = \frac{1}{2M}\sum_p \int \frac{1}{z} H_{\mu\nu}^{(\wp)}\ (p = zP, p' = p+q) f_p(z)\delta((p+q)^2 - m_\wp^2)dz \quad (9.163)$$

Now $\delta((p+q)^2 - m_\wp^2) = \delta(2p\cdot q + q^2) = \delta(2zP\cdot q + q^2) = \frac{1}{2M\nu}\delta(z-x)$

$$\nu = \frac{p\cdot q}{M},\ Q^2 = -q^2,\ x = \frac{Q^2}{2M\nu} \quad (9.164)$$

Using (9.164) we can carry out the z-integration in (9.163) and get

$$\begin{aligned} W_{\mu\nu}^{(\wp)} &= \frac{1}{4M^2\nu}\sum_\wp \frac{1}{x} H_{\mu\nu}^{(\wp)}(p = xP, p' = p+q) f_\wp(x) \\ &= \frac{1}{M^2}\sum_\wp \frac{1}{x}[xP_\mu(xP+q)_\nu + (xP+q)_\mu P_\nu \\ &\qquad - g_{\mu\nu}(P\cdot q) \mp i \in_{\mu\nu\alpha\beta} P^\alpha q^\beta] f_\wp(x) \\ &= \frac{1}{M^2\nu}\sum_\wp f_\wp(x)[2xP_\mu P_\nu - g_{\mu\nu}(M\nu) \\ &\qquad + (P_\mu q_\nu + q_\nu P_\nu) \mp i_{\in_{\mu\nu\alpha\beta}} P^\alpha q^\beta] \end{aligned} \quad (9.165)$$

comparing (9.165) with (9.147) we get

$$W_2(Q^2,\nu) = \frac{1}{\nu}\sum_\wp 2x f_\wp(x)$$

$$W_1(Q^2,\nu) = \frac{1}{M}\sum_\wp f_\wp(x)$$

$$\frac{1}{2}W_3(Q^2,\nu) = \mp\frac{1}{\nu}\sum_\wp f_\wp(x)$$

We can rewrite these relations as

$$\nu W_2(Q^2, \nu) = \sum_{\wp} 2s f_{\wp}(x) = F_2(x)$$

$$2MW_1(Q^2, \nu) = \sum_{\wp} 2f_{\wp}(x) = F_1(x)$$

and $$\nu W_3(Q^2, \nu) = \sum_{\wp} 2 \in_{\wp} f_{\wp}(x) = F_3(x) \tag{9.166}$$

where $\in_{\wp} = +1$, if $\wp$ is a particle

-1, if $\wp$ is an antiparticle

As in the case of DI eN scattering we see that $F_1(x)$ and $F_2(x)$ are exactly related in the parton model consistent with experimental data

$$F_1(x) = \frac{1}{x} F_2(x) \tag{9.167}$$

Identifying the charged partons with quarks and antiquarks and restricting to two flavor case we see that the parton level processes are

$$\nu_\mu + d \to \mu^- + u, \quad \nu_\mu + \overline{u} \to \mu^- + \overline{d}, \tag{9.168}$$

$$\overline{\nu}_\mu + u \to \mu^+ + d, \quad \overline{\nu}_\mu + \overline{d} \to \mu^+ + \overline{u} \tag{9.169}$$

Hence $F_2^{\nu p}(x) = 2x[d(x) + \overline{u}(x)]$, $F_3^{\nu p}(x) = 2[d(x) - \overline{u}(x)]$,

$$F_2^{\overline{\nu} p}(x) = 2x[u(x) + \overline{d}(x)], \; F_3^{\nu p}(x) = 2[u(x) - \overline{d}(x)], \tag{9.170}$$

Structure functions for $\nu n(\overline{\nu} n)$ scattering are obtained from those for $\nu p(\overline{\nu} p)$ scattering by interchanging $u \longleftrightarrow d$. Hence

$$F_2^{\nu n}(x) = 2x[u(x) + \overline{d}(x)], \; F_3^{\nu n}(x) = 2[u(x) - \overline{d}(x)],$$

$$F_2^{\overline{\nu} n}(x) = 2x[d(x) + \overline{u}(x)], \; F_3^{\nu n}(x) = 2[d(x) - \overline{u}(x)], \tag{9.171}$$

Therefore, $$F_i^{\nu N} = \frac{1}{2}(F_i^{\nu p} + F_i^{\nu p}) = \frac{1}{2}(F_i^{\overline{\nu} n} + F_i^{\overline{\nu} p}) = F_i^{\overline{\nu} N} \tag{9.172}$$

Thus $$F_2^{\nu N}(x) = F_2^{\overline{\nu} N}(x) = x[d(x) + u(x) + \overline{u}(x) + \overline{d}(x)] \tag{9.173}$$

$$F_3^{\nu N}(x) = F_3^{\nu N}(x) = [d(x) + u(x) - \overline{u}(x) - \overline{d}(x)] \tag{9.174}$$

Substituting parton model results (9.167), (9.172)–(9.174) in equation (9.153) and neglecting the $M/2E$ term, we get

$$\frac{d\sigma^{I,\nu N}}{dxdy} = \frac{G_F^2}{2\pi} s[xq(x) + x\overline{q}(x)(1-y)^2] \tag{9.175}$$

$$\frac{d\sigma^{I,\bar{\nu}N}}{dxdy} = \frac{G_F^2}{2\pi} s[x\bar{q}(x) + xq(x)(1-y)^2] \tag{9.176}$$

where $q(x) = d(x) + u(x),\ \bar{q}(x) = \bar{u}(x) + \bar{d}(x),$ (9.177)

From the above results, we can summarize the quark-parton model predictions for DI lepton-nucleon scattering

(1) $\sigma_{\text{tot}}^{\nu,\bar{\nu}} = \alpha E \quad \alpha = \text{constant}$

(2) $a = \dfrac{F_2^{eN(x)}}{F_2^{\nu N}(x)} = \langle e_p^2 \rangle = \dfrac{1}{2}(e_u^2 + e_d^2) = \dfrac{5}{18} = 0.278$

(3) $b = \displaystyle\int_0^1 F_2^{\nu N}(x)dx = \int_0^1 x[u(x) + d(x) + \bar{u}(x) + \bar{d}(x)]dx = 1 - \in_g$

(4) $R = \dfrac{\sigma_{\text{tot}}^{\bar{\nu}}(s)}{\sigma_{\text{tot}}^{\nu}(s)} = \dfrac{\frac{1}{3}\in_q + \in_{\bar{q}}}{\in_q + \frac{1}{3}\in_{\bar{q}}} \approx \dfrac{1}{3},$

$$(5)\ B = \frac{\displaystyle\int_0^1 xF_3^{\nu N}(x)dx}{\displaystyle\int_0^1 F_2^{\nu N}(x)dx} = \frac{\in_q - \in_{\bar{q}}}{\in_q + \in_{\bar{q}}} \approx 1 \tag{9.178}$$

Experimental data agree with all the predictions.

$$a = 0.29 \pm 0.03, \qquad b = 0.48 \pm 0.02$$
$$R = 0.38 \pm 0.02, \qquad b = 0.8$$

The experimental value of b implies that neutral partons which can be identified with the gluons, carry about 50% of the proton momentum. Moreover, the experimental value of B implies that $\in_{\bar{q}} / (\in_q + \in_{\bar{q}}) \sim 0.1$. This means that antiquarks carry about 5% of the proton momentum.

9.5.4 More Sum Rules

We have talked about the Gottfried sum rule while discussing DI electron-nucleon scattering in the framework of quark-parton model. There are two similar sum rules that concern DI neutrino-nucleon scattering.

Adler Sum Rule:

$$\begin{aligned}\int_0^1 [F_2^{\nu n}(x) - F_2^{\nu p}(x)]dx &= 2\int_0^1 [\{u(x) + \bar{d}(x)\} - \{d(x) + \bar{u}(x)\}] \\ &= 2\int_0^1 [\{u(x) - \bar{u}(x)\} - \{d(x) - \bar{d}(x)\}]dx = 4I_3^{(p)} = 2\end{aligned} \tag{9.179}$$

Gross-Llewellyn Smith Sum Rule:

$$\begin{aligned}\int_0^1 F_3^{\nu N}(x)dx &= \frac{1}{2}\int_0^1 [F_3^{\nu p}(x) + F_3^{\nu n}(x)]dx \\ &= \frac{1}{2}\int_0^1 2[\{d(x) - \overline{u}(x)\} + \{u(x) - \overline{d}(x)\}]dx \\ &= \frac{1}{2}\int_0^1 2[\{u(x) - \overline{u}(x)\} + \{d(x) - \overline{d}(x)\}]dx \\ &= 3B(p) = 3 \end{aligned} \quad (9.180)$$

The values of the integrals on the lhs of equation (9.179) and (9.180) can be calculated from the experimental data. According to the quark-parton model their values correspond to the third component of *i*-spin and the baryon number of the proton, respectively.

Problems

1. Show that an exponential charge distribution of the form $\rho(r) = \rho_0 e^{-\mu r}$ leads to a form factor

$$F(q^2) = \frac{1}{(1 + q^2/\mu^2)^2}$$

   Find the value of μ that leads to an rms charge radius of 0.8 fm.

2. Consider the elastic scattering of a relativistic electron from a point-like spinless charged particle (electron-pion scattering). Show that to lowest order in electromagnetism and in the high energy approximation the y-distribution has the form

$$\frac{d\sigma}{dy} = \frac{2\pi\alpha^2}{s} \cdot \frac{1-y}{y^2}$$

   where $y = \nu/E$, $\nu = E - E'$. Here E and E' are the energies of the incident electron and the scattered electron, respectively, in the Lab frame (fixed target experiment)

3. Show that the quark-parton model together with the V-A theory of weak interactions predicts that

$$\frac{\sigma(\overline{\nu}N)}{\sigma(\nu N)} \approx \frac{1}{3}$$

if one neglects the contributions of the antiquarks. Here $\sigma(\nu N)$ and $\sigma(\overline{\nu} N)$ are the total DI inclusive cross sections for neutrino-nucleon and antineutrino-nucleon scattering, respectively.

4. Consider inclusive photoproduction, $\gamma N \to X$. Assuming the photon to be virtual space-like, show that the ratio of total inclusive cross sections for longitudinal and transverse photons, σ_1, and σ_T, respectively can be expressed in the form,

$$\frac{\sigma_L}{\sigma_T} = \frac{\left(1 - \frac{v^2}{q^2}\right) W_2(v, Q^2) - W_1(v, Q^2)}{W_1(v, Q^2)}$$

where W_2 and W_1 are nucleon structure functions, $Q^2 = -q^2$ and $\nu = P \cdot q/M$, P and q being the nucleon and photon 4-momenta and M is the nucleon mass.

Show that the parton model predicts that in the DI limit

$(\nu \to \infty,\ Q^2 \to \infty,\ x = Q^2/2M\nu = \text{fixed})$

$$\frac{\sigma_L}{\sigma_T} \to 0$$

If the charged partons are spin $-\frac{1}{2}$ particles, and

$$\frac{\sigma_T}{\sigma_L} \to 0$$

If the charged partons are spin-0 particles.

Chapter 10

Electron-positron Annihilation into Hadrons ($e^-e^+ \to$ hadrons)

10.1 e^-e^+ Annihilation Physics

We now consider another process involving leptons and hadrons, namely, the process $e^-e^+ \to$ hadrons at high energies, which has provided rich information about the physics of hadrons, their structure and interactions. The fact that the process can at all occur is a consequence of the RQFT. The electron interaction being predominantly electromagnetic in nature, QED is applicable. Because of the smallness of the electromagnetic coupling strength ($\alpha = e^2/4\pi = 1/137$) one needs consider only the lowest order non-vanishing term in the perturbative expansion of the S-matrix for the process. Energy-momentum conservation requires that $W^2 \geq \left(\sum_{i=1}^{n_f} m_i\right)^2$, n_f being the number of particles in the final state. Since the dominant contribution comes from the diagram with 1-photon intermediate state, the final state must have the quantum numbers of the photon ($J^{PC} = 1^{--}$, $Q = B = L = S = 0$).

An important observable for the process is $\sigma(s)$, the total cross section as a function of $s = W^2$, $W =$ total CM energy. In order to have more possibilities for the final state and thus to probe any possible new physics, one must go for higher and higher energies. This means that stationary-target expts are less preferable to colliding beam expts. This is because for the process $a+b \to f$, $s \propto E$ ($E \gg m$) in the stationary target expt, whereas $s \propto E^2$ in the colliding beam expt, where E is the beam energy. However, the luminosity (and hence the event rate) is much smaller in collider expts and one has to take special steps to increase the luminosity to a reasonable value.

The first e^-e^+ collider was built in the early 1960's at FRASCATI in Italy with $\sqrt{3} \leq 3$ Gev. The SPEAR (SLAC, USA) with 3 GeV $\leq \sqrt{s} \leq$ 7.5 GeV started doing expts. in early 1970's. These expts led to the discovery of the charmonium states $(J/\psi, \psi', \ldots)$ and the τ-lepton. Other important e^-e^+ colliders to be built were PETRA (DESY, Hamburg, Germany) with 36 GeV $\leq \sqrt{s} \leq$ 46 GeV, CESR (Cornell, USA) and TRISTAN (Japan). The highest energy e^-e^+ collider to be built so far is the LEP machine (CERN, Europe), LEPI with $\sqrt{s} \leq$ 100 GeV and LEP II with $\sqrt{s} \leq$ 200 GeV.

At very low energies one can have only elastic scattering, $e^+e^- \to e^-e^+$ (Bhaba Scattering). The non-vanishing lowest order contribution comes from two Feynman diagrams, called the exchange diagram and the annihilation diagram, respectively.

$e^+e^- \to e^+e^-$:

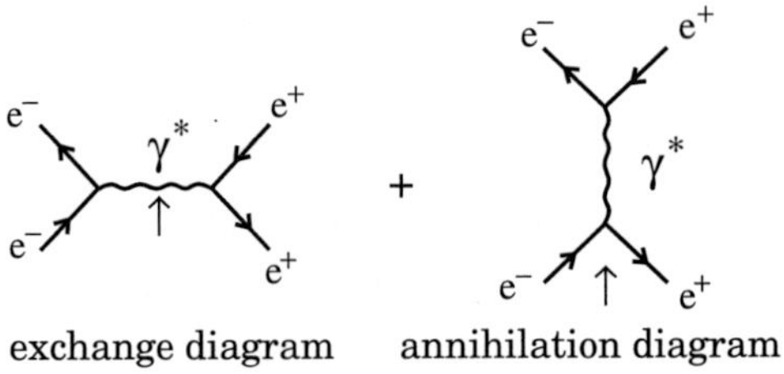

Fig. 10.1

As the energy increases particle production starts taking place. For $\sqrt{s} \geq 2m_\mu$ ($\approx$ 200 MeV) one sees the process $e^+e^- \to \mu^+\mu^-$ for which only the annihilation diagram contributes.

$e^+e^- \to \mu^+\mu^-$: $q = p_1' + p_2' = p_1 + p_2$, $s = q^2 = (p_1 + p_2)^2$, $s \geq 4m_\mu^2$

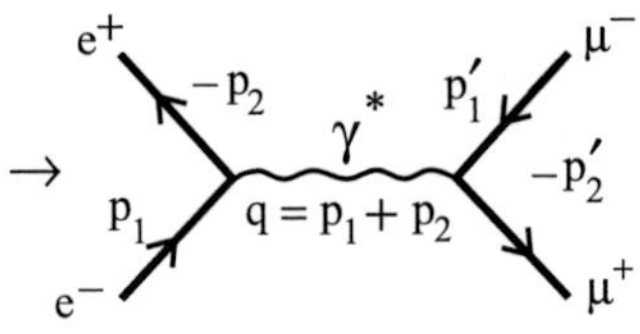

Fig. 10.2

As the energy increases further, hadron production starts taking place for $\sqrt{s} \geq 2m_\pi$ ($\approx$ 300 MeV) for which also only the annihilation diagram contributes in the lowest order (second order)

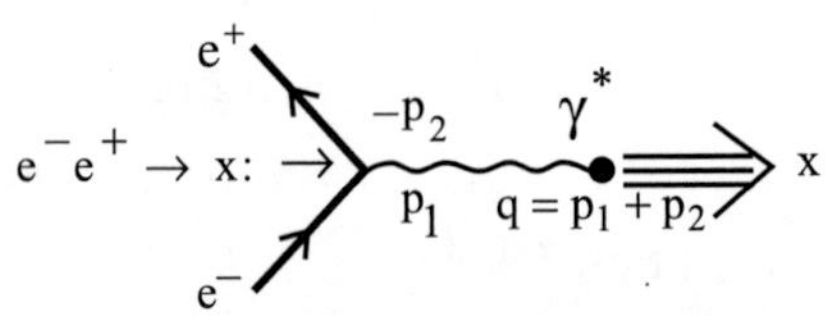

Fig. 10.3

$X =$ anything (hadrons subject to conservation laws)

$= \pi^+\pi^-,\ \pi^+\pi^-\pi^0,\ K^+K^-,\ \bar{p}p,\ \pi^+\pi^-\pi^+\pi^-$ etc

One measures

$$\sigma(e^+e^- \to \text{hadrons}) = \sum_X \int d\sigma(e^-e^+ - X),\ \sigma = \sigma(s) \tag{10.1}$$

As one plots $\sigma(s)$ as a function of $W = \sqrt{s}$ one first sees a number of peaks or "resonances". Any such resonance is interpreted as the formation of an intermediate state with mass M_i, width Γ_i and with quantum numbers of the photon ($J^{PC} = 1^{--}$, $Q = B = 0$, $I = 0, 1$). These are identified with neutral vector mesons $\rho^0(773)$, $\omega(785)$, $\phi(1020)$ and their excited states. For $\sqrt{s} \geq 2$ GeV the behavior of $\sigma(s)$ as a function of $\sqrt{s} = W$ becomes flat (plateau). This behavior continues till $\sqrt{s} = 3$ GeV. (For $\sqrt{s} > 3$ GeV one again starts seeing peaks). The behavior of $\sigma(s)$ as a function of s away from the resonance region can be understood in the framework of the quark-parton model. For this one considers the ratio of $\sigma(e^+e^- \to \text{hadrons})$ and $\sigma(e^-e^+ \to \mu^-\mu^+)$ at the same s,

$$R(s) = \frac{\sigma(e^-e^+ \to \text{hadrons})}{\sigma(e^-e^+ \to \mu^+\mu^-)} \tag{10.2}$$

10.2 $e^-e^+ \to \mu^-\mu^+$

We first consider the pure QED process $e^-e^+ \to \mu^-\mu^+$ at high energies ($s \gg 4m_\mu^2$). Because of the smallness of electromagnetic coupling strength ($\alpha = e^2/4\pi = 1/137$), the dominant contribution comes from the 1-photon annihilation diagram (Fig. 10.2). From the standard invariant cross section formula, we have

$$\sigma = \int d\sigma(e^-e^+ \to \mu^-\mu^+) = \frac{1}{L}\int |\overline{A}|^2 \cdot (2\pi)^4 \delta_4(P_i - P_f) d\phi_f \tag{10.3}$$

Here $L = 4\sqrt{(p_1 \cdot p_2)^2 - m_e^2},\ \approx 4p_1 \cdot p_2$

$$P_i = p_1 + p_2,\ P_f = p_1' + p_2'$$

$$d\phi_f = \frac{d^3p_1'}{(2\pi)^3(2E_i')} \cdot \frac{d^3p_2'}{(2\pi)^3(2E_2')}$$

$$|\overline{A}|^2 = \frac{1}{4}\sum_{\text{spin}} |A|^2$$

$$A \approx A^{(2)} = \frac{4\pi\alpha}{q^2} l^\mu l'_\mu$$
$$l^\mu = \langle o|J^\mu_{\rm em}(o)|p_1 p_2\rangle_{\rm electron} = \overline{v}_e(p_2)\gamma^\mu u_e(p_1)$$
$$l'_\mu = \langle p'_1, p'_2|J^{\rm em}_\mu(o)|o\rangle_{\rm muon} = \overline{u}(p'_1)\gamma_\mu v(p'_2) \tag{10.4}$$

The Mandelstam variables for the process are, in the high energy approximation,

$$s = (p_1 + p_2)^2 \approx 2p_1p_2$$
$$t = (p_1 - p'_1)^2 \approx -2p_1p'_1$$
$$u = (p_1 - p'_2)^2 \approx -2p_1p'_2$$

In the CM frame, $\vec{p_1} + \vec{p_2} = 0$, $P_i = p_1 + p_2 = (\sqrt{s}, 0)$ and at high energies, we can write

$$p_1 \approx \frac{\sqrt{s}}{2}(1, \hat{z}),\ p_2 \approx \frac{\sqrt{s}}{2}(1, -\hat{z})$$
$$p'_1 \approx \frac{\sqrt{s}}{2}(1, \hat{n}),\ p'_2 \approx \frac{\sqrt{s}}{2}(1, -\hat{n}) \tag{10.5}$$
$$\hat{z}\cdot\hat{n} = \cos\theta$$

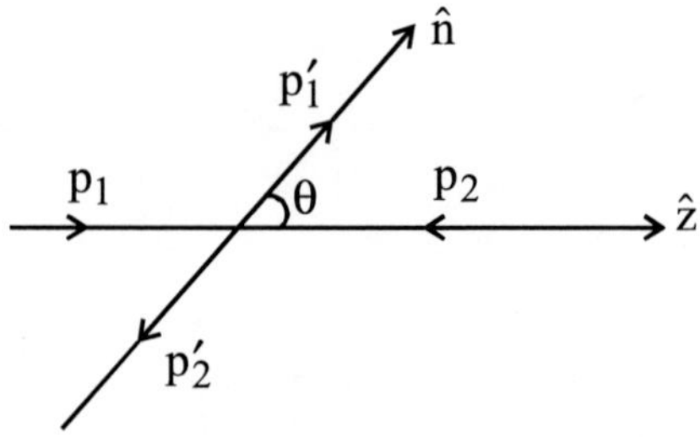

Fig. 10.4

We evaluate the phase space integral in the CM frame. We have from (10.3), (10.4) and (10.5)

$$\sigma \approx \frac{1}{2s}\int |\overline{A}|^2 (2\pi)^4 \delta_3(\vec{p}'_1 + \vec{p}'_2)\delta(E'_1 + E'_2 - \sqrt{s})\frac{d^3p'_1 d^3p'_2}{4(2\pi)^6 E'_1 E'_2}$$
$$= \frac{1}{2s}\int |\overline{A}|^2 \cdot \frac{1}{4(2\pi)^2}\delta(E'_1 + E'_2 - \sqrt{s})\frac{d^3p'_1}{E'_1 E'_2} \tag{10.6}$$

We can write

$$\frac{d^3p'_1}{E'_1} = \frac{|\vec{p}'_1|^2 d|p'_1| d\Omega}{E'_1} \approx E'_1 dE'_1 d\Omega$$

Also, because of δ-functions, we have, $|\vec{p}'_2| = |\vec{p}'_1| = |\vec{p}'|$, and hence $E'_1 = E'_2 = E'$

Therefore, (10.6) becomes

$$\left(\frac{d\sigma}{d\Omega}\right)_{\text{CM}} = \frac{1}{2s}\cdot\frac{1}{4(2\pi)^2}\int\overline{|A|^2}\delta(2E'-\sqrt{s})dE' = \frac{1}{64\pi^2 s}\overline{|A|^2} \tag{10.7}$$

(Although we have used high energy approximation, the result (10.7) is exact for any $2\to 2$ process.)

We next compute $\overline{|A|^2}$. We have

$$|\overline{A}|^2 = \frac{1}{4}\sum_{\text{spin}}|A|^2 = \left(\frac{4\pi\alpha}{q^2}\right)^2 L^{\mu\nu}L'_{\mu\nu} \tag{10.8}$$

where $L^{\mu\nu} = \frac{1}{2}\sum\limits_{\text{spin}} l^{\mu}l^{\nu+} = \frac{1}{2}\text{Tr}[\gamma^{\mu}\not{p}_1\gamma^{\nu}\not{p}_2]$

$$= 2[p_1^{\mu}p_2^{\nu} + p_2^{\mu}p_1^{\mu} - g^{\mu\nu}(p_1\cdot p_2)]$$

and $\quad L'_{\mu\nu} = \frac{1}{2}\sum\limits_{\text{spin}} l'_{\mu}(l'_{\nu})^{+} = \frac{1}{2}\text{Tr}[\gamma_{\mu}\not{p}'_2\gamma_{\nu}\not{p}'_1]$

$$= 2[p'_{2\mu}p'_{1\nu} + p'_{1\mu}p'_{2\nu} - g_{\mu\nu}(p'_1\cdot p'_2)] \tag{10.9}$$

Therefore (10.8) becomes

$$\overline{|A|^2} = \frac{(4\pi\alpha)^2}{s^2} = 8[(p_1\cdot p'_1)(p_2\cdot p'_2) + (p_1\cdot p'_2)(p_2\cdot p'_1)]$$

$$= \frac{(4\pi\alpha)^2}{s^2}\cdot 2(t^2+u^2) \tag{10.10}$$

Comparing $\overline{|A|^2}_{e^-e^+\to\mu^-\mu^+}$ given by Eq. (10.10) with $\overline{|A|^2}_{e\mu\to e\mu}$ given by Eq. (9.41) we see that the 2 amplitudes are related. One is obtained from the other by interchanging the Mandelstam variables s and $t(s\to t)$ for the corresponding processes. This is an example of what is known as “crossing symmetry” (The amplitudes for 2 processes, one obtained from the other, by interchanging initial state particle by a final state antiparticle with the sign of the 4-momenta reversed are related).

Substituting (10.10) in (10.7) and using (10.5) We get

$$\left(\frac{d\sigma}{d\pi}\right)_{\text{CM}}^{e^-e^+\to\mu^-\mu^+} = \frac{\alpha^2}{4s}(1+\cos^2\theta) \tag{10.11}$$

where θ is the production angle in the CM frame. Integrating over all angles we get from (10.11)

$$\sigma_{\text{tot}}(e^-e^+ \to \mu^-\mu^+) = \frac{4\pi\alpha^2}{3s} \tag{10.12}$$

Clearly, the result (10.12) is Lorentz invariant.

10.3 $e^-e^+ \to$ hadrons

According to the quark-parton model, at very high energies and away from the resonance regions, $\sigma(e^+e^- \to \text{hadrons})$ is given by

$$\begin{aligned}\sigma(e^+e^- \to \text{hadrons}) &= \sum_q \sigma(e^+e^- \to q\bar{q}) \cdot P(q\bar{q} \to \text{hadrons}) \\ &= \sum_q \sigma(e^-e^+ \to q\bar{q}) \end{aligned} \tag{10.13}$$

i.e. e^-e^+ first annihilate into a $q\bar{q}$ pair and this $q\bar{q}$ pair then hadronise. Due to confinement, quarks cannot exist in free states and they must eventually form hadrons and hence the probability P of hadronisation is taken to be 1.

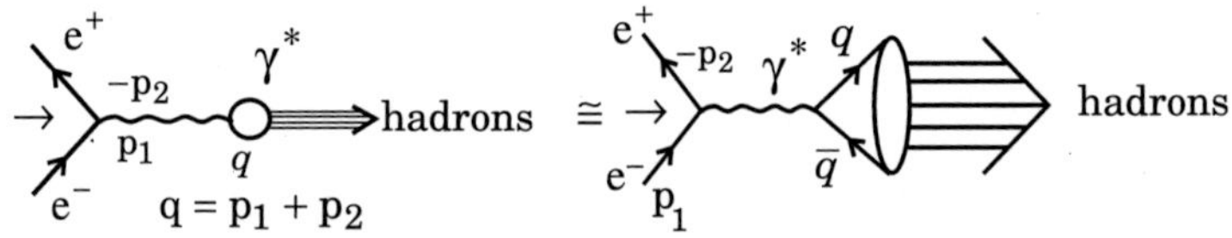

Fig. 10.5

The process $e^-e^+ \to q\bar{q}$ is similar to $e^-e^+ \to \mu^+\mu^-$ except for the fact that quark charge is e_q in unit of e $\left(e_q = \frac{+2}{3}, \frac{-1}{3}\right)$. Thus for the 1-photon annihilation diagram the em coupling factor e^2 is replaced by $e^2 e_q$. Thus the results for $e^-e^+ \to q\bar{q}$ are obtained from those of $e^-e^+ \to \mu^-\mu^+$ by making the above change. Therefore, we get

$$\left(\frac{d\sigma}{d\pi}\right)_{\text{CM}}^{e^+e^-\to q\bar{q}} = \frac{\alpha^2 e_q^2}{4s}(1 + \cos^2\theta) \tag{10.14}$$

and $$\sigma_{\text{tot}}(e^+e^- \to q\bar{q}) = \frac{4\pi\alpha^2}{3s} e_q^2 \tag{10.15}$$

Experimentally one measures the quantity $R(s)$ defined in Eq. (10.2) which in the parton model becomes

$$R(s) = \frac{\sigma(e^+e^- \to \text{ hadrons})}{\sigma(e^+e^- \to \mu^+\mu^-)} = \sum_q e_q^2 \theta(s - 4m_q^2) \tag{10.16}$$

The θ-function in (10.16) indicates that only those quarks whose mass satisfies the inequality $4m_q^2 < s$ contribute. In (10.16) the sum is over both flavor and color quantum numbers, since color has nothing to do with em interactions, (10.16) becomes

$$R(s) = \frac{\sigma(e^+e^- \to \text{ hadrons})}{\sigma(e^+e^- \to \mu^+\mu^-)} = N_c \sum_{q(\text{flavors})} e_q^2 \theta(s - 4m_q^2) \tag{10.17}$$

For 2 GeV $< \sqrt{s} <$ 3 GeV, all the light quark degrees of freedom contribute and we get,

$$R(s) = N_c\left[e_u^2 = e_d^2 + e_s^2\right] = N_c\left[\left(\frac{2}{3}\right)^2 + \left(-\frac{1}{3}\right)^2 + \left(-\frac{1}{3}\right)^2\right] = N_c\left(\frac{2}{3}\right)$$
$$= 2 \text{ if } N_c = 3 \tag{10.18}$$

The value agrees with the experimental result.

As the total CM energy $w = \sqrt{s}$ increases beyond 3 Gev, 2 resonances appear in $R(s)$, at $\sqrt{s}$ = 3.1 GeV and 3.7 GeV. Since they must have the quantum numbers of the photon ($J^{PC} = 1^{--}$, $Q = 0$, $B = 0$) they are interpreted as vector mesons formed of new heavy quarks with mass $\geq$ 1.5 GeV.

$$\psi, \psi' = Q\overline{Q} \tag{10.19}$$

One would like to identity this new heavy quark with the charm quark c introduced previously to explain certain features of weak interactions. This interpretation is corroborated by further exptl. results regarding the decay products of the excited states of $c\bar{c}$ bound system (charmonium).

Above $\sqrt{s} > 5$ GeV, one is again away from the resonance regions but now the charm degree of freedom is also excited and we have for 5 GeV $<$ $\sqrt{s} < 9$ GeV

$$R(s) = N_c\left[e_u^2 + e_d^2 + e_s^2 + e_c^2\right] = 3\left[\frac{4}{9} + \frac{1}{9} + \frac{1}{9} + \frac{4}{9}\right] = 10/3 \tag{10.20}$$

Above $\sqrt{s} < 9$ GeV, one enters a new resonance region with 3 narrow resonances $\upsilon(9.46)$, $\upsilon(10.02)$, and $\upsilon(10.35)$ which are interpreted as bound states of a new heavier b quark and its antiparticle: $\upsilon = b\bar{b}$ with $e_b = -\frac{1}{3}$ and $m_b \approx 4.5$ GeV. Above this energy $\sigma(e^+e^- \to \text{hadrons})$ again becomes a smooth function of s. Thus for $\sqrt{s} > 11$ GeV we have

$$R(s) = N_c[e_u^2 + e_d^2 + e_s^2 + e_c^2 + e_b^2]$$
$$= 3\left[\left(\frac{2}{3}\right)^2 + \left(\frac{-1}{3}\right)^2 + \left(\frac{-1}{3}\right)^2 + \left(\frac{2}{3}\right)^2 + \left(\frac{-1}{3}\right)^2\right]$$
$$= 11/3 \tag{10.21}$$

The sixth quark, the t-quark is much heavier ($m_t = 175$ GeV). No $t\bar{t}$ bound state is expected. Moreover, e^+e^- colliders have yet to reach $\sqrt{s} \geq 2m_t = 350$ GeV. The t quark was discovered in 1995 in $p\bar{p}$ collider experiments at the FNAL.

10.4 $e^+e^- \to hX$

The inclusive process $e^+e^- \to hX$ is related to DI electron scattering $e^- + \bar{h} \to e^- + X$ by crossing symmetry and therefore one expects a scaling behavior. According to the quark-parton model

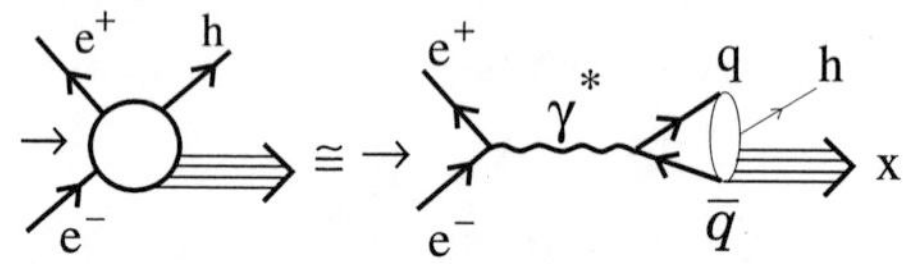

Fig. 10.6

and $$\frac{d\sigma}{dz}(e^+e^- \to hX) \approx \sum_q \sigma(e^+e^- \to q\bar{q})[D_q^h(z) + D_{\bar{q}}^h(z)] \tag{10.22}$$

$D_q^h(z), D_{\bar{q}}^h =$ Fragmentation functions

$D_q^h(z)dz =$ probability of a quark fragmenting into a hadron h carrying a fraction of quark momentum between z and $z + dz (0 \leq z \leq 1)$ we define

$$\sigma_{\text{tot}} = \sum_q \sigma(e^-e^+ \to q\bar{q}) \tag{10.23}$$

Normalization gives

$$\sum_h \int_0^1 zD_q^h(z)dz = 1 = \sum_h \int_0^1 zD_{\bar{q}}^h(z)dz \tag{10.24}$$

Also $\sum_q \int_{z_m}^1 [D_q^h(z) + D_{\bar{q}}^h(z)] = dz = n_h =$ multiplicity of the hadron h.

$$z_m = \frac{2m_h}{\sqrt{s}} = \text{the minimum value of } z. \tag{10.25}$$

with $\sigma(e^+e^- \to q\bar{q})$ given by (10.15) we get

$$\frac{1}{\sigma_{\text{tot}}}\frac{d\sigma}{dz}(e^+e^- \to hX) = \frac{\sum_q e_q^2[D_q^h(z) + D_{\bar{q}}^h(z)]}{\sum_q e_q^2} = f(z) \tag{10.26}$$

Thus $\frac{1}{\sigma_{\text{tot}}}\frac{d\sigma}{dz}(e^+e^- \to hX)$ scales.

with $D_q^h(z) \approx N\dfrac{(1-z)^n}{z}$, N, n constant (10.27)

one gets $n_h \approx \log\left(\dfrac{\sqrt{s}}{2m_h}\right)$ (10.28)

Thus one has logarithmic growth of hadron multiplicity, $n_h \propto \log s$

10.5 Hadron Jets

According to the quark-parton model, the final state hadrons in the process $e^+e^- \to$ hadrons, when viewed in the CM frame, are emitted in two back-to-back jets. This is because the hadrons are formed by the fragmentation of quarks and/or antiquarks with limited transverse momentum. The jet axis is thus oriented along the direction of the parent q and $\overline{q}$ in the CM frame.

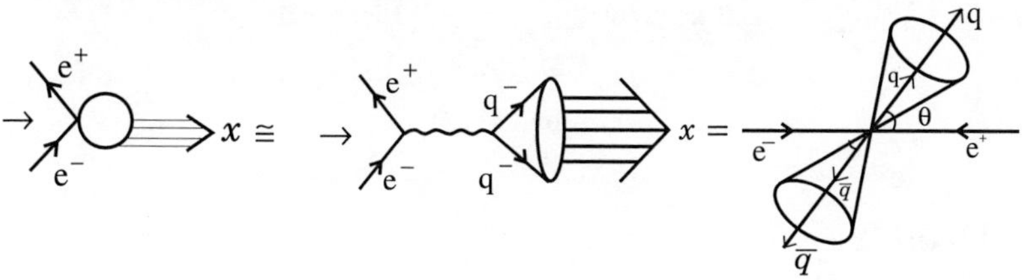

Fig. 10.7

The angular distribution of the hadron jet axis is therefore, the same as that for the process $e^-e^+ \to q\overline{q}$ in the CM frame and according to eqn. (10.14) is $\propto (1 + \cos^2\theta)$ where $\theta =$ angle between the e^+e^- momentum and $q\overline{q}$ momentum directions in the CM frame. Such 2-jet events have actually been seen both at SPEAR (SLAC, USA) and at PETRA (DESY, Germany). For $\sqrt{s} > 7$ GeV, 3-jet events have also been seen. They correspond to the emission of a hard gluon from either q or $\overline{q}$, ie the process $e^-e^+ \to q\overline{q}g \cdot$ 3-jet events were first observed at PETRA for $\sqrt{s} > 10$ GeV.

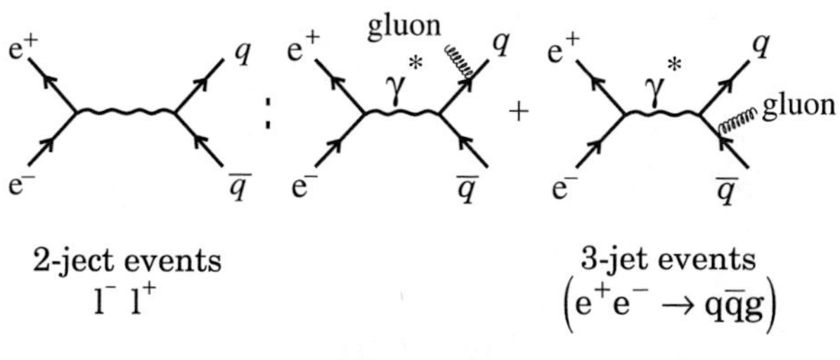

Fig. 10.8

One expects

$$\frac{\sigma(3\text{jets})}{\sigma(2\text{jets})} = 0(\alpha_s) \tag{10.29}$$

Experimentally

$$\frac{\text{No. of 3-jet events}}{\text{No. of 2-jet events}} \approx 10\text{–}15\% \tag{10.30}$$

This implies $\alpha_s \approx 0.12, \alpha_s = \dfrac{g_s^2}{4\pi}$ = strong coupling fine structure constant.

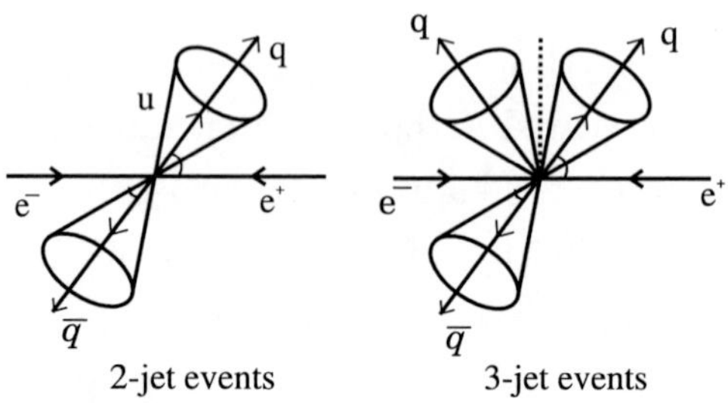

CM frame

Fig. 10.9

10.6 Drell-Yan Process

We now consider another DI process involving leptons and hadrons, in which "scaling" is observed and which also can be understood in the frame-work of the quark-parton model. This process is popularly known as the Drell-Yan process:

$$h_1 + h_2 \to \ell^+\ell^- + X \tag{10.31}$$

It corresponds to production of a large invariant-mass lepton-pair (plus anything) in high-energy hadron collisions.

In (10.31), $\ell = e, \mu$ and $(h_1, h_2) = (p, p), (\overline{p}, p), (\pi^{\pm}, N)$ with $N = p, n$. In the quark-parton model, the process is approximated by

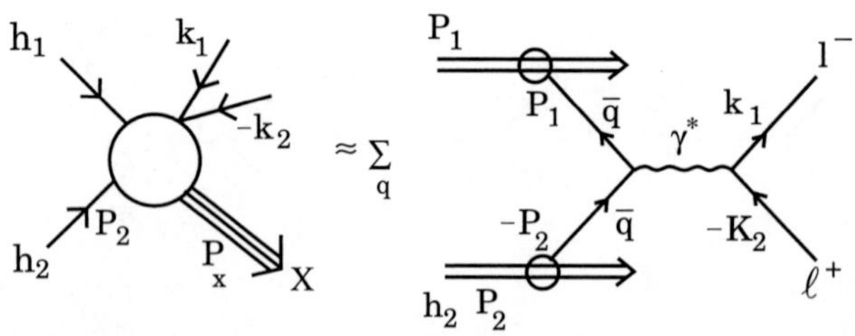

Fig. 10.10

observed process	parton-level process
$h_1 + h_2 \rightarrow \ell^+ + \ell^- + X$	$q + \overline{q} \rightarrow \gamma^* \rightarrow \ell^- + \ell^+$
$P_1 + P_2 = k_1 + k_2 + P_X$	$p_1 + p_2 = q = k_1 + k_2$

$$p_1 = x_1 P_1, \; p_2 = x_2 P_2$$

$$0 \leq x_1, x_2 \leq 1 \tag{10.32}$$

$$s = (P_1 + P_2)^2 \approx 2P_1 P_2$$

$$\hat{s} = (p_1 + p_2)^2 = (x_1 P_1 + x^2 P_2)^2 \approx x_1 x_2 s$$

$$M^2 = (k_1 + k_2)^2 \geq 4m_\ell^2, \; (m_\ell = \text{lepton mass}) \tag{10.33}$$

In the above, x_1 and x_2 are the fractions of the parent hadron momenta carried away by the constituent partons that take part in the subprocess. The remaining partons recombine to form the final-state hadrons (X). M is known as the invariant mass of the outgoing lepton pair. It is clearly, a Lorentz-invariant quantity.

According to the quark-parton model,

$$\sigma(h_1 + h_2 \rightarrow \ell^+ + \ell^- + X) = \sum_q \int\int d\sigma(q + \overline{q} \rightarrow \ell^+ + \ell^-)$$
$$[f_q^{h_1}(x_1) f_{\overline{q}}^{h_2}(x_2) + (1 \leftrightarrow 2)] dx_1 dx_2 \tag{10.34}$$

In (10.34) there are integration over final lepton pair phase space as well as the two (Lorentz-invariant) momentum fraction variables x_1, x_2.

$$\sigma(q + \overline{q} \rightarrow \ell^+\ell^-) = \int d\sigma(q + \overline{q} \rightarrow \ell^+\ell^-) = \frac{4\pi\alpha^2}{3\hat{s}} e_q^2 \tag{10.35}$$

Inserting in (10.34) a factor $\int \delta(M^2 - \hat{s}) dM^2 = 1$, and using eq. (10.35) we get

$$\frac{d}{dM^2}\sigma(h_1 + h_2 \rightarrow \ell^+\ell^- + X) = \sum_q \frac{4\pi\alpha^2}{3\hat{s}} e_q^2 \left[f_q^{h_1}(x_1) f_{\overline{q}}^{h_2}(x_2) + (1 \leftrightarrow 2)\right]$$
$$\times \, \delta(M^2 - \hat{s}) dx_1 dx_2 \tag{10.36}$$

Now $\delta(M^2 - \hat{s}) = \delta(M^2 - x_1 x_2 s)$

$$= \frac{1}{s}\delta(x_1 x_2 - \tau)\frac{1}{M^2} x_1 x_2 \delta(x_1 x_2 - \tau) \tag{10.37}$$

where $\tau = M^2/s$

We can, therefore, write (10.36) in the form

$$\frac{d}{dM^2}\sigma(h_1 + h_2 \to \ell^+ + \ell^- + X)$$
$$= \frac{4\pi\alpha^2}{3M^4}\sum_q e_q^2[x_1 f_q^{h_1}(x_1)\cdot x_2 f_{\bar{q}}^{h_2}(x_2) + (1 \leftrightarrow 2)]\delta(x_1x_2 - \tau)dx_1dx_2 \quad (10.38)$$

Eq. (10.38) implies that

$$M^4\frac{d\sigma}{dM^2} = f(\tau) = f(M^2/s) \quad (10.39)$$

which means $M^4 d\sigma/dM^2$ "scales". Experimental data show the "scaling" behavior predicted by the quark-parton model.

A test of the quark-parton model is the prediction of the ratio $\sigma(\pi^- + 12_C \to \mu^+\mu^- X)/\sigma(\pi^+ + 12_C \to \mu^+\mu^- X)$ at high energies which agrees with experimental data. This agreement is also consistent with the identification of charged partons with fractionally charged quarks.

One finds that

$$R = \frac{\sigma(\pi^- + 12_C \to \mu^+\mu^- X)}{\sigma(\pi^+ + 12_C \to \mu^+\mu^- X)} = \frac{\sum_{q=u}\sigma(q+\bar{q} \to \mu^+\mu^-)}{\sum_{q=d}\sigma(q+\bar{q} \to \mu^+\mu^-)}$$
$$= \frac{18e_u^2}{18e_d^2} = 4 \quad (10.40)$$

Problems

1. Draw a Feynman diagram to demonstrate that the process $e^-e^+ \to \pi^-\pi^+$ measures the electromagnetic form factor $F_\pi(q^2)$ of the pion. If the process is dominated by the ρ-meson intermediate state ($e^-e^+ \to \gamma^* \to \rho^0 \to \pi^-\pi^+$), show that the rms electromagnetic charge radius of the pion is 0.64 fm ($m_\rho = 769$ MeV/c^2).

2. Assuming that the massive lepton-pair production in hadron-hadron collisions at high energies occurs through the Drell-Yan process, $q\bar{q} \to \gamma^* \to \ell^-\ell^+$, show that the quark-parton model predicts that

$$\frac{\sigma(\pi^+ + 12_C \to \mu^-\mu^+ X)}{\sigma(\pi^- + 12_C \to \mu^-\mu^+ X)} = \frac{1}{4}$$

3. Assume that the valence quark momentum distribution in the nucleon is of the form $A(1-x)^3$ and that the valance quark/antiquark momentum distribution in the pion is of the form $B(1-x)$, where A and B are constants and x is the fraction of hadron momentum carried by a constituent quark/antiquark. Find an expression for $d\sigma/dM^2$ for the production of massive muon pairs in a pion-proton collision (assuming it goes through the Drell-Yan process) as a function $\tau = M^2/s$, where $\sqrt{s} = W =$ pion-proton CM energy and $M^2 =$ invariant mass squared of the muon pair.

Chapter 11

Quantum Chromo Dynamics (QCD)

The success of the quark-parton model seems to indicate that at large momentum transfer, that is at short distance, the strong interaction between quarks becomes sufficiently weak so that quarks may be regarded as effectively non-interacting with one another. On the other hand, at low-momentum transfer i.e. at large distance ($\geq$ hadron size) the strong interaction coupling strength must be sufficiently large so that quarks remain permanently bound (confined) within hadrons. This property of the coupling strength is characteristic of non-abelian gange theories, i.e. field theories which are invariant under local gauge transformations with the underlying transformation group being non-abelian instead of the simple $U(1)$ as in the case of quantum electrodynamics.

11.1 $U(1)$ Gauge Theories

Let us consider the Lagrangian density of a single free Dirac field,

$$\pounds_0 = \overline{\psi}(x)(i\gamma^\mu\partial_\mu - m)\psi(x) \tag{11.1}$$

$\pounds_0$ is clearly invariant under the transformation,

$$\psi(x) \to \psi'(x) = e^{-i\theta}\psi(x)$$

$$\overline{\psi}(x) \to \overline{\psi'}(x) = \overline{\psi}(x)e^{+i\theta} \tag{11.2}$$

where θ is a real parameter independent of x. This transformation is known as global gauge transformation (global because it is same for all $\psi(x)$). If we consider an infinitesimal transformation $\theta = \delta\theta =$ small, then

$$\psi(x) \to \psi'(x) = e^{-i\delta\theta}\psi(x) \approx (1 - i\delta\theta + \cdots)\psi(x) = \psi(x) + \delta\psi(x)$$

$$\delta\psi = -i\delta\theta\psi(x)$$

$$\text{similarly, } \delta\overline{\psi} = \overline{\psi}(x)(i\delta\theta) \tag{11.3}$$

under the infinitesimal global gauge transformation,

$$\pounds_0 \to \pounds_0 + \delta\pounds_0$$

$$\delta\pounds_0 = \frac{\partial\pounds}{\partial\psi}\delta\psi + \frac{\partial\pounds}{\partial(\partial_\mu\psi)}\delta(\partial_\mu\psi) = \frac{\partial\pounds}{\partial\psi}\delta\psi + \frac{\partial\pounds}{\partial(\partial_\mu\psi)}\partial_\mu(\delta\psi)$$

$$= \partial_\mu \left(\frac{\partial \pounds}{\partial(\partial_\mu \psi)} \right) \delta\psi + \frac{\partial \pounds}{\partial(\partial_\mu \psi)} \partial_\mu(\delta\psi) = \partial_\mu \left[\frac{\partial \pounds}{\partial(\partial_\mu \psi)} \delta\psi \right]$$

$$= \partial_\mu [\overline{\psi}(x) i\gamma^\mu(-i\delta\theta)\psi(x)] = \delta\theta \partial_\mu [\overline{\psi}(x)\gamma^\mu \psi(x)] = \delta\theta \partial_\mu J^\mu$$

where $J^\mu = \overline{\psi}(x)\gamma^\mu\psi(x)$ (11.4)

Thus invariance of the Lagragian under arbitrary infinitesimal global gauge transformation implies that $\delta L_0 = 0$ which in turn implies

$$\partial_\mu J^\mu = 0 \tag{11.5}$$

i.e. there exists a conserved current J^μ and the corresponding charge

$$Q = \int J^0(x) d^3x$$

is a constant of motion, $dQ/dt = 0$

Next we consider the transformation

$$\psi(x) \to \psi'(x) = e^{-i\theta(x)}\psi(x) \tag{11.6}$$

where $\theta(x)$ is still real but now a function of x; i.e. the transformation is different at different x. Such a transformation is said to be a local gauge transformation. L_0 is not invariant under such transformations because

$$\partial_\mu \psi(x) \to \partial_\mu [e^{-i\theta(x)}\psi(x)] = e^{-i\theta}\partial_\mu \psi(x) + [\partial_\mu(e^{-i\theta})]\psi(x) \neq e^{-i\theta}\partial_\mu \psi$$

However, we can obtain a Lagrangian which is invariant under local gauge transformations, if we replace the ordinary derivative by gauge covariant derivative

$$\partial_\mu \to D_\mu = \partial_\mu + ieA_\mu(x) \tag{11.7}$$

where $A_\mu(x)$ is a vector field (called the gauge field) with the transformation property

$$A_\mu(x) \to A_\mu(x) = A_\mu(x) + \frac{1}{e}\partial_\mu \theta \tag{11.8}$$

and e is a constant called the charge. Thus $\pounds_0$ is replaced by $\pounds'_0$ where

$$\pounds'_0 = \overline{\psi}(x)(i\gamma^\mu D_\mu - m)\psi(x) \tag{11.9}$$

We can easily check that

$$D_\mu \psi \to (D_\mu \psi)' = D'_\mu \psi' = (\partial_\mu + ieA'_\mu)\psi' = e^{-i\theta}(D_\mu \psi) \tag{11.10}$$

and therefore, $\pounds'_0$ is invariant under local gauge transformation defined by Eqs. (11.6), (11.7) and (11.8). To get the full Lagrangian we have to add to L'_0, the Lagrangian density due to the gauge field alone. Because of the gauge transformation property (11.8) of A_μ,

$$F_{\mu\nu} = \partial_\mu A_\nu - \partial_\nu A_\mu \tag{11.11}$$

is invariant under local gauge transformation. Thus a suitable choice for gauge field Lagrangian is

$$\mathcal{L}_{\text{gauge}} = -\frac{1}{4} F_{\mu\nu} F^{\mu\nu} \tag{11.12}$$

There can not be a term of the form $A_\mu A^\mu$ (mass term) in the Lagrangian because it is not invariant under transformation (11.8). This implies that the gauge field can not have a mass 'Thus the "minimal" full Lagragian density is given by

$$\begin{aligned} \mathcal{L} &= \mathcal{L}'_0 + \mathcal{L}_{\text{gauge}} = \overline{\psi}(x)(i\gamma^\mu D_\mu - m)\psi(x) - \frac{1}{4} F_{\mu\nu} F^{\mu\nu} \\ &= \mathcal{L}_0 + \mathcal{L}_{\text{int}} + \mathcal{L}_{\text{gauge}} \end{aligned} \tag{11.13}$$

where $\mathcal{L}_0 = \overline{\psi}(x)(i\gamma^\mu \partial_\mu - m)\psi(x)$

$$\mathcal{L}_{\text{int}} = e\overline{\psi}(x)\gamma^\mu A_\mu(x)\psi(x) = eJ^\mu(x) A_\mu(x) \tag{11.14}$$

and $$\mathcal{L}_{\text{gauge}} = -\frac{1}{4} F_{\mu\nu} F^{\mu\nu} \tag{11.12}$$

$\mathcal{L}_{\text{int}}$ gives the interaction between the matter field and the gauge field. It may be noted that the gauge field couples only to conserved currents.

The gauge theory we have described above is known as a $U(1)$ or Abelian gauge theory, since it involves a 1-parameter gauge transformation and hence two such transformations commute. The best known example of $U(1)$ gauge theories is Quantum Electrodynamics (QED) with the gauge boson identified with the photon, the quantum of the electromagnatic field.

11.2 Nonabelian Theories

We again consider a free Dirac Lagrangian

$$\mathcal{L}_0 = \overline{\psi}(x)(i\gamma^\mu \partial_\mu - m)\psi(x) \tag{11.15}$$

where, $\psi(x)$ is now a n-component field corresponding to an n-dim representation of an internal symmetry group. $\mathcal{L}_0$ is invariant under global gauge transformation.

$$\psi(x) \xrightarrow{U(\theta)} \psi'(x) = U(\theta)\psi(x)$$

$$U(\theta) = e^{-i\theta},\ \theta = \sum_{a=1}^{r} \theta_a \frac{t_a}{2} \tag{11.16}$$

It is a r-parameter transformation, t_a's are $n \times n$ matrices being the n-dim representation of the generators of the symmetry group being considered. θ_a's are independent of x and t_a satisfy the commutation relations

$$\left[\frac{t_a}{2}, \frac{t_b}{2}\right] = \text{i}f_{abc}\frac{t_c}{2} \tag{11.17}$$

f'_{abc}s $(a, b, c = 1, \ldots, r)$ are known as the structure constants of the group. They are independent of representations used to compute them and are characteristic of the group.

As a consequence of the global gauge invariance, by Noether's theorem, there exists a conserved current

$$J^{\mu,a}(x) = \overline{\psi}(x)\gamma^{\mu}\frac{t_a}{2}\psi(x) \tag{11.18}$$

such that $\partial_\mu J^{\mu,a} = 0$

We now consider the same transformation (11.16) but now $\theta_a = \theta_a(x) \cdot$ $(a = 1, \ldots r)$. This is known as local gauge transformation, i.e. the transformation is different at different x.

Now $\partial_\mu \psi \xrightarrow{U} (\partial_\mu \psi) = \partial_\mu \psi' = \partial_\mu(U\psi) = U(\partial_\mu \psi) + (\partial_\mu U)\psi \neq U(\partial_\mu \psi)$

Thus $\overline{\psi}\partial_\mu \psi$ is not invariant under local gauge transformation, although $\overline{\psi}\psi$ is still invariant. To implement local gauge invariance we replace ∂_μ by D_μ, where

$$D_\mu = \partial_\mu - igA_\mu(x) \tag{11.19}$$

where $A_\mu(x) = \sum_{a=1}^{r} \frac{t_a}{2} A_\mu^a(x)$ is a matrix-valued field ($n \times n$ matrix)

$A_\mu^a(x)(a = i \cdots r)$ are a set of r fields transforming as the adjoint representation of the symmetry group and like a 4-vector under Lorentz transformation. Now,

$$D_\mu \psi \xrightarrow{U} (D_\mu \psi)' = D'_\mu \psi' = (\partial_\mu - igA'_\mu)(U\psi) = \partial_\mu(U\psi) - igA'_\mu(U\psi)$$

$$= U(\partial_\mu \psi) + (\partial_\mu U)\psi - igA'_\mu(U\psi) \tag{11.20}$$

We want $(D_\mu \psi)' = U(D_\mu \psi) = U(\partial_\mu - igA_\mu)\psi$ (11.21) from (11.20) and (11.21)

$$U(\partial_\mu\psi) + (\partial_\mu U)\psi - igA'_\mu(U\psi) = U(\partial_\mu\psi) - igU(A_\mu\psi)$$

or $$A'_\mu(U\psi) = UA_\mu\psi + \frac{1}{ig}(\partial_\mu U)\psi$$

or $$A'_\mu(U\psi) = UA_\mu U^{-1}(U\psi) + \frac{1}{ig}(\partial_\mu U)U^{-1}(U\psi)$$

This means under local gauge transformation, $A_\mu(x)$ must transform in the following way,

$$A_\mu(x) \xrightarrow{U} A'_\mu(x) = UA_\mu(x)U^{-1} + \tfrac{1}{ig}(\partial_\mu U)U^{-1} \tag{11.22}$$

We have $(D_\mu\psi)' = U(D_\mu\psi)$ [eq. (11.21)]

or $D'_\mu\psi' = U(D_\mu\psi)$

or, $D'_\mu(U\psi) = UD_\mu U^{-1}(U\psi)$

This means under gauge transformation,

$$D_\mu \xrightarrow{U} D'_\mu = UD_\mu U^{-1} \tag{11.23}$$

Equation (11.22) and (11.23) give the transformation properties of the gauge field $A_\mu(x)$ and the covariant derivative D_μ. Under an infinitesimal local gauge transformation,

$$\begin{aligned} U = e^{-i\delta\theta(x)} &\approx 1 - i\delta\theta(x) \\ &\approx 1 - i\sum_a \delta\theta_a\frac{t_a}{2}, \quad \delta\theta_a\text{'s are infinitesimal} \end{aligned} \tag{11.24}$$

The individual gauge fields transform in the following way

$$A^a_\mu(x) \xrightarrow{U} A'^a_\mu(x) \approx A^a_\mu(x) - \frac{1}{g}\partial_\mu\delta\theta_a(x) - f_{abc}\delta\theta_b A^c_\mu(x) \tag{11.25}$$

as can be seen by substituting (11.24) in (11.22) and using the definition of $A_\mu(x)$ and the CR's of t_a's (Eq. 11.7).

It is now easily seen that

$$\mathcal{L}'_0 = \overline{\psi}(x)(i\gamma^\mu D_\mu - m)\psi(x) \tag{11.26}$$

is invariant under local gauge transformations.

We have next to construct $\mathcal{L}_{\text{gauge}}$, the Lagrangian density involving gauge fields alone. We first define a rank-2 Lorentz-tensor

$$G_{\mu\nu} = \frac{1}{ig}[D_\mu, D_\nu] = \frac{1}{ig}[\partial_\mu - igA_\mu, \partial_\nu - igA_\nu]$$

$$= -\partial_\mu A_\nu + \partial_\nu A_\mu + ig[A_\mu, A_\nu] \tag{11.27}$$

From (11.23) it is easily seen that

$$G_{\mu\nu} \xrightarrow{U} G'_{\mu\nu} = UG_{\mu\nu}U^{-1} \tag{11.28}$$

From (11.28) it follows that

$$\pounds_{\text{gauge}} = -\frac{1}{2}\text{Tr}\,(G_{\mu\nu}G^{\mu\nu}) \tag{11.29}$$

is both Lorentz-invariant and gauge-invariant

$$\text{Writing } G_{\mu\nu} = \sum_{a=1}^{r} G^a_{\mu\nu}\frac{t_a}{2} \tag{11.30}$$

$$\text{we get} \quad G^a_{\mu\nu} = -\partial_\mu A^a_\nu + \partial_\nu A^a_\mu - \sum_{b,c} g f_{abc} A^b_\mu A^c_\nu \tag{11.31}$$

$$\text{and} \qquad \pounds_{\text{gauge}} = -\frac{1}{2}\text{Tr}\,(G_{\mu\nu}G^{\mu\nu}) = -\frac{1}{4}\sum_{a=1}^{r} G^a_{\mu\nu}G^{\mu\nu,a} \tag{11.32}$$

As in the case of Abelian gauge theories we can not have a term of the form $\text{Tr}(A_\mu A^\mu)$ in $\pounds_{\text{gauge}}$ since it is not gauge invariant. Thus the gauge fields in non-abelian gauge theories are also massless. The full gauge-invariant Lagrangian is then given by

$$\pounds = \pounds'_0 + \pounds_{\text{gauge}} = \overline{\psi}(x)(i\gamma^\mu D_\mu - m)\psi(x) - \frac{1}{2}\text{Tr}(G_{\mu\nu}G^{\mu\nu})$$

$$= \pounds_0 + \pounds_{\text{int}} + \pounds_{\text{gauge}} \tag{11.33}$$

where $\pounds_0$ and $\pounds_{\text{gauge}}$ are given by equation (11.25) and (11.34) and

$$\pounds_{\text{int}} = g\overline{\psi}(x)\gamma^\mu A_\mu(x)\phi(x) = g\sum_{a=1}^{r} J^{\mu,a}(x)A^a_\mu(x) \tag{11.34}$$

where $J^{\mu,a}(x) = \overline{\psi}(x)\gamma^\mu \frac{t_a}{2}\psi(x)$

is the conserved current (Eq. 11.18)

One striking feature of non-abelian gauge theories is the appearance of self-interaction terms among the gauge bosons because the generators of the gauge group do not commute. Because of the form of the gauge field tensor $G_{\mu\nu}$, there are 3-boson and 4-boson interaction terms in $\pounds_{\text{gauge}}$,

$$g f_{abc}(\partial_\mu A^a_\nu) A^{\mu,b} A^{\nu,c} \tag{11.36a}$$

and $\quad g^2 f_{abc} f_{ab'c} A^b_\mu A^c_\nu, A^{\mu,b'} A^{\nu,c'}$ (11.36)

Non-abelian gauge theories as possible candidates for strong interaction were first studied by C.N. Yang and R.L. Mills in 1954 and these are now known as Yang-Mills theories. They chose i-spin symmetry group $SU(2)_I$ as the gauge group and the fermions were identified with the nucleons belonging to the fundamental representation $\left(I = \frac{1}{2}\right)$ of $SU(2)_I$ and the gauge fields were identified with the vector mesons belonging to the adjoint representation $(I = 1)$ of $SU(2)_I$. However, these theories as possible candidates of strong interaction were abandoned since experimentally no massless vector meson was discovered.

Interest in non-abelian gauge theories as possible candidates of strong interaction was revived after the discovery of scaling in deep-inelastic lepton-hadron scattering. In 1973 Gross and Wilczek and independently Politzer proved that non-abelian gauge theories have the remarkable property of asymptotic freedom which could explain the success of the parton model in explaining scaling phenomena by assuming the partons to be weakly coupled.

Quarks are distinguished from leptons in having nonzero color quantum number.Moreover,quarks participate in strong interaction while leptons do not.Thus it was natural to choose "color" as the strong-interaction "charge" and the gauge group identified with $SU(3)_c$. The gauge bosons were identified with the gluons, the quanta presumed to mediate the strong interaction among quarks. The gluons must belong to the adjoint representation 8 of $SU(3)_c$ (and not the other choice 1). There was no problem in the gluons to be massless (as required by local gauge invariance) because they can not be observed in free state for the same reason that quarks do not exist in free states. They both belong to non-singlet representation of $SU(3)_c$ and according to the color hypothesis only color-singlet states are physical, that is observable. Thus quarks and gluons are both permanently confined within hadrons. This confinement property is another characteristic of non-abelian gauge theories which unfortunately has not yet been proved.

The theory of strong interaction among quarks mediated by gluons with $SU(3)_c$ as the underlying gauge group is known as Quantum Chromodynamics (QCD) in analogy with Quantum Electrodynamics (QED).

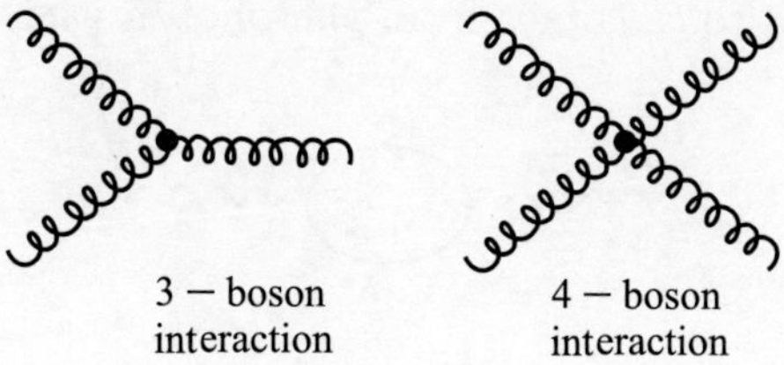

Fig. 11.1

11.3 Running Coupling Constant-Asymptotic Freedom

Define $g = g(Q^2)$ (11.37)

g satisfies the Renormalization group (RG)equation

$$\frac{d}{dt}g(t) = \beta(g), \; t = \frac{1}{2}\ell n(Q^2/\mu^2) \tag{11.38}$$

In perturbation theory

$$\beta(g) = -b_0 g^3 + b_1 g^5 + \cdots \tag{11.39}$$

To lowest order in g

$$\frac{d}{dt}g(t) = -b_0 g^3 \tag{11.40}$$

Integrating, we get

$$\int_{g_i}^{g_f} \frac{dg}{g^3} = -b_0 \int_0^1 dt, \; g_i = g(\mu^2), \; g_f = g(Q^2)$$

or $$\frac{1}{g_f^2} - \frac{1}{g_i^2} = 2b_0 t$$

or $$\frac{1}{\alpha(Q^2)} - \frac{1}{\alpha(\mu^2)} = +8\pi b_0 t, \; \alpha = \frac{g^2}{4\pi} \tag{11.41}$$

or $$\alpha(Q^2) = \frac{\alpha(\mu^2)}{1 + 8\pi b_0 \alpha(\mu^2) t} \tag{11.42}$$

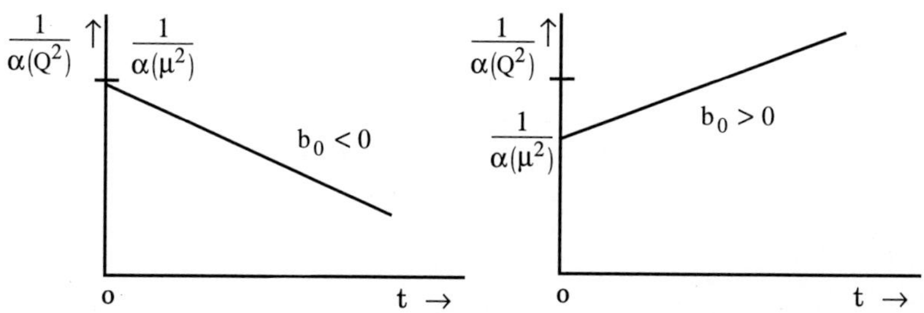

Fig. 11.2

QED: Contribution to b_0 comes from the photon vacuum polarization diagram

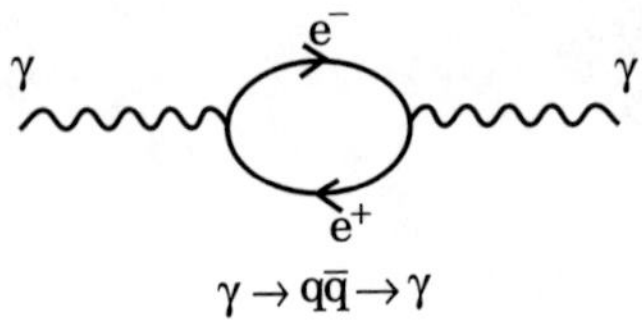

Fig. 11.3

$$\text{and} \quad b_0 = -\frac{1}{12\pi^2} \tag{11.43}$$

$$\text{This gives } \alpha(Q^2) = \frac{\alpha(\mu^2)}{1 - \frac{\alpha(\mu^2)}{3\pi}\ell n(Q^2/\mu^2)} \tag{11.44}$$

One can choose the renormalization point $\mu = m_e =$ electron mass

QCD: Contribution to b_0 comes from two diagrams in Fig. 11.3

$$\text{and} \quad b_0 = \frac{1}{16\pi^2}\left(\frac{11}{3}N_c - \frac{2}{3}N_f\right) \tag{11.45}$$

where N_c = number of quark colors = 3 for $SU(3)_c$ and N_f = number of quark flavors which can contribute.

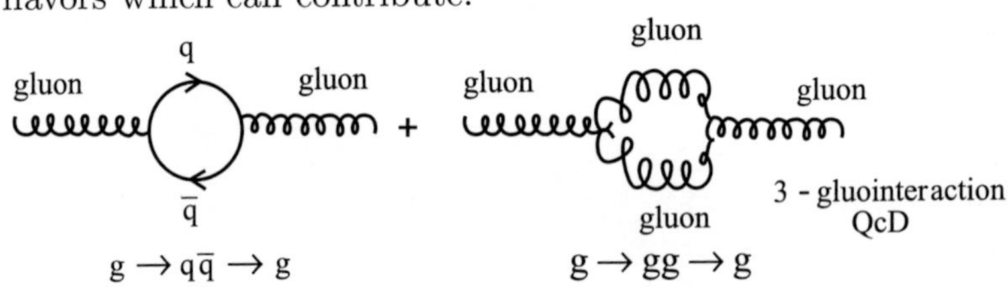

Fig. 11.4

Since the physical quark mass is not defined, there is no mass parameter in QCD. So one chooses a parameter $\wedge$ called the "scale parameter" such that

$$\frac{1}{\alpha_s(\wedge^2)} = 0 \tag{11.46}$$

From (11.41) and (11.38)

$$\begin{aligned}\frac{1}{\alpha_s(Q^2)} &= \frac{1}{\alpha_s(\mu^2)} + 4\pi b_0 \ell n\left(\frac{Q^2}{\mu^2}\right)\\ &= \frac{1}{\alpha_s(\mu^2)} + 4\pi b_0\left[\ell n\left(\frac{Q^2}{\wedge^2}\right) + \ell n\left(\frac{\wedge^2}{\mu^2}\right)\right]\\ &= \frac{1}{\alpha_s(\wedge^2)} + 4\pi b_0 \ln(Q^2/\wedge^2)\\ &= 4\pi b_0 \ell n\left(\frac{Q^2}{\wedge^2}\right)\end{aligned} \tag{11.47}$$

In getting (11.47) we have used (11.46)

we can write (11.47) in the form

$$\alpha_s(Q^2) = \frac{1}{4\pi b_0 \ln(Q^2/\wedge^2)} = \frac{4\pi}{b \ln(Q^2/\wedge^2)} \tag{11.48}$$

where $b = \frac{11}{3}N_c - \frac{2}{3}N_f = 11 - \frac{2}{3}N_f$, for $SU(3)_c$ (11.49)

Thus for $N_f < 16$, $b_0 > 0$

Equation (11.48) implies that for $Q^2 \gg \wedge^2, \alpha_s(Q^2) \to 0$ and one has what is known as "asymptotic freedom" i.e. quark-gluon coupling becomes very small at large Q^2 (short distance) and one can regard the quarks as essentially non-interacting in this limit. This explains the success of the quark-parton model in describing DI scattering. At the other extreme, i.e. $Q^2 \sim \wedge^2$, small Q^2 or large distance $\alpha_s(Q^2)$ becomes very large and quarks remain strongly bound or confined within hadrons. However, in this limit perturbation theory is not applicable. This is the reason why quark confinement has not yet been proved, although it seems to be true since quarks have so far not been seen in free states.

11.4 Comparison of QED and QCD

Both QED and QCD are gauge theories, i.e. the underlying Lagrangian is invariant under local gauge transformations. As a consequence the mediators (the quanta of gauge field) are, in both cases, massless vector particles (the photon and the gluons respectively) and the gauge fields couple only to conserved currents. However, in QED the gauge group $U(1)_Q$ is Abelian, where as the QCD gauge group $SU(3)_c$ is non-Abelian. The gauge field (i.e. the photon) in QED does not have self interaction since it is electrically neutral, i.e. does not carry the "charge" of electromagnetic interaction. On the other hand, the gauge fields (i.e. the gluons) in QCD have self-interactions (3-gluon and 4-gluon vertices). The gluons belong to the adjoint representation 8 (non-singlet) of $SU(3)_c$. Hence they have non-zero color, i.e. they carry "charge" of strong interaction. In fact, the gluon self-interactions are responsible for most of the differences between QED and QCD. QCD is asymptotically free, i.e. $\alpha_s(Q^2) \to 0$ as $Q^2 \to \infty$, whereas QED is not asymptotically free. This means $\alpha(Q^2)$ is small at small Q^2 (i.e. at large distances). Thus there is "screening" of electric charge. On the other hand, $\alpha_s(Q^2)$ is large at small Q^2 (i.e. at large distances) and therefore there is "anti-scrcening" of "color". This behavior of $\alpha_s(Q^2)$ is responsible for confinement of "color" i.e. confinement of quarks and gluons within hadrons since both belong to non-singlet representation of $SU(3)_c$.

11.5 QCD Corrections to the Parton Model Results

The parton model may be regarded as the zeroth approximation of QCD. In the parton model the strong interactions between quarks at large Q^2 are neglected and the quarks are treated as free particles. This agrees with the fact that in QCD $\alpha_s(Q^2) \to 0$ as $Q^2 \to \infty$. Since $\alpha_s(Q^2)$ is small at large Q^2, perturbation theory is applicable in the DI domain of $Q^2 \gg \wedge^2$. Perturbative QCD has been quite successful in explaining the observed deviations from parton-model results. These deviations are logarithmic being $0(1/\ell n Q^2)$. This agrees with the behavior of $\alpha_s(Q^2)$ at large Q^2 as given by equation (11.48).

***ep* $\to$ *eX*:** The lowest-order QCD corrections come from the 1-gluon diagrams

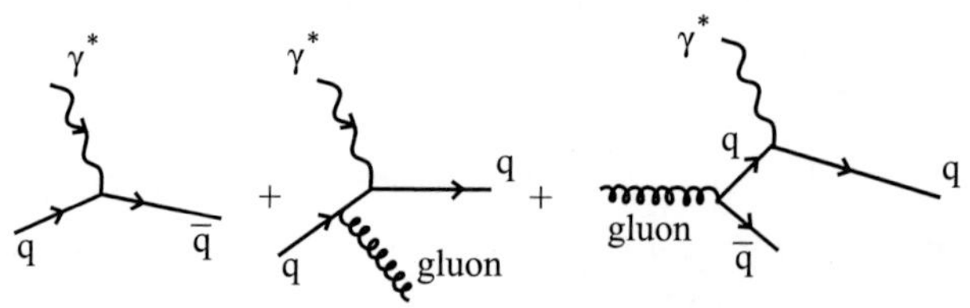

Fig. 11.5

$e^-e^+ \to$ hadrons: The lowest-order QCD correction comes from the diagrams

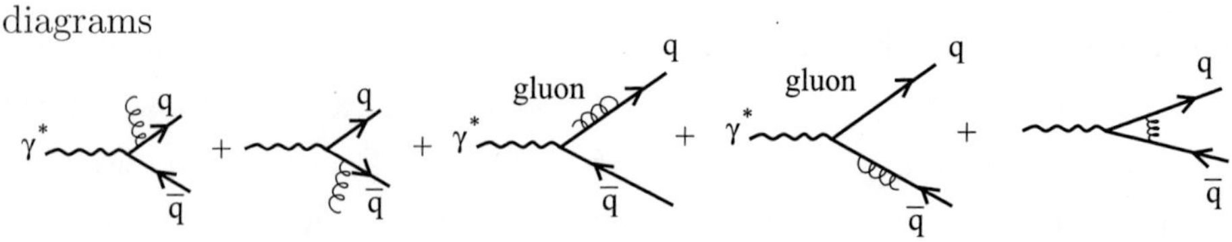

Fig. 11.6

One finds

$$R(s) = \frac{\sigma(e^+e^- \to \text{ hadrons})}{\sigma(e^+e^- \to \mu^+\mu^-)} = 3\sum_q e_q^2 \left[1 + \frac{\alpha_s(Q^2)}{\pi}\right] \tag{10.50}$$

where $s = Q^2$, $\sqrt{s} = W$ = total CM energy. Comparison with experimental data gives $\alpha_s(10 \text{ GeV}) \approx 0.15$ and $\alpha_s(90 \text{ GeV}) \sim 0.12$.

11.6 Evolution of Structure Functions

QCD makes quantitative predictions about the Q^2 evolution of the quark distribution functions in the limit $\alpha_s(Q^2) \ll 1$ which requires $Q^2 \gg \wedge^2$. Adding $\gamma^* q \to qg$ to the parton model process $\gamma^* q \to q$ (first 2 diagrams of Fig. 11.4), one finds that to lowest order in QCD.

$$f(x) \to f(x, Q^2) = f(x) + \Delta f(x, Q^2)$$

$$\text{where } \Delta f(x, Q^2) = \frac{\alpha_s}{2\pi} \ell n \left(\frac{Q^2}{2\pi}\right) \int_x^1 \frac{dy}{y} P_{qq}\left(\frac{x}{y}\right) f(y) \tag{11.51}$$

one can write (11.51) as an integro-differential equation for $f(x, Q^2)$

$$\frac{d}{d\ell n Q^2} f(x, Q^2) = \frac{\alpha_s}{2\pi} \int_x^1 \frac{dy}{y} P_{qq}\left(\frac{x}{y}\right) f(y, Q^2) \tag{11.52}$$

Eq. (11.52) is known as the Altarelli-Parisi equation. The equation mathematically expresses the fact that a quark with momentum fraction x could have come from parent quark with a larger momentum fraction $y(x = zy, 0 \leq z \leq 1)$ which has radiated a gluon. The probability of its happening is proportional to $\alpha_s P_{qq}(x/y)$, where $P(z)$ is the probability of radiation of a spin-1 boson from a spin $-\frac{1}{2}$ fermion. $P(z)$ is of the form

$$P(z) = c\left(\frac{1 + z^2}{1 - z}\right) \tag{11.53}$$

where c is a group theoretic factor $c = 1$ for QED and $c = 4/3$ for QCD. The $z = 1$ singularity is associated with the emission of a "soft" boson. The integral in (11.52) is a sum over all possible momentum fractions of the parent quark ($y > x$).

Given the quark distribution function at some reference point, $f(x, Q_0^2)$ one can compute it in QCD for any value Q^2 by using Altarelli-Parisi equation (Eq. 11.52). The study of systematics of the Q^2 dependence of structure functions shows that around $x = 0.25$ the structure function F_2 scales, ie, there is no Q^2 dependence at this particular x value. But for $x < 0.25$, $F_2(x, Q^2)$ increases with Q^2, while for $x > 0.25$, $F_2(x, Q^2)$ decreases with Q^2.

To order α_s we should also include the contribution from the process $\gamma^* g \to q\overline{q}$, i.e. the process in which a gluon in the initial proton produces a $q\overline{q}$ pair, to which the virtual photon then comples (third diagram of Fig. 11.5). If we include the pair production contribution to the evolution of the quark distribution function the Altarelli-Parisi equation (eq. 11.52) becomes

$$\frac{d}{d\ell nq^2} f(x, Q^2) = \frac{\alpha_s}{2\pi} \int_x^1 \frac{dy}{y} \left[P_{qq}\left(\frac{x}{y}\right) f(y, Q^2) + P_{qg}\left(\frac{x}{y}\right) g(y, Q^2) \right] \quad (11.54)$$

In (11.54), $g(y, Q^2)$ is the gluon distribution function and $P_{qg}(z)$ represents the probability that a gluon annihilates into a $q\overline{q}$ pair such that the quark has a fraction z of its momentum which in turn has a fraction y of the momentum of the initial proton ($x = zy, 0 \leq z \leq 1$). One finds that

$$P_{qg}(z) = c'[z^2 + (1 - z)^2] \quad (11.55)$$

where c' is a group theoretic factor. $c' = 1/2$ for QCD. For completeness, one also needs an equation for the evolution of the gluon distribution function in the proton, which in analogy with the evolution of the quark distribution function is given by

$$\frac{d}{d\ell nQ^2} g(y, Q^2) = \frac{\alpha_s}{2\pi} \int_x^1 \frac{dy}{y} \left[\sum_i f_i(y, Q^2) P_{gq}\left(\frac{x}{y}\right) \right.$$
$$\left. + g(y, Q^2) P_{gg}\left(\frac{x}{y}\right) \right] \quad (11.56)$$

In (11.56) the sum runs over all quarks and antiquarks of all flavors. P_{gq} is independent of quark flavor if the quark masses are negligibly small. It is found that

$$P_{gq}(z) = \frac{4}{3} \left[\frac{1 + (1 - z)^2}{z} \right]$$

and $$P_{gg}(z) = 6 \left[\frac{1 - z}{z} + \frac{z}{1 - z} + z(1 - z) + \left(\frac{11}{2} - \frac{n_f}{3} \right) \delta(1 - z) \right] \quad (11.57)$$

In the above, $P_{gq}(z)$ is the probability for the process $q\overline{q} \to g$ and $P_{gg}(z)$ is the probability for the process $g \to gg$ such that final state gluon carries a fraction z of the initial quark or gluon momentum.

Problems

1. Draw the Feynman diagrams and write down the corresponding invariant amplitudes for the following processes to the lowest order in QCD:

$$qq \to qq, \; q\overline{q} \to gg, \; gg \to gg$$

2. Compute $|\overline{A}|^2$, the |invariant amplitude|^2, averaged over the initial state and summed over the final state spins and colours for the process $qq' \to qq'$, where q and q' are quarks of different flavor.

3. Compute the color factor (a) in the 1-gluon exchange potential $V(q\bar{q})$ in mesons and (b) in the 1-gluon exchange potential $V(qq)$ in a baryon by considering the following Feynman diagrams respectively.

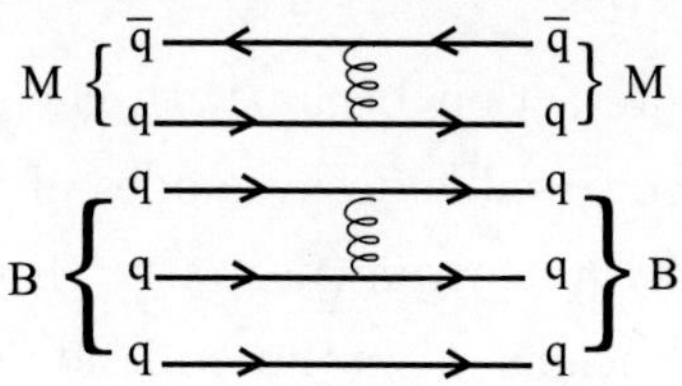

Chapter 12

Weak Interaction and the V-A Theory

12.1 Historical Survey

The study of weak interaction began with the attempt to understand β-decay, one of the three possible decay modes of radioactive nuclei. In electron (positron) β-decay a neutron (proton)-rich nucleus decays into a more stable neighbouring nucleus with the emission of an electron (positron). Two facts were established from the early studies of β-active nuclei. First, the energy spectrum of the emitted electron (positron) was a continuous one with an end-point (maximum) energy, equivalent to the mass difference of the parent and daughter nuclei. Thus, either energy is not conserved in β-decay or it could not be a two-body decay. Secondly, the half-lives of β-active nuclei were much longer than the electromagnetic decay time. This implied that the β-decay is mediated by a force much weaker than the electromagnetic force. Since the β-decay was a nuclear phenomenon, it was named "weak nuclear force" in order to distinguish it from the other known (strong) nuclear force responsible for nuclear binding.

In order to save the energy-momentum conservation principle in β-decay, Pauli in 1930 postulated the existence of a massless, electrically neutral particle which was emitted along with the electron and carried away the missing energy. The particle was subsequently chirstened 'neutrino' by Fermi. The neutrino can participate only in weak interactions. However, in order to conserve angular momentum in β-decay, the neutrino must have a half-integral spin because the angular momentum transfer between the parent and daughter nuclei is always integral and the electron has spin-$\frac{1}{2}$. Furthermore, since there is no lepton in the initial state of the decay process, the neutrino emitted along with the electron in the decay must be an antilepton. Thus an antineutrino (neutrino) is emitted in electron (positron) β-decay.

Making use of Pauli's neutrino hypothesis, Fermi in 1932 gave a theory of β-decay based on an analogy with the electromagnetic interaction. The interaction Lagrangian suggested by Fermi for the decay $a \to b + c + \overline{d}$ (all spin-$\frac{1}{2}$fermions) was of the form

$$L_{\text{int}} \sim G_F[(\overline{\psi}_b \gamma^\mu \psi_a)(\overline{\psi}_c \gamma_\mu \psi_d) + h \cdot c.] \tag{12.1}$$

in which the electromagnetic field A_μ is replaced by the Dirac bilinear covariant transforming as a 4-vector under Lorentz transformation, i.e. $J^\mu A_\mu$ is replaced by $J^\mu J_\mu^+ + h \cdot c.$ Equation (12.1) is a special case of what is now known as the 4-fermion interaction.

The existence of the neutrino was first convincingly proved some 20 years later (in 1953) by Reines and Cowan in an ingenious experiment making use of a beam of antineutrinos ($\overline{\nu}_e$) from a nuclear fission reactor.

In the meantime, many subatomic particles were discovered and it was found that most of them decay with a long lifetime which implied that their decays were also mediated by a force similar to the weak nuclear force. All these forces were thus part of what is now known as weak interaction or weak force.

In order to resolve the $\tau - \theta$ puzzle in the charged kaon decay, Lee and Yang in 1957 suggested that parity may not be conserved in weak interactions. Parity violation was subsequently experimentally detected by C. S. Wu in the β-decay of polarized 60_{CO} nucleus. Studies of weak decays convincingly proved that not only parity but also many of the conservation laws respected by strong and electromagnetic interactions are not obeyed by weak interactions.

Another interesting fact about weak interactions is that the neutrinos associated with different charged leptons are different. This was first postulated by Ponte carvo for the electron and muon case and was experimentally confirmed by Steinberger and others in mid 1960's using accelerator neutrinos (mostly associated with the muon). After the discovery of the τ-lepton in 1975, it was found that it has its own neutrino. Thus $\nu_e \neq \nu_\mu \neq \nu_\tau$.

Although the Fermi theory of β-decay subsequently modified and extended was reasonably successful in explaining weak decays which are all low-energy processes it could not be applied to high energy weak processes (neutrino scattering and production). This is because for them one needs higher-order (loop) corrections and the Fermi theory is not renormalizable.

It was established in 1960's that in order to have a renormalizable theory of weak interaction,it has to unified with the electromagnetic interaction. The unified theory of weak and electromagnetic interactions is known as the electroweak theory.

12.2 Nuclear β-Decay

The nuclear β-decay can be subdivided into two types—electron β-decay and the positron β-decay. There is an associated process known as the electron capture.

electron β-decay: ${}^A_ZX \rightarrow {}^A_{Z+1}Y + e^- + \overline{\nu}_e$

positron β-decay: ${}^A_ZX \rightarrow {}^A_{Z-1}Y + e^+ + \nu_e$

electron capture: $e^- + {}^A_ZX \rightarrow {}^A_{Z-1}Y + \nu_e$ (12.2)

The effective processes in (12.2) are as follows

electron β-decay: $n \rightarrow p + e^- + \overline{\nu}_e$

positron β-decay: $p \rightarrow n + e^+ + \nu_e$

electron capture: $e^- + p \rightarrow n + \nu_e$ (12.3)

Although the free neutron decays, the free proton is stable. The second process in (12.3) can occur only in a nucleus where the difference in the binding energies of the parent and the daughter nuclei more than compensates the negative Q-value of the process. (This process occurs as the first step in the thermonuclear reactions leading to the fusion of H-atoms to form He-atoms which is responsible for the source of energy of the main sequence stars like the sun.)

Nuclear β-transitions are classified into allowed and forbidden transitions (the latter are further classified into first forbidden, second forbidden and so on). The lifetimes of forbidden transitions are much longer than those of allowed transitions. There is no orbital angular momentum transfer and no change of parity between the parent and daughter nuclei in allowed transitions ($\Delta l = 0$, $\Delta\pi = NO$) whereas in forbidden transitions, ($\Delta l \geq 1$ and $\Delta\pi = (-)^{\Delta l}$. In allowed transitions, $\Delta l = 0$ and hence $\Delta J = \Delta S = |\vec{S}_e + \vec{S}_e| = 0, 1$. Allowed transitions with $\Delta J = 0$ are called pure Fermi transitions, and those with $\Delta J = 1$ are called pure Gamow-Teller transitions, Mixed transitions have $\Delta J = 0$ and 1.

The original Fermi theory can account for pure Fermi transitions reasonably well but can not explain pure Gamow-Teller or mixed transitions. However, all β-transition can be taken care of by a generalization of the 4-fermion interaction proposed by Fermi (eq. 12.1)

$$L_f^W = G_F \sum_i c_i[(\overline{\psi}_p \Gamma_i \psi_n)(\overline{\psi}_e \Gamma^i \psi_\nu) + h \cdot c], \; i = \text{S, V, T, A, P} \tag{12.4}$$

In (12.4) one has sum over all 16 possible 4×4 Dirac matrices, classified as scalar (S), vector (V), Tensor (T). Axial vector (A) and pseudoscalar (P) based on Lorentz transformation properties.

The electron β-decay matrix element then becomes, to first order in G_F,

$$\text{A} = G_F \sum_a c_a[\overline{u}_p(p_f)\Gamma_a u_n(p_i)][\overline{u}_e(p_e)\Gamma^a v_\nu(p_\nu)] \tag{12.5}$$

Here $u(p)$ and $v(p)'$'s are 4-component Dirac spinors and p_i, p_f, p_e and p_ν are respectively, the 4-momenta of the parent nucleus, daughter nucleus, decay electron and antineutrino. If one uses nonrelativistic approximation for nuclear motions, one gets

$$\begin{aligned}
&\text{S} : \overline{u}_p 1 u_n \approx \chi_p^+ \chi_n \\
&\text{V} : \overline{u}_p \gamma^\mu u_n \approx \chi_p^+ \chi_n \delta_{\mu 0} \\
&\text{A} : \overline{u}_p \gamma^\mu \gamma_5 u_n \approx \chi_p^+ \sigma^j \chi_n \delta_{\mu j} \\
&\text{T} : \overline{u}_p \sigma^{\mu\nu} u_n \approx \in_{jkl} \chi_p^+ \sigma^l \chi_n \delta_{\mu j} \delta_{\nu k} \\
&\text{P} : \overline{u}_p \gamma_5 u_n \approx 0
\end{aligned} \tag{12.6}$$

Here χ_p and χ_n are 2-component Pauli spinors of the Proton and the neutron, respectively and j, k, $l = 1, 2, 3$. Thus S and V can contribute to Fermi transitions, while T and A can contribute to GT transitions. ($i =$ P does not contribute in nuclear β-decay.)

The decay probability is given by ($\hbar = c = 1$)

$$\frac{1}{\tau} = \Gamma = \int |\overline{A}|^2 \cdot (2\pi)^4 \delta_4(p_f + p_e + p_\nu - p_i) d\phi_f \tag{12.7}$$

where $d\phi_f =$ final state phase space volume element

$$d\phi_f = \frac{d^3 p_e}{(2\pi)^3 (2E_e)} \cdot \frac{d^3 p_\nu}{(2\pi)^3 (2E_\nu)} \cdot \frac{d^3 p_f}{(2\pi)^3}$$

$$|\overline{A}|^2 = \frac{1}{2} \sum_{\text{spin}} |A|^2 \tag{12.8}$$

In the rest frame of the parent nucleus, $p_i = (M_i \vec{0})$.

We have $\delta_4(p_f + p_e + p_\nu - p_i) = \delta_3(\vec{p}_f + \vec{p}_e + \vec{p}_\nu)\delta_3(E_f + E_e + E_\nu - M_i)$

$$\approx \delta_3(\vec{p}_f + \vec{p}_e + \vec{p}_\nu)\delta_3(E_e + E_\nu - E_0)$$

where $E_0 = M_i - E_f \approx M_i - M_f$

Also $d^3p_\nu = |\vec{p}_\nu|^2 d|\vec{p}_\nu| d\Omega_\nu = E_\nu^2 dE_\nu d\Omega_0$ ($\because$ neutrino is massless)

and $d^3p_e = |\vec{p}_e|^2 d|^2 d|\vec{p}_e| d\Omega_e = |\vec{p}_e| E_e dE_e d\Omega_e$ (12.9)

Carrying out the momentum integration in (12.7) with the help of energy-momentum conserving δ-functions, we have

$$\frac{1}{\tau} = \Gamma = \frac{1}{4(2\pi)^5} \int \frac{|\overline{A}|^2 |\vec{p}_e| E_e (E_0 - E_e)^2 dE_e d\Omega_e d\Omega_\nu}{E_e E_\nu} \tag{12.10}$$

Using nonrelativistic approximation for nuclear motion, we have

$$|\overline{A}|^2 = G_F^2 |\overline{M}|^2 \cdot 4E_e E_\nu \left[A + B\vec{\beta}_e \cdot \vec{\beta}_\nu + C\frac{m_e}{E_e}\right] \tag{12.11}$$

where $|\overline{M}|^2 = \dfrac{1}{2}\sum\limits_{\text{spin}} |M|^2$, $\vec{\beta}_i = \vec{p}_i / E_i$

From (12.6) we have for Fermi transitions:

$$M = \chi_p^+ \chi_n,\ A = |c_s|^2 + |c_v|^2,\ B = |c_v|^2 - |c_s|^2,\ C = 2Re(c_s c_v^*)$$

and for GT transitions:

$$\begin{aligned} &M = \chi_p^+ \vec{\sigma} \chi_n, \quad A = |c_A|^2 + |c_T|^2, \\ &B = \frac{1}{3}(|c_A|^2 - |c_T|^2), \quad C = 2Re(c_T c_A^*) \end{aligned} \tag{12.12}$$

(a) Electron energy spectrum: Substituting (12.11) in (12.10) and integrating over angles one gets

$$\frac{d\Gamma}{dE_e} = \frac{G_F^2}{2\pi^3} |\overline{M}|^2 |\vec{p}_e| E_e (E_0 - E_e)^2 \left[A + C\left(\frac{m_e}{E_e}\right)\right] \tag{12.13}$$

An analysis of the electron spectrum shapes in pure Fermi and pure GT transitions indicates $C \approx 0$. Thus either c_s or c_v and either c_T or c_A are zero, i.e. there are only two non-zero effective couplings.

(b) Electron-neutrino correlations: The correlations between the electron and neutrino momentum $\vec{p}_e$ and $\vec{p}_\nu$ are given by the second term in (12.11) (the $\frac{1}{3}$ factor in the case of GT transitions is due to the three possible orientations for the $J = 1$ total angular momentum of the leptons, thus reducing the correlations). The experiments measure the electron and the nuclear recoil directions from which the neutrino direction can be inferred. The results provide unambiguous evidence for pure V and A coupling ($c_{\rm S} = c_{\rm T} = 0$)

(c) $f\tau$ Values and coupling constants: The mean life time τ for the β-decay is obtained by integrating (12.13) over electron energy.

$$\frac{1}{\tau} = \Gamma = \frac{G_F^2}{2\pi^3}|\overline{M}|^2 A \int_{m_e}^{E_0} |\vec{p}_e| E_e (E_0 - E_e)^2 dE_e \tag{12.14}$$

The integral is conveniently denoted by $m_e^5 f$ so that f is dimension less. the '$f\tau$ value' is then given by

$$(f\tau)^{-1} = \frac{G_F^2 m_e^5}{2\pi^3}|\overline{M}|^2 A \tag{12.15}$$

$$\text{where } f = \int_1^{x_0} \sqrt{x^2 - 1} \cdot x(x_0 - x)^2 dx \cdot \left(x = \frac{E_e}{m_e},\ x_0 = \frac{E_0}{m_e}\right) \tag{12.16}$$

The value of the coupling constants may be obtained from measurements of $(f\tau)$. It is conventional to use the half life $t_{\frac{1}{2}}$ rather than mean life τ. Typical $ft_{\frac{1}{2}}$ values for pure Fermi transitions are of the order of 3000 which yields a value of $G_F C_V \approx 10^{-5}/M_p^2$ ($\hbar = c = 1$). By comparing the $ft_{\frac{1}{2}}$ value for $14_{\rm O}$. β-decay (pure Fermi transition) one gets

$$\frac{(ft_{1/2})_{14_{\rm O}}}{(ft_{\frac{1}{2}})_n} = \frac{3100 \pm 20}{1080 \pm 16} = \frac{c_V^2 + 3c_A^2}{2c_V^2} \tag{12.17}$$

(the factor 2 comes for the fact that contributions can come from either of the 2 protons outside the $12_{\rm C}$ core in $14_{\rm O}$).

$$\text{This gives } \left|\frac{c_A}{c_V}\right| = 1.25 \pm 0.02 \tag{12.18}$$

12.3 Parity Non-conservation in β-decay

There is unambiguous experimental evidence for parity non-conservation in β-decay. Lee and Yang (1956) had pointed out before the experiment on polarized $60_{\rm Co}$ decay that the interaction Lagrangian (12.4), which is a scalar, would require modification by inclusion of a pseudoscalar term to take into account parity nonconservation in the weak interaction, so that c_i in equation (12.4) is replaced by $\frac{1}{\sqrt{2}}(c_i + c_i'\gamma_5)$

The modified interaction Lagrangian thus becomes

$$\begin{aligned} L^w_{\text{int}} &= \frac{G_F}{\sqrt{2}} \sum_i [(\overline{\psi}_p \Gamma_i \psi_n)\overline{\psi}_e \Gamma^i (c_i + c_i'\gamma_5)\psi_\nu + h.c.] \\ &= \frac{G_F}{\sqrt{2}} \sum_i \left[\left(\frac{c_i + c_i'}{2}\right) (\overline{\psi}_p \Gamma_i \psi_n)\overline{\psi}_e \Gamma^i (1+\gamma_5)\psi_\nu \right. \\ &\qquad \left. + \left(\frac{c_i - c_i'}{2}\right) (\overline{\psi}_p \Gamma_i \psi_n)\overline{\psi}_e \Gamma^i (1-\gamma_5)\psi_\nu + h.c. \right] \end{aligned} \tag{12.19}$$

For maximum parity violation, $c_i' = \pm c_i$ (assuming c_i and c_i's to be real). Only one of the 2 terms in (12.19) survives, the second one if the neutrino is left-handed (antineutrino is right handed) and the first one if the neutrino is right-handed (antineutrino is left-handed). The actual helicity of the neutrino was unambiguously determined to be negative (left-handed) in an experiment by Goldhaber etal. Thus, finally, the β-decay interaction Lagrangian can be written in the form

$$\begin{aligned} L^w_{\text{int}} &= \frac{G_F}{\sqrt{2}} \sum_{i=\text{V, A}} [c_i(\overline{\psi}_p \Gamma_i \psi_n)\overline{\psi}_e \Gamma_i (1-\gamma_5)\psi_\nu + h.c.] \\ &= \frac{G_F}{\sqrt{2}} [\{\overline{\psi}_p \gamma_\mu (c_V - c_A\gamma_5)\psi_n\}\{\overline{\psi}_e \gamma^\mu (1-\gamma_5)\psi_\nu\} + h.c.] \end{aligned} \tag{12.20}$$

The relative phase of the vector and axial vector interaction was determined by studying the asymmetry in the angular distribution of the electrons emitted in the decay of polarized neutrons and was found to be negative. Since the V and A interaction terms occur in L^W_{int} with a relative negative sign, it is known as the V-A theory.

As consequences of parity nonconservation in weak interaction, one can have pseudoscalar (parity-nonconserving) terms in $|A|^2$ (square of the modulus of the matrix element), i.e. terms of the form $\langle \vec{J}_N \rangle \cdot \vec{p}_e$ and $\langle \vec{\sigma}_e \rangle \cdot \vec{p}_e$. The first one will lead to forward-backward asymmetry in the angular distribution of the decay electrons and the second one to longitudinal polarization of the electron.

12.4 Helicity of the Neutrino

The remarkable fact about the neutrino, which interacts only via weak interaction, is that only left-handed neutrinos (ν_L) and right-handed anti-neutrinos ($\overline{\nu}_R$) are observed in nature. This obviously violates C-invariance which requires ν and $\overline{\nu}$ to have identical weak interactions. It also violates parity-invariance, which requires the states ν_L and ν_R to have identical weak interactions (since the helicity operator $h = \vec{\sigma} \cdot \vec{p}/|\vec{p}|$ is a pseudoscalar, the parity operator converts ν_L to ν_R). However, the observed fact is consistent with CP-invariance, since

$$\nu_L \xrightarrow{\text{CP}} \overline{\nu}_R \tag{12.21}$$

The helicity of the neutrino was first unambiguously determined in an ingenious experiment by Goldhaber etal. The process studied was electron capture in 152_{Eu}.

$$e^- + {}^{152}Eu \rightarrow {}^{152}Sm^* + \nu_e \tag{12.22}$$

The excited state of ^{152}Sm which is formed decays to the ground state by γ-emission

$$^{152}Sm^* \rightarrow {}^{152}Sm + \gamma \tag{12.23}$$

The electron is captured from the K-shell and the initial state has zero momentum, so that the ν_e and $^{152}Sm^*$ nucleus recoil in opposite directions. The experiment selected events in which the γ was emitted in the direction of motion of the decaying $^{152}Sm^*$ nucleus. This was done by resonant scattering from a second Sm target. It relies on the fact that the γ's travelling in the direction opposite to the ν_e have slightly more energy than those emitted in the same direction as ν_e, and only the former have enough energy to excite the resonance level. Thus the overall observed reaction was

$$\begin{array}{ccc} e^- + {}^{152}Eu & \longrightarrow & {}^{152}Sm \;\; + \nu_e + \gamma \\ (J=0) & & (J=0) \end{array} \tag{12.24}$$

where all three final-state particles were collinear. The helicity of ν_e can then be deduced from the measured helicity of the γ by applying angular momentum conservation along the event axis to the overall reaction. In doing this, no orbital angular momentum is involved, because the initial electron is captured from the K-shell (L = 0) and all the final state particles move along the event axis. Hence the spin components of the ν_e $\left(\pm\frac{1}{2}\right)$

and the spin-components of the photon (± 1) must add up to give the spin components ($\pm\frac{1}{2}$) of the initial electron. This gives two possible spin configurations of the three particles as shown in Fig. 12.1 below and in each case the helicity of the γ and ν_e are the same (both left-handed or both right-handed).

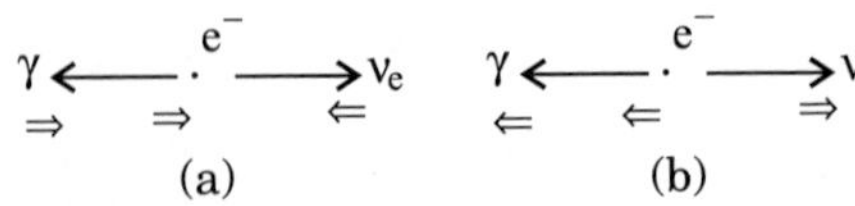

Fig. 12.1

The γ helicities were measured by passing the photons through iron which could be magnetized either parallel or anti parallel to the γ-direction. The scattering of the γ's in the iron depends on the relative orientation of the γ-helicity and the magnetisation direction of the iron, since electrons with spin opposite to the γ-angular momentum can absorb this angular momentum by spin-flip, whereas for an electron with spin parallel to the γ angular momentum there is no equivalent process. The absorbtion of the γ's in the iron are therefore greater when the iron magnetisation and γ-helicity are opposite and this phenomenon allows a unique determination of the γ-helicity. After allowing for known depolarising effects the data were found to be consistent with the occurrence of completely left-handed neutrinos (ν_L) corresponding to the possibility (a) of the spin configurations in Fig. 12.1.

A similar experiment for the $\overline{\nu}_e$ was carried out involving the emission of γ-ray following the positron decay of ^{203}Hg nucleus and the measured photon polarization (helicity) in this case was found to be consistent with the occurrence of right-handed antineutrinos ($\overline{\nu}_R$) only.

12.5 Muon Decay

In modern terminology the muon is known to decay in the following way

$$\begin{aligned} \mu^- &\to e^- + \overline{\nu}_e + \nu_\mu \\ \mu^+ &\to e^+ + \nu_e + \overline{\nu}_\mu \end{aligned} \tag{12.25}$$

since both the muon number and electron number must be separately conserved. Thus the muon decay is similar to the neutron decay in the sense that both decays are 3-body decay and involve two fermion pairs. However, while in the neutron decay (n, p) pair can be treated nonrelativistically, (μ, ν_μ) pair in μ-decay has to be treated relativistically. Energy-momentum conservation requires

$$P = p + k + k'$$

$$\text{with } P^2 = m_\mu^2,\ p^2 = m_e^2,\ k^2 = k'^2 = 0 \tag{12.26}$$

In the rest frame of the muon $P = (m_\mu, \vec{0})$. The maximum energy of the decay electron in the rest frame of muon is given by

$$(E_e)_{\max} = \frac{m_\mu^2 - m_e^2}{2m_\mu} \approx \frac{m_\mu}{2} \tag{12.26}$$

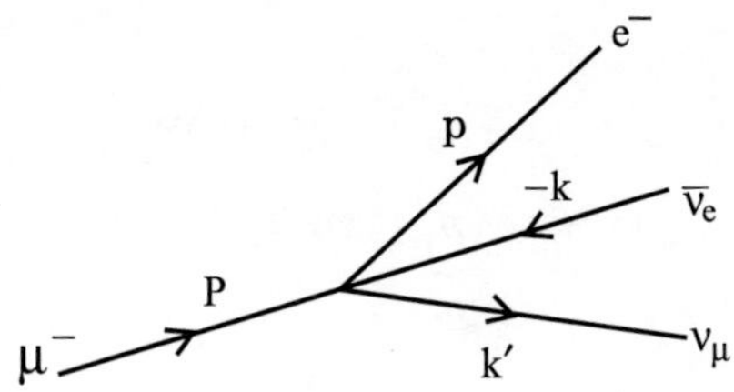

Fig. 12.2

The S-matrix element for transition $i \to f$ (initial state to final state) is written in the form

$$S_{fi} = \delta_{fi} - i(2\pi)^4 \delta_4(P_f - P_i)A \tag{12.27}$$

where A is the invariant amplitude for the transition. In the present problem $P_i = P_\mu = P$ and $P_f = P_e + P_{\overline{\nu}_e} + P_{\overline{\nu}_\mu} = p + k + k'$. Also we work in the rest frame of the muon in which case $P = (m_\mu, \vec{o})$. The invariant amplitude for μ^--decay in the V-A theory is given by

$$A = \frac{G_F}{\sqrt{2}}[\overline{u}_{(\nu_\mu)}(k')\gamma^\mu(1-\gamma_5)u_{(\mu)}(P)\overline{u}_e(p)\gamma_\mu(1-\gamma_5)v_{(\nu_\mu)}(k)] \tag{12.29}$$

$$\frac{1}{\tau} = \Gamma = \frac{1}{2m_\mu}\int |\overline{A}|^2 (2\pi)^4 \delta_4(P_f - P_i)d\phi_f \tag{12.30}$$

where $d\phi_f$ = final state Lorentz invariant phase space volume element

$$= \frac{d^3p}{(2\pi)^3(2E_p)}\,\frac{d^3k}{(2\pi)^3(2E_k)}\,\frac{d^3k'}{(2\pi)^3(2E_{k'})}$$

$$|\overline{A}|^2 = \frac{1}{2}\sum_{\text{spin}} |A|^2 \tag{12.31}$$

Substituting (12.29) in (12.31) we get

$$|\overline{A}|^2 = \frac{G_F^2}{2}\frac{1}{2}L^{\mu\nu}_{(e)}L^{(\mu)}_{\mu\nu} \tag{12.32}$$

where
$$\begin{aligned}
L^{\mu\nu}_{(e)} &= \sum_{\text{spin}}[\overline{u}_e(p)\gamma^\mu(1-\gamma_5)v_{\nu_e}(k)][\overline{u}_e(p)\gamma^\nu(1-\gamma_5)v_{\nu_e}(k)]^+ \\
&= \sum_{\text{spin}}[\overline{u}_e(p)\gamma^\mu(1-\gamma_5)v_{\nu_e}(k)v^+_{\nu_e}(k)(1-\gamma_5)^+\gamma^{\nu^+}\gamma^{o^+}u_e(p)] \\
&= \sum_{\text{spin}}[\overline{u}_e(p)\gamma^\mu(1-\gamma_5)v_{\nu_e}(k)\overline{v}_{\nu_e}(k)\gamma^o(1-\gamma_5)\gamma^{\nu^+}\gamma^o u_e(p)] \\
&= \sum_{\text{spin}}[\overline{u}_e(p)\gamma^\mu(1-\gamma_5)v_{\nu_e}(k)\overline{v}_{\nu_e}(k)(1+\gamma_5)\gamma^\nu u_e(p)] \\
&= \mathrm{Tr}[\gamma^\mu(1-\gamma_5)\not{k}(1+\gamma_5)\gamma^\nu(\not{p} + m_e)] \\
&= 2\mathrm{Tr}[(1+\gamma_5)\gamma^\mu\not{k}\gamma^\nu(\not{p} + m_e)] \\
&= 2\mathrm{Tr}[\gamma^\mu\not{k}\gamma^\nu\not{p}+\gamma^5\gamma^\mu\not{k}\gamma^\nu\not{p}] \\
&= 8[\{K^\mu p^\nu - g^{\mu\nu}(K\cdot p) + p^\mu K^\nu\} + i\in^{\mu\alpha\nu\beta}K_\alpha p_\beta] \\
&= 8[\{K^\mu p^\nu + p^\mu K^\nu - g^{\mu\nu}(K\cdot p) - i\in^{\mu\alpha\nu\beta}K_\alpha p_\beta]
\end{aligned}$$

and similarly

$$\begin{aligned}
L^{(\mu)}_{\mu\nu} &= \sum_{\text{spin}}\overline{u}_{\nu_\mu}(k')\gamma_\mu(1-\gamma_5)u_\mu(P)][\overline{u}_{\nu_\mu}(k')\gamma_\nu(1-\gamma_5)u_\mu(P)]^\dagger \\
&= 8[P_\mu k'_\nu + k'_\mu P_\nu - g_{\mu\nu}(P\cdot k') - i\in_{\mu\nu\rho\sigma}P^\rho k'^\sigma]
\end{aligned}$$

This gives

$$\begin{aligned}
L^{\mu\nu}_{(e)}L^\mu_{\mu\nu} &= 128[\{(k\cdot P)(k'\cdot p) + (k\cdot k')(p\cdot P)\} \\
&\qquad + \{(k\cdot P)(k'\cdot p) - (k\cdot k')(p\cdot P)\}] \\
&= 128[\{(k\cdot P)(k'\cdot p) + k\cdot k')(p\cdot P)\} \\
&\qquad + \{(k\cdot P)(k'\cdot p) - (k\cdot k')(p\cdot P)\}]256(k\cdot P)(k'\cdot p) \\
&= 256(k\cdot P)(k'\cdot p)
\end{aligned} \tag{12.33}$$

In deriving (12.33) we have used

$$\in^{\mu\nu\alpha\beta}\in_{\mu\nu\rho\sigma} = -2(g^\alpha_\rho g^\beta_\sigma - g^\alpha_\sigma g^\beta_\rho)$$

Substituting (12.33) in (12.32) we get

$$|\overline{A}|^2 = 64G_F^2(k\cdot P)(k'\cdot p) \tag{12.34}$$

To proceed further, we integrate over the phase space of the unobserved neutrinos in (12.30). Since $|\overline{A}|^2$ is linear in (k, k'), we are led evaluate the integral

$$I^{\alpha\beta} = \int \frac{d^3k}{2E_k}\frac{d^3k'}{2E_{k'}} k^\alpha k'^\beta \delta_4(Q - k - k') \tag{12.35}$$

with $Q = P - p$

The integral transforms under Lorentz transformation like a rank-2 tensor and is to be constructed out of $g^{\alpha\beta}$ and the 4-vector Q^α (only 4-vector available). Thus

$$I^{\alpha\beta} = (Ag^{\alpha\beta} + BQ^\alpha Q^\beta) \tag{12.36}$$

where A and B are Lorentz scalars (function of Q^2).The values of A and B can be determined by evaluating the integral in the CM frame of the neutrions in which Q^α is time-like $[Q^2 = (k + k')^2 \geq o]$. In this frame $Q = (\sqrt{Q^2}, \vec{o})$

$$I^{\alpha\beta}_{\text{CM}} = \int \frac{d^3k}{2E_k}\frac{d^3k'}{2E_{k'}} k^\alpha k'^\beta \delta(Q_o - E_k - E_{k'})\delta_3(\vec{k} + \vec{k'}), \quad Q_o = \sqrt{Q^2}$$

$$= \int \frac{d^3k}{4|\vec{k}|^2} k^\alpha k'^\beta \delta(Q_o - 2(|\vec{k}|)$$

$$= \frac{1}{4}\int d|\vec{k}|d\Omega_k \quad k^\alpha k'^\beta \delta(Q_o - 2(|\vec{k}|) \tag{12.37}$$

In (12.37) $k = |\vec{k}|(1, \hat{n}), \quad k' = |\vec{k}|(1, -\hat{n}), \quad \hat{n} = (\theta, \phi)$ (12.38)

From (12.37) and (12.38) we get

$$I^{\alpha\beta}_{\text{CM}} = \frac{1}{4}\cdot 4\pi \cdot \frac{1}{2}\left(\frac{Q_0}{2}\right)^2 \quad \alpha = \beta = 0$$

$$= \frac{1}{4}\left(\frac{-4\pi}{3}\right)\cdot\frac{1}{2}\left(\frac{Q_0}{2}\right)^2, \quad \alpha = \beta = j \quad (j = 1, 2, 3) \tag{12.39}$$

Thus we get

$$\frac{\pi Q^2}{8} = A + BQ^2, \quad \frac{-\pi Q^2}{24} = -A$$

or $$A = \frac{\pi Q^2}{24}, \quad B = \frac{1}{12} \tag{12.40}$$

Therefore (12.36) becomes

$$I^{\alpha\beta} = \frac{\pi}{24}(Q^2 g^{\alpha\beta} + 2Q^\alpha Q^\beta) \tag{12.41}$$

Substituting (12.34), (12.35) and (12.41) in (12.30) we get

$$\frac{1}{\tau} = \Gamma = \frac{1}{2m_\mu}\int 64G_F^2 P_\alpha p_\beta \frac{\pi}{24}(Q^2 g^{\alpha\beta} + 2Q^\alpha Q^\beta)\frac{d^3p}{2E_p}\frac{1}{(2\pi)^5}$$

$$= \frac{1}{2m_\mu}\cdot 64G_F^2 \cdot \frac{\pi}{24}\cdot\frac{1}{2(2\pi)^5}\int [(Q^2 P\cdot p) + 2(Q\cdot P)(Q\cdot p)]\cdot\frac{d^3p}{E_p} \tag{12.42}$$

where $Q = P - p$

From (12.42) we get, neglecting the electron mass

$$\frac{d\Gamma}{dE_e} = \frac{G_F^2 m_\mu^2}{12\pi^3}E_e^2\left(3 - \frac{4E_e}{m_\mu}\right) \tag{12.43}$$

where we have written $E_p = E_e =$ electron energy. From (12.43) we get

$$\frac{1}{\tau} = \Gamma = \frac{G_F^2 m_\mu^2}{12\pi^3}\int_0^{\frac{1}{2}m_\mu} E_e^2\left(3 - \frac{4E_e}{m_\mu}\right)dE_e$$

$$= \frac{G_F^2 m_\mu^5}{12\pi^3}\int_0^{\frac{1}{2}} x^2(3-4x)dx = \frac{G_F^2 m_\mu^5}{192\pi^3}$$

Thus $$\frac{1}{\tau} = \Gamma = \frac{G_F^2 m_\mu^5}{192\pi^3} \tag{12.44}$$

The value of G_F determined from the observed muon lifetime

$$\tau_\mu = (2.21 \pm 0.003)\times 10^{-6} \text{ sec.}$$

and (12.44) is

$$G_F = 1.015 \pm 0.003)\times 10^{-5}/M_p^2 \tag{12.45}$$

12.6 Pion Decay

The charged pion decays dominantly into muons

$$\pi^- \to \mu^- + \overline{\nu}_\mu, \quad \pi^+ \to \mu^+ + \overline{\nu}_\mu \tag{12.46}$$

We apply the V-A theory to study the pion decay,

$$L_{\text{int}} = \frac{G_F}{\sqrt{2}}(J_\mu^W J^{\mu,W+} + h\cdot c) \tag{12.47}$$

with $J_\mu^W = J_\mu^V - J_\mu^A$ we are strongly motivated to make this approach by two experimental facts.

(1) The muon in the pion decay is longitudinally polarized. (μ^- in π^- decay is right-handed and μ^+ in π^+ decay is left-handed)

(2) The electronic decay mode of the charged pion, although energetically more favored is highly suppressed

$$R = \frac{\Gamma(\pi^- \to e^- + \overline{\nu}_e)}{\Gamma(\pi^- \to \mu^- + \overline{\nu}_\mu)} = 1.3 \times 10^{-4} \tag{12.48}$$

Both features of the pion decay can be understood in terms of the V-A theory

(1) In the pion rest frame the muon and neutrino emerge with equal and opposite 3-momenta, since energy and momentum must be conserved. Therefore, both must be right-handed or both must be left-handed if their angular momenta along the direction of motion are to cancel as required by angular momentum conservation, since the pion spin is zero. In V-A theory the neutrino is left-handed and the antineutrino is right-handed and hence the μ^- in π^- decay must be right-handed and the μ^+ in π^+ decay must be left-handed.

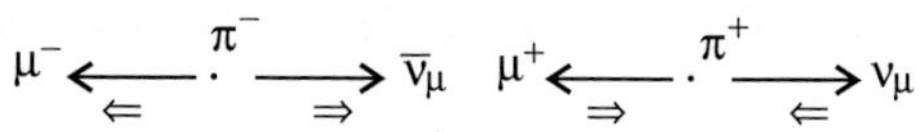

Fig. 12.3

(2) In β-decay, the electron has negative longitudinal polarization ($P_e^L = -\beta_e$). If $m_e = 0$, $\beta = 1$ and $P_e^L = 1$, i.e. the electron will be completely left-handed (like the neutrino). Thus the probability P_l of the lepton $l(l = e^-, \mu^-)$ being right-handed is proportional to $1 - \beta_l^2$

$$P_l \alpha 1 - \beta_l^2 = \frac{1}{\gamma_1^2} = \left(\frac{m_l}{E_l}\right)^2 \tag{12.49}$$

In the rest frame of the pion,

$$E_l = \frac{m_\pi}{2}\left(1 + \frac{m_l^2}{m_\pi^2}\right) \quad (p_\pi = p_\nu + p_l)$$

Therefore,

$$R = \frac{\Gamma(\pi^- \to e^- + \overline{\nu}_e)}{\Gamma(\pi^- \to \mu^- + \overline{\nu}_\phi)} \propto \left(\frac{m_e}{m_\mu}\right)^2 \times \text{factor of order 1}$$

Since $\frac{m_e}{m_\mu} = \frac{1}{207}$, $R \ll 1$

We consider the decay

$$\pi^- \to l^- + \overline{\nu}_l \quad (l = e, \mu)$$

$$S_{fi} = \delta_{fi} - i(2\pi)^4 \delta_4(P_f - P_i)A \tag{12.50}$$

$$P_i = p_\pi = P, \quad P_f = p_l + p_\nu = p + k$$

In the rest frame of the pion $[P = (m_\pi, \vec{o})]$

$$\frac{1}{\tau} = \Gamma = \frac{1}{2m_\pi} \int |\overline{A}|^2 \cdot (2\pi)^4 \delta_4(p + k - P) d\phi_f \tag{12.51}$$

where $d\phi_f = \dfrac{d^3p}{(2\pi)^3(2E_p)} \dfrac{d^3k}{(2\pi)^3(2E_k)}$

$$|\overline{A}|^2 = \sum_{\text{spin}} |A|^2$$

In V-A theory with the interaction given by (12.47)

$$A = \frac{G_F}{\sqrt{e}} l^\mu h_\mu \tag{12.52}$$

where $l^\mu = \langle l(p)\overline{\nu}_l(k)|J^{\mu,W^+}(o)|o\rangle = \overline{u}_l(p)\gamma^\mu(1 - \gamma_5)v_{\nu_l}(k)$ (12.53)

and $h_\mu = \langle o|J_\mu^W(o)|\pi^-(P)\rangle = -\langle o|J_\mu^A(o)|\pi^-(P)\rangle = -if_\pi P_\mu$ (12.54)

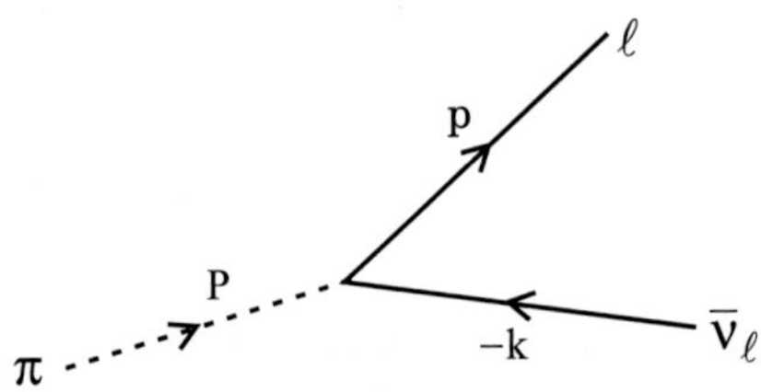

Fig. 12.4

Since the pion is pseudoscalar, only the axial vector part of the weak current contributes in the pion decay. The last equality follows from general considerations. f_π is a constant known as the pion decay constant. Therefore

$$A = \frac{G_F}{\sqrt{2}}(-if_\pi P_\mu)\overline{u}_l(p)\gamma^\mu(1 - \gamma_5)v_{\nu_l}(k)$$

$$= -i\frac{G_F}{\sqrt{2}} f_\pi m_l \overline{u}_l(p)(1 - \gamma_5)v_{\nu_l}(k) \tag{12.55}$$

In obtaining (12.55) Dirac equations for lepton spinors have been used

$$|\overline{A}|^2 = \sum_{\text{spins}} |A^2| = G_F^2 \cdot f_\pi^2 m_l^2 \cdot \text{Tr}[(1 - \gamma_5)\not{k}(\not{p} + m_l)]$$

$$= 4G_F^2 f_\pi^2 m_l^2 (k \cdot p) \tag{12.56}$$

Evaluating the phase space integral in the rest frame of the pion, we get

$$\Gamma = \frac{1}{\tau} = \frac{1}{2m_\pi}|\overline{A}|^2 \cdot \frac{1}{4\pi} \cdot \frac{|\vec{p}|}{m_\pi} \tag{12.57}$$

Substituting (12.56) for $|\overline{A}|^2$ in (12.57), we finally obtain

$$\frac{1}{\tau} = \Gamma = \frac{G_F^2}{8\pi} f_\pi^2 m_\pi m_l^2 \left(1 - \frac{m_l^2}{m_l^2}\right) \tag{12.58}$$

In (12.58) we see the explicit dependence of Γ on m_l

From eq. (12.58) we get

$$R = \frac{\Gamma(\pi^- \to e^-\overline{\nu}_e)}{\Gamma(\pi^- \to \mu^-\overline{\nu}_\mu)} = \left(\frac{m_e}{m_\mu}\right)^2 \left[\frac{1 - m_e^2/m_\pi^2}{1 - m_\mu^2/m_\pi^2}\right]^2 \approx 1.23 \times 10^{-4} \tag{12.59}$$

which agrees quite well with the experimental value given in (12.48). Moreover, from the observed value of the charged pion lifetime

$\tau_\pi = (2.55 \pm 0.03)10^{-8}$ sec., one finds $f_\pi = 0.93 m_\pi$

12.7 The Conserved Vector Current (CVC) Hypothesis

In V-A theory of weak interactions,

$$L_{\text{int}}^{\text{weak}} = \frac{G_F}{\sqrt{w}}[J_\mu^W J^{\mu \cdot W^+} + h \cdot c] \tag{12.60}$$

where $J_\mu^W = J_\mu^V - J_\mu^A$.

Furthermore, $J_\mu^W = (J_\mu^W)_{\text{hadron}} + J_\mu^W)_{\text{lepton}}$

In neutron decay,

$$n \to p + e^- + \overline{\nu}_e$$

the hadronic part of the matrix element is given by

$$\langle p(p')|(J_\mu^W)_h|n(p)\rangle = \overline{u}_p(p')\gamma_\mu(c_V - c_A\gamma_5)u_n(p) \tag{12.61}$$

$$c_V = c_V(q^2), \quad c_A = c_A(q^2), \quad q = p - p'$$

$$m_e^2 \le q^2 \le (m_n - m_p) \Rightarrow q^2 \approx 0$$

The weak vector current in the (p, n) case is given by

$$J_\mu^{W,V} = \overline{\psi}_p\gamma_\mu\psi_n = \overline{\psi}\gamma_\mu\tau_+\psi \tag{12.61}$$

where $\psi = \begin{pmatrix} \psi_p \\ \psi_n \end{pmatrix}$, $\tau_\pm = \frac{1}{2}(\tau_1 \pm i\tau_2)$

ψ transforms like a doublet under $\text{SU}(2)_I$ transformation

$$J_\mu^{em} = \overline{\psi}_p \gamma_\mu \psi_p = \overline{\psi} \cdot \frac{1}{2}(1+\tau_3)\psi = \frac{1}{2}[(J_\mu^{em})_{I=0} + (J_\mu^{em})_{I=1}] \quad (12.62)$$

where $(J_\mu^{em})_{I=0} = \psi \cdot 1 \cdot \psi = \overline{\psi}_p \psi_p + \overline{\psi}_n \psi_n$

and $(J_\mu^{em})_{I=1} = \psi \cdot \tau_3 \cdot \psi = \overline{\psi}_p \psi_p - \overline{\psi}_n \psi_n$

$J_\mu^{W,V}$, $(J_\mu^{W,V})^+$ and $(J_\mu^{em})_{I=1}$ form a SU(2) Lie algebra. The corresponding "charges" are the generators of the SU(2) symmetry transformation $(Q = \int J^0 d^3x)$. This property has the following consequences:

(1) $J_\mu^{W,V}$ is conserved since J_μ^{em} is conserved.

$$\partial_\mu J^{\mu,W,V} = 0 \quad (12.63)$$

(2) The matrix elements of J_μ^{em} and $J_\mu^{W,V}$ are related.

The result (1) is known as the CVC hypothesis.

Form factors: In order to illustrate the second consequence of CVC, we first introduce the notion of form factors. From general considerations, the matrix element of the electromagnetic current J_μ^{em} between nucleon states is given by

$$\begin{aligned} h_\mu &= \langle p'|J_\mu^{em}(0)|p\rangle_N \\ &= \overline{u}(p')[\gamma_\mu f_1(q^2) + p_\mu f_2(q^2) + p'_\mu f_3(q^2)]u(p) \quad [N = p_1 n] \\ q &= p' - p, \;\; p^2 = p'^2 = M^2 \end{aligned}$$

$\partial^\mu J_\mu^{em}(x) = 0$ gives $q^\mu h_\mu = 0$ which means $f_2 = f_3$. Thus we have

$$\langle p'|J_\mu^{em}(0)|p\rangle_N = \overline{u}(p')[\gamma_\mu f_1(q^2) + (p+p')_\mu f_2(q^2)]u(p) \quad (12.64)$$

Using Gordon's decomposition of electromagnetic currents, we can write (12.64) in the form

$$\langle p'|J_\mu^{em}(0)|p\rangle_N = \overline{u}(p')\left[\gamma_\mu F_1(q^2) + \frac{i\sigma_{\mu\nu}q^\nu}{2M}F_2(q^2)\right]u(p) \quad (12.65)$$

where $F_2 = 2Mf_2$, $F_1 = f_2 + 2Mf_2$

$F_1(q^2)$ and $F_2(q^2)$ are known as Dirac form factors. Their normalizations are $F_1^N(0) = e_N$, $F_2^N(0) = \kappa_N$ = anomalous part of the magnetic moment (in unit of $e/2M$)

Thus $F_1^p(0) = 1$, $F_1^n(0) = 0$, $F_2^p(0) = 1.79$, $F_2^n(0) = -1.91$ (12.66)

Because of (12.62) we can write

$$F_i^N(q^2) = \frac{1}{2}[F_1^S(q^2) + \tau_3 F_1^V(q^2)] \quad (i = 1, 2)$$

This gives

$$F_i^p(q^2) = \frac{1}{2}[F_1^S(q^2) + F_i^V(q^2)] \quad \text{and} \quad F_i^n(q^2) = \frac{1}{2}[F_1^S(q^2) + F_i^V(q^2)]$$

Alternatively,

$$F_i^S(q^2) = [F_i^p(q^2) + F_i^n(q^2)] \quad \text{and} \quad F_i^V(q^2) = [F_i^p(q^2) - F_i^n(q^2)]$$

This gives

$$F_i^S(0) = e_p = 1, \quad F_1^V(0) = e_p = 1$$

$$F_2^S(0) = \kappa_p + \kappa_n = -0.12, \quad F_2^V(0) = \kappa_p - \kappa_n = 3.70$$

The CVC hypothesis then gives

$$\langle p(p')|J_\mu^{W,V}(0)|n(p)\rangle = \overline{u}_p(p')\left[\gamma_\mu F_1^V(q^2) + i\frac{\sigma_{\mu\nu}q^\nu}{2M}F_2^V(q^2)\right]u_n(p) \tag{12.68}$$

For small $q^2 (q^2 \approx 0)$ eqn. (12.68) gives

$$\langle p(p')|J_\mu^{w,V}(0)|n(p)\rangle \approx \overline{u}_p(p')\gamma_\mu F_1^V(o)u_n(p) \tag{12.69}$$

Comparing with eqn. (12.60) we get

$$c_V \approx F_1^V(0) = 1 \tag{12.70}$$

As a further consequence of the CVC hypothesis, we consider the rare decay, $\pi^- \to \pi^0 + e^- + \overline{\nu}_e$. Since the pion is a pseudoscalar we have

$$\langle \pi^0(p')|J_\mu^W(0)|\pi^-(p)\rangle = \langle \pi^0(p')|J_\mu^{W,V}(0)|\pi^-(p)\rangle \tag{12.71}$$

Because of the CVC hypothesis, the rhs is related to the pion electromagnetic form factor

$$\begin{aligned}\langle \pi^0(p')|J_\mu^{W,V}(0)|\pi^-(p)\rangle &= \langle \pi^0|I_+|\pi^-\rangle\langle \pi^-(p')|J_\mu^{em}(p)|\pi^-(p)\rangle \\ &= \sqrt{2}\cdot(p+p')_\mu F_\pi(q^2)\end{aligned} \tag{12.72}$$

From (12.71) and (12.72) we have

$$\langle \pi^0(p')|J_\mu^W(0)|\pi^-\rangle = \sqrt{2}(p+p')_\mu F_\pi(q^2) \tag{12.73}$$

This then predicts

$$R = \frac{\Gamma(\pi^- \to \pi^0 e^- \overline{\nu}_e)}{\Gamma(\pi^- \to \mu^- e^- \overline{\nu}_e)} = 1.07 \times 10^{-8} \tag{12.74}$$

The predicted value agrees well with $R_{\text{expt}} = (1.023 \pm 0.069) \times 10^{-8}$

12.8 PCAC Hypothesis

Consider the charged pion decay in V-A theory. From eq. (12.54) one gets

$$\langle 0|\partial_\mu J^{\mu,A}(0)|\pi^-(p)\rangle = f_\pi m_\pi^2 \tag{12.75}$$

Thus $\partial_\mu J^{\mu,A}(x) = 0 \Rightarrow$ either f_π or m_π^2 or both are zero, none of which are true. Thus the weak axial vector current is not conserved. However, the smallness of the rhs of (12.75) indicates that deviation from conservation is not very large. This has led to invoking what is known as the partially conserved axial vector current hypothesis (PCAC):

$$\lim_{m_\pi \to 0} \partial_\mu J^{\mu,A}(x) = 0 \tag{12.76}$$

i.e. the axial vector current is conserved in the limit of pion mass being zero. One sometimes uses the stronger hypothesis,

$$\partial_\mu J^{\mu,A}(x) = C\phi_\pi(x) \tag{12.77}$$

where $\phi_\pi(x)$ is the pion field and $C = f_\pi m_\pi^2 =$ constant.

12.8.1 Callan-Treiman Relation

A consequence of the PCAC hypothesis is the Callan-Treiman relation which relates the axial vector coupling constant to the pion-nucleon strong coupling constant. The matrix element of the weak (V-A) current between the nucleon states can be written as

$$\langle p(p')|J_\mu^W(0)|n(p)\rangle \approx \overline{u}_p(p')\gamma_\mu(c_V - c_A\gamma_5)u_n(p) \tag{12.78}$$

where $J_\mu^W = J_\mu^{W,V} - J_\mu^{W,A}$, $c_V = c_V(q^2)$, $c_A = c_A(q^2)$, $q = p - p'(q^2 \approx 0)$

The CVC hypothesis implies

$$\langle p(p')|J_\mu^{W,V}(o)|n(p)\rangle = \overline{u}_p(p')\left[\gamma_\mu F_1^V(q^2) + i\frac{\sigma_{\mu\nu}q^\nu}{2M}F_2^V(q^2)\right]u_n(p) \tag{12.68}$$

with $F_1^V(o) = 1$, $F_2^V(o) = 3.70$

Comparison of (12.78) with (12.68) gives

$$c_V = c_V(o) = F_1^V(o) = 1$$

i.e. weak vector coupling constant is unrenormalized by strong interaction effects.

The most general form of weak axial vector current between the nucleon states can similarly be written as

$$\langle p(p')|J_\mu^{W,A}(o)|n(p)\rangle = \overline{u}_p(p')[\gamma_\mu\gamma_5 F_1^A(q^2) + q_\mu\gamma_5 F_2^A(q^2) + (p+p')_\mu\gamma_5 F_3^A(q^2)]u_n(p) \quad (12.79)$$

G-parity considerations imply $F_3^A(q^2) = 0$. Comparison of (12.78) and (12.79) gives

$$F_1^A(o) = c_A = 1.21 \quad (12.80)$$

The second term in (12.79) is known as the "induced pseudoscalar" term. $F_2^A(q^2)$ receives contribution from the 1-pion exchange diagram, Fig. 12.5(b)

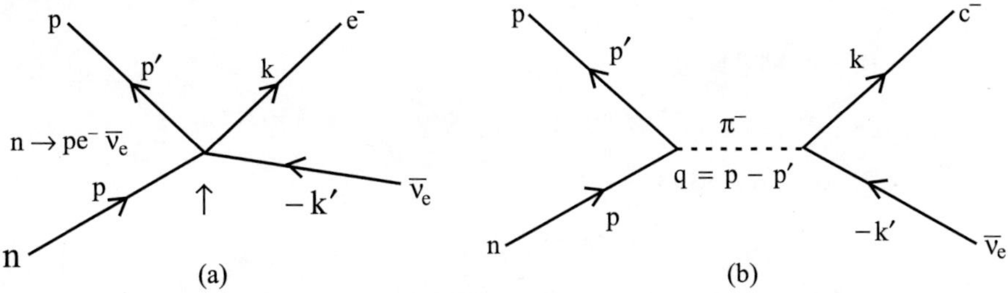

Fig. 12.5

Consider the decay $n \to pe^-\overline{\nu}_e$

Contribution of diagram (b) in Fig. 12.5 to the decay is given by

Amplitude $A = (n \to p\pi^-$ vertex$) \times ($pion propagator$) \times (\pi^- \to e^-\overline{\nu}_e$ vertex$)$

$$= \{g\sqrt{2}\overline{u}_p(p')\gamma_5 u_n(p)\} \cdot \left(\frac{i}{q^2 - m_\pi^2}\right) \cdot \frac{G_F}{\sqrt{2}}(-if_\pi q_\mu)l^\mu \quad (12.81)$$

where $l^\mu = \overline{u}_e(k)\gamma^\mu(1-\gamma_5)v_{\nu_e}(k')$

According to V-A theory,

$$A = \frac{G_F}{\sqrt{2}} < p(p')|J_\mu^{W,A}(o)|n(p) > l^\mu \quad (12.82)$$

Comparison of (12.82), (12.79) and (12.81) gives

$$F_2^A(q^2) = \frac{(g\sqrt{2})f_\pi}{q^2 - m_\pi^2} \quad (12.83)$$

$$\text{Now } \langle p(p')|\partial_\mu J^{\mu,W,A}(o)|n(p)\rangle = \frac{1}{i}\langle p(p')\left|\left[\hat{P}_\mu, J^{\mu,W,A}(o)\right]\right|n(p)$$

$$= \frac{1}{i}(p'-p)_\mu\langle p(p')|J^{\mu,W,A}(o)|n(p)\rangle$$

$$= iq^\mu\langle p(p')|J_\mu^{W,A}(o)|n(p)\rangle \quad (12.84)$$

Substituting (12.79) on the rhs. of (12.84) we get

$$\begin{aligned}\langle p(p')|\partial_\mu J^{\mu,W,A}(o)|n(p)\rangle &= i\overline{u}_p(p')[\not{q}\gamma_5 F_1^A(q^2) + q^2\gamma_5 F_1^A(q^2) \\ &\quad + q^2\gamma_5 F_2^A(q^2)]u_n(p) \\ &= i\overline{u}_p(p')[-2MF_1^A(q^2) + q^2 F_2^A(q^2)]\gamma_5 u_n(p) \quad (12.85)\end{aligned}$$

where we have used Dirac eqn. with $m_n \approx m_p = M$

Using the PCAC hypothesis, [Eq. (12.76)] we get

$$\lim_{m_\pi \to 0} -2MF_1^A(q^2) + q^2 F_2^A(q^2) = 0 \quad (12.86)$$

Substituting for $F_2^A(q^2)$ from (12.83) into (12.86)

$$2MF_1^A(q^2) = f_\pi g\sqrt{2} \quad (12.87)$$

Since $q^2 \approx 0$, (12.87) becomes

$$2MF_1^A(o) \approx f_\pi g\sqrt{2} \quad (12.88)$$

From (12.80) and (12.88) we get

$$2Mc_A \approx f_\pi g\sqrt{2} \quad (12.89)$$

Eqn. (12.89) is known as the Callan-Treiman relation $c_A = 1.21$, $\frac{g^2}{4\pi} = 14$ gives

$$f_\pi = 0.87m_\pi \quad (12.90)$$

Experimental data on pion lifetime gives $f_\pi = 0.93m_\pi$ The two values agree reasonably well justifying the use of the PCAC hypothesis.

12.9 Weak Decay of Strange Particles: The Cabibbo Theory

Weak decay of strange particles (kaons and hyperons) are of three types:

(1) purely leptonic: $K^- \to \mu^-\overline{\nu}_\mu$, $K^+ \to \mu^+\nu_\mu$ (12.91)

(2) semi-leptonic: $\sum^- \to ne^-\overline{\nu}_e$, $K^- \to \pi^0 l^- \overline{\nu}_l$, $K_L^0 \to \pi^- l^+ \nu_l$ (12.92)

(3) non-leptonic: $\Lambda \to N\pi$, $\sum^\pm \to N\pi$, $\Xi \to \Lambda\pi$

$\Omega^- \to \Xi\,\pi$, $K^\pm \to \pi^\pm\pi^0$, $\pi^\pm\pi^+\pi^-$, $\pi^\pm\pi^0\pi^0$, $\Omega^- \to \Lambda K^-$

$K_s^0 \to \pi\pi$, $K_L^0 \to 3\pi$ (12.93)

Semileptonic decays of strange particles are characterized by the selection rule $\Delta Q = \Delta S$, where ΔQ and ΔS are the changes in electric charge and strangenes of the hadrons. $\Delta Q = \Delta S = 1$ is expected if $\Delta I_3 = \frac{1}{2}$, from the relation $Q = I_3 + \frac{1}{2}(B + S)$. As examples of the application of the rule, the decay $\sum^- \to ne^-\overline{\nu}_e$ is observed, with branching ratio 1.08×10^{-3} but not the decay $\sum^+ \to ne^+\nu_e$ for which the branching ratio is $< 5 \times 10^{-6}$.

Non-leptonic decay of strange particles is characterized by the selection rule $|\Delta S| = 1$, $\Delta I = \frac{1}{2}$.

The rule predicts

$$\frac{\Gamma(\Lambda \to n\pi^0)}{\Gamma(\Lambda \to n\pi^0) + \Gamma(\Lambda \to p\pi^-)} = \frac{(\sqrt{1/3})^2}{(\sqrt{1/3})^2 + (\sqrt{2/3})^2} \times 1.036$$

$$= \frac{1}{3} \times 1.036 = 0.345 \tag{12.94}$$

where the correction factor 1.036 accounts for the slightly different phase space for $n\pi^0$ and $p\pi^-$ states. The observed value is 0.358 ± 0.005 in agreement with this prediction, when account is taken of small electromagnetic correction. As another example, $\Delta I = \frac{1}{2}$ rule predicts.

$$\frac{\Gamma(K_s^0 \to \pi^0\pi^0)}{\Gamma(K_s^0 \to \pi^0\pi^0) + \Gamma(K_s^0 \to \pi^+\pi^-)} = \frac{1}{3} \tag{12.95}$$

while the experimental value is 0.313.

Deviation from the $\Delta I = \frac{1}{2}$ rule is observed in cases where the final state can have both $I = \frac{1}{2}$ and $I = \frac{3}{2}$. For example, the observed branching ratio

$$\frac{\Gamma(\Omega^- \to \Xi^0 \pi^-)}{\Gamma(\Omega^- \to \Xi^- \pi^0)} = 2.94 \pm 0.03 \tag{12.96}$$

while the $\Delta I = \frac{1}{2}$ rule predicts the ratio to be 2. Thus the $I = \frac{3}{2}$ amplitude is also present, although highly suppressed.

Of more interest are the absolute decay rates of the $|\Delta S| = 1$ transition, compared with $|\Delta S| = 1$ transition. It has been found that the $|\Delta S| = 1$ transitions are suppressed by a factor ~ 20 relative to the $\Delta S = 0$ transition. The reduced decay rate for the $|\Delta S| = 1$ transition is accounted for by the Cabibbo theory. In this theory,

$$H_{\text{int}}^W = \frac{G_F}{\sqrt{2}}[J_\mu^W J^{W,\mu^+} + h \cdot c.]$$

with $J_\mu^W = (J_\mu^W)_l + (J_\mu^W)_h$

$$(J_\mu^W)_l = \sum_{l=e,\mu,\tau} \bar{l}\gamma(1 - \gamma_5)\nu_l$$

$$(J^W_\mu)_h = (J^W_\mu)_{h,\Delta s=0} \cos\theta_c + (J^W_\mu)_{h,\Delta s=1} \sin\theta_c \tag{12.97}$$

where θ_c = Cabibbo angle.

Comparison of weak coupling constants involved in muon decay ($\mu^- \to e^-\overline{\nu}_e\nu_\mu$), neutron decay ($n \to pe^-\overline{\nu}_e$) and Λ semileptonic decay ($\Lambda^- \to pe^-\overline{\nu}_e$) in Cabibbo theory gives

$$G_\mu = G_F,\ G_\beta = G_F \cos\theta_c,\ G_\Lambda = G_F \sin\theta_c \tag{12.98}$$

Comparison of ^{14}O decay rate with that for muon decay gives $\cos^2\theta_c$ and hence θ_c (vector) $= 0.210 \pm 0.025$ (both pure vector transitions) while comparison of decay $\pi^- \to \pi^0 e^-\overline{\nu}_e$ and $K^- \to \pi^0 e^-\overline{\nu}_e$ gives $\tan^2\theta_c$ and hence θ_c (vector) $= 0.25 \pm 0.05$.

An estimate of θ_c can also be obtained by comparing the pure axial vector transitions $K^+ \to \mu^+\nu_\mu$ and $\pi^+ \to \mu^+\nu_\mu$

$$R_{\text{th}} = \frac{\Gamma(K^+ \to \mu^+\nu_\mu)}{\Gamma(\pi^+ \to \mu^+\nu_\mu)}$$

$$= \tan^2\theta_c \cdot \left(\frac{f_K}{f_\pi}\right)^2 \left(\frac{m_K}{m_\pi}\right) \cdot \left(\frac{1 - m_\mu^2/m_K^2}{1 - m_\mu^2/m_\pi^2}\right)^2 \tag{12.99a}$$

and $$R_{\text{expt}} = \frac{(0.6377)(1/\tau_k)}{(1/\tau_\pi)} \approx 1.34 \tag{12.99b}$$

SU(3) symmetry gives $f_K \approx f_\pi$. One than gets by equating R_{th} with R_{expt}

$$\theta_c \text{ (axial)} = 0.269 \pm 0.001$$

One does not get precisely same values in the three cases but the reasons for small discrepancies can be understood. This would take us onto subjects like non conservation of axial vector currents and quark mass effects. When these are taken into consideration, the relative rates of $|\Delta S| = 1$ and $|\Delta S| = 0$ transitions are consistent with a unique value of the Cabibbo angle $\theta_c \approx 0.25$ radian. Cabibbo theory also explains the small discrepancy between the β-decay coupling constant and the muon decay coupling constant ($G_\beta/G_\mu = \cos\theta_c$).

In the quark model description

$$\underset{(udd)}{n} \to \underset{(uud)}{p}e^-\overline{\nu}_e \quad \text{is } d \to ue^-\overline{\nu}_e$$

$$\text{and} \quad \underset{(uds)}{\Lambda} \to \underset{(uud)}{p} e^- \overline{\nu}_e \quad \text{is } s \to u e^- \overline{\nu}_e \tag{12.100}$$

Therefore, in the quark model description, the hadronic weak current can be written as

$$\begin{aligned}(J_\mu^W)_h &= (\overline{u}\Gamma_\mu d)\cos\theta_c + (\overline{u}\Gamma_\mu s)\sin\theta_c, \quad \Gamma_\mu = \gamma_\mu(1-\gamma_5) \\ &= \overline{u}\Gamma_\mu d'\end{aligned}$$

$$\text{where } d' = d\cos\theta_c + s\sin\theta_c \tag{12.101}$$

One can say that in the Cabibbo theory, d and s quark states participating in weak transitions are "rotated" by an angle θ_c, known as the Cabibbo angle.

12.10 Absence of $|\Delta S| = 1$ Weak Neutral Currents: GIM Mechanism

With the development of high energy neutrino beam facilities (mostly ν_μ and $\overline{\nu}_\mu$) one observed "neutral current" (muonless) events that are ascribed to processes of the form $\nu_\mu e^- \to \nu_\mu e^-$ and $\nu_\mu N \to \nu_\mu N$ (also with $\nu_\mu \longleftrightarrow \overline{\nu}_\mu$). All neutral current processes observed so far are characterized by the selection rule $\Delta S = 0$. No $|\Delta S| = 1$ neutral current (flavour changing neutral current, FCNC) processes ($K^0 \to \mu^+\mu^-$, $\Lambda \to ne^+e^-$, $K^+ \to \pi^+\nu\overline{\nu}$ etc) have so far been seen. For example, the ratio of neutral current to charged current ratio in kaon decay is

$$\frac{\Gamma(K^+ \to \pi^+\nu\overline{\nu})}{\Gamma(K^+ \to \pi^0\mu^+\nu_\mu)} \langle 10^{-5} \tag{12.102}$$

In V-A theory the process $K^0 \to \mu^+\mu^-$ can go through the "loop" diagram (Fig. 12.6) with the amplitude $G_F^2 \sin\theta_c\cos\theta_c$ which although small is still observable. The problem is more serious in models (GSW model) which predict the existence of weak neutral currents. If

$$(J^W)_h^{ch} = \overline{u}\Gamma d', \quad (d' = d\cos\theta_c + s\sin\theta_c)$$

$$\begin{aligned}\text{then } (J^W)_h^{\text{neutral}} &= \overline{u}\Gamma u + \overline{d'}\Gamma d' \\ &= \overline{u}\Gamma u + (\overline{d}\Gamma d\cos^2\theta_c + \overline{s}\Gamma s\sin^2\theta_c) \\ &\qquad + (\overline{d}\Gamma s + \overline{s}\Gamma d)\sin\theta_c\cos\theta_c\end{aligned} \tag{12.103}$$

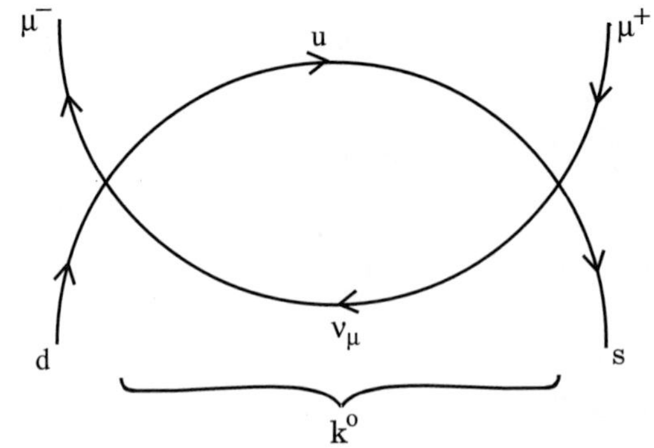

$k^0 \to \mu^- \mu^+$

Fig. 12.6

The first 2 terms on the rhs correspond to $\Delta S = 0$ neutral current, while the last term corresponds to $\Delta S = 1$ neutral current ($d \leftrightarrow s$). Thus $|\Delta S| = 1$ neutral current process should be possible since $\sin\theta_c \neq 0$. Glashow, Iliopoulos and Maiani (GIM) proposed in 1976 the existence of a fourth quark, called the 'charm' quark carrying a new flavor called 'charm' C ($C = 1$ for c-quark and $C = 0$ for all other quarks) and electric charge $+2/3$ (same as the u quark) and a modification of weak charged current. In GIM model

$$(J^W)_h^{ch} = \overline{u}\Gamma d' + \overline{c}\Gamma s'$$

with $d' = d\cos\theta_c + s\sin\theta_c$ (same as in cabibbo theory)

and $s' = -d\sin\theta_c + s\cos\theta_c$ (orthogonal to d') (12.104)

As a result the neutral current in the GIM model becomes

$$\begin{aligned}(J^W)_h^{\text{neutral}} &= \overline{u}\Gamma u + \overline{c}\Gamma c + \overline{d}'\Gamma d' + \overline{s}'\Gamma s' \\ &= \overline{u}\Gamma u + \overline{c}\Gamma c + \overline{d}\Gamma d + \overline{s}\Gamma s \end{aligned} \tag{12.105}$$

The unwanted FCNC currents ($|\Delta s| = 1$ current) have got cancelled because of the property of the orthogonal transformation,

$$\begin{pmatrix} d' \\ s' \end{pmatrix} = \begin{pmatrix} \cos\theta_c & \sin\theta_c \\ -\sin\theta_c & \cos\theta_c \end{pmatrix} \begin{pmatrix} d \\ s \end{pmatrix} \tag{12.106}$$

This is known as the GIM mechanism. It may be noted that the existence of the c-quark gives a second contribution to $K^0 \to \mu^+\mu^-$ in the V-A theory, with the u quark in the loop replaced by the c quark. The amplitude in this case is $\propto -G_F^2 \sin\theta_c \cos\theta_c$. Thus the contribution of the 2 diagrams cancel out if m_c is not too different from m_u. The c-quark was subsequently discovered in 1974 ($J/\psi = c\bar{c}$) with $m_c \sim 1.5$ GeV and it was a great triumph for the theory.

12.11 The $K^0 - \overline{K}^0$ Problem

As first suggested by Gell-Mann and Pais and later experimentally confirmed, there exist two neutral K-meons, K^0 and $\overline{K}^0$ in order to get the correct multiplet structure of hadrons. They belong to different i-spin multiplets.

$$K = \begin{pmatrix} K^+ \\ K^0 \end{pmatrix} \quad \text{and} \quad \overline{K} = \begin{pmatrix} \overline{K}^0 \\ K^- \end{pmatrix}$$

$$I = \frac{1}{2}, \; Y = +1(S = +1), \quad I = \frac{1}{2}, \; Y = -1(S = -1) \tag{12.107}$$

They are produced via strong interaction in hadron-hadron collisions:

$$\begin{aligned} \pi^- + p &\to \Lambda + K^0 \\ &\to \Sigma^- + K^+ \\ &\to n + K^0 + \overline{K}^0 \end{aligned} \qquad \begin{aligned} \pi^+ + p &\to \Sigma^+ + K^+ \\ &\to p + K^+ + \overline{K}^0 \end{aligned} \tag{12.108}$$

The threshold energy for $\overline{K}^0$ production is higher than that for K^0 production because of the necessity to conserve strangeness. K^0 and $\overline{K}^0$ are antiparticles of each other just like K^+ and K^-. The CPT theorem states that particles and anti particles must have the same mass and lifetime. However, the neutral Kaon decay modes are associated with two distinct lifetimes. They are denoted by K_S and K_L with lifetimes $\tau_S = 0.9 \times 10^{-10}$ sec. and $\tau_L = 0.5 \times 10^{-7}$ sec. respectively ($K_S \to 2\pi$, $K_L \to 3\pi$, $\pi\mu\nu$). Thus weak interaction (mass) eigen states (K_S, K_L) are not the same as strong interaction (flavor) eigenstates ($K^0, \overline{K}^0$). This is because of the possibility of $K^0 - \overline{K}^0$ mixing since strangeness is not conserved in weak interactions. Although neither charge conjugation invariance (C) nor parity (P) invariance is respected by weak interaction, CP is known to be conserved in most weak decays. Thus K_S and K_L are expected to be CP eigenstates.

Let us consider 2π and 3π decay modes,

(a) **2π system**—For $\pi^+\pi^-$, $C = P = (-)^l$ so that $CP = (-)^{2l} = +1$. For $\pi^0\pi^0$, $C = P = +$ so that $CP = +1$

(b) **3π system**—The small Q value (70 MeV) of the 3π decay mode suggests $l = 0$, i.e. the 3 pions are in a relative s-state. By the previous argument $CP(\pi^+\pi^-) = +$

The π^0 has $C = +$ and intrinsic parity $\eta_{\pi^0} = -$ and therefore $CP = -1$ so $CP(\pi^+\pi^-\pi^0) = -$. For $l > 0$, both +ve and −ve CP eigenvalues can result, but such decays are strongly suppressed by angular momentum barrier effects. For $3\pi^0$, $CP = (+)^3(-1)^3 = -1$. Thus K_S and K_L are expected to have $CP = +1$ and $CP = -1$ respectively. The decay $K_S \to 2\pi$ is expected to be faster than the decay $K_L \to 3\pi$ because of the availability of larger phase space in the former case. Since K^0 and $\overline{K}^0$ are antiparticles of each other.

$$CP|K^0\rangle = \eta|\overline{K}^0\rangle \quad \text{and} \quad CP|\overline{K}^0\rangle = \eta'|\overline{K}^0\rangle \tag{12.109}$$

where η, η' are arbitrary phase factors ($|\eta| = |\eta'| = 1$). We choose $\eta = \eta' = +1$. Thus $|K^0\rangle$ and $|\overline{K}^0\rangle$ are not CP eigenstates, but we can form the linear combinations

$$|K_1\rangle = \frac{1}{\sqrt{2}}(|K^0\rangle + |\overline{K}^0\rangle) \quad \text{and} \quad |K_2\rangle = \frac{1}{\sqrt{2}}(|K^0\rangle - |\overline{K}^0\rangle) \tag{12.110}$$

so that $CP|K_1\rangle = +|K_1\rangle$, $CP|K_2\rangle = -|K_2\rangle$ (12.111)

Therefore assuming CP invariance we can identify K_S and K_L with K_1 and K_2, respectively,

$|K_S\rangle = |K_1\rangle,\ |K_L\rangle = |K_2\rangle$

Strangeness Oscillations

One of the consequences of mixing due to weak interaction effects is strangeness oscillation. The general solution of the Schrodinger equation

$$i\hbar\frac{\partial}{\partial t}|\psi(t)\rangle = \hat{H}|\psi(t)\rangle \tag{12.112}$$

can be written as

$$|\psi(t)\rangle = \sum_n c_n e^{-iE_n t/\hbar}|\psi_n\rangle = \sum_n c_n|\psi_n(t)$$

where $|\psi_n\rangle's$ are eigenstates of the Hamiltomian operator $\hat{H}$

$$\hat{H}|\psi_n\rangle = E_n|\psi_n\rangle \tag{12.113}$$

If $\hat{H}$ is hermitean, then E_n's are real and $\psi_n(t)\rangle$'s are known as stationary solutions of the Schrodinger eqn. $\langle\psi_n(t)|\psi_n(t)\rangle = \langle\psi_n(0)|\psi_n(0)\rangle$. For decaying states one can write the solution of (12.113) in the same form but with $E_n \to E_n - i\frac{\Gamma}{2}$, $\frac{\Gamma}{\hbar} = \frac{1}{\tau}$, τ ~lifetime of the state. Thus any general state (time-dependent) of the $K^0 - \overline{K}^0$ system can be written as

$$|\psi(t)\rangle = \sum_{n=1}^{2} c_n |K_n\rangle e^{-i\left(E_n - \frac{i\Gamma_n}{2}\right)t} \quad (\hbar = c = 1) \tag{12.114}$$

In the rest system of the particles. $E_n = m_n$ and $\tau_n = \frac{1}{\Gamma_n}$ is the proper lifetime. Now suppose one starts with a pure K^0 beam. That is

$$|\psi(0)\rangle = |K^0\rangle = c_1|K_1\rangle + c_2|K_2\rangle \tag{12.115}$$

Since CP invariance has been assumed, [CP, H_W] = 0, eigenstates of CP are also eigenstates of the total Hamiltonian $\hat{H} = \hat{H}_s + \hat{H}_{em} + \hat{H}_W$. From (12.110)

$$|K^0\rangle = \frac{1}{\sqrt{2}}(|K_1\rangle + |K_2\rangle), \ |\overline{K}^0\rangle = \frac{1}{\sqrt{2}}(|K_1\rangle - |K_2\rangle) \tag{12.116}$$

From (12.115) and (12.116) $c_1 = c_2 = \frac{1}{\sqrt{2}}$ Thus with $|\psi(0)\rangle$ given by (12.115) we can write the state at later time as

$$\begin{aligned}
|\psi(t)\rangle = |K^0(t)\rangle &= \frac{1}{\sqrt{2}}\left\{|K_1\rangle_e^{-i\left(m_1 - \frac{i}{2\tau_1}\right)t} + |K_2\rangle_e^{-i\left(m_2 - \frac{l}{2\tau_2}\right)t}\right\} \\
&= \frac{1}{\sqrt{2}}\cdot\frac{1}{\sqrt{2}}\left\{(|K^0\rangle + |\overline{K}^0\rangle)_e^{-i\left(m_1 - \frac{i}{2\tau_1}\right)t} + (|K^0\rangle - |\overline{K}^0\rangle)_e^{-i\left(m_2 - \frac{i}{2\tau_2}\right)t}\right\} \\
&= \frac{1}{2}\left[|K^0\rangle\left\{e^{-i\left(m_1 - \frac{i}{2\tau_1}\right)t} + e^{-i\left(m_2 - \frac{i}{2\tau_2}\right)t}\right\}\right. \\
&\qquad \left. + |\overline{K}^0\rangle\left\{e^{-i\left(m_1 - \frac{i}{2\tau_1}\right)t} - e^{-i\left(m_2 - \frac{i}{2\tau_2}\right)t}\right\}\right]
\end{aligned} \tag{12.117}$$

After a time for free decay in vacuo the K^0 intensity will be

$$\begin{aligned}
I(K^0) &= \left|\frac{1}{2}\left\{e^{-i\left(m_1 - \frac{i}{2\tau_1}\right)t} + e^{-i\left(m_2 - \frac{i}{2\tau_2}\right)t}\right\}\right|^2 \\
&= \frac{1}{4}\left[e^{-t/\tau_1} + e^{-t/\tau_2} + 2e^{-\frac{t}{2}\left(\frac{1}{\tau_1} + \frac{1}{\tau_2}\right)}\cos(\Delta m)t\right]
\end{aligned} \tag{12.118}$$

where $\Delta m = |m_1 - m_2|$

Similarly the $\overline{K}^0$ intensity will be

$$I(\overline{K}^0) = \left|\frac{1}{2}\left\{e^{-i\left(m_1 - \frac{i}{2\tau_1}\right)t} - e^{-i\left(m_2 - \frac{i}{2\tau_2}\right)t}\right\}\right|^2$$

$$= \frac{1}{4}\left[e^{-t/\tau_1} + e^{-t/\tau_2} - 2e^{-\frac{t}{2}\left(\frac{1}{\tau_1}+\frac{1}{\tau_2}\right)t}\cos(\Delta m)t\right] \quad (12.119)$$

Thus the K^0 and $\overline{K}^0$ intensities oscillate with the frequency Δm.

If one measures the number of $\overline{K}^0$ interaction events (i.e. hyperon yield) as a function of distance from the K^0 source, one can therefrom deduce Δm. The present value is given by $\Delta m \tau_1 = 0.477 \pm 0.002$. Going over to ordinary units, we get $\Delta m = 3.52 \times 10^{-6}$ ev/c^2 and a fractional mass difference

$$\frac{\Delta m}{m} = 0.7 \times 10^{-14} \quad (12.120)$$

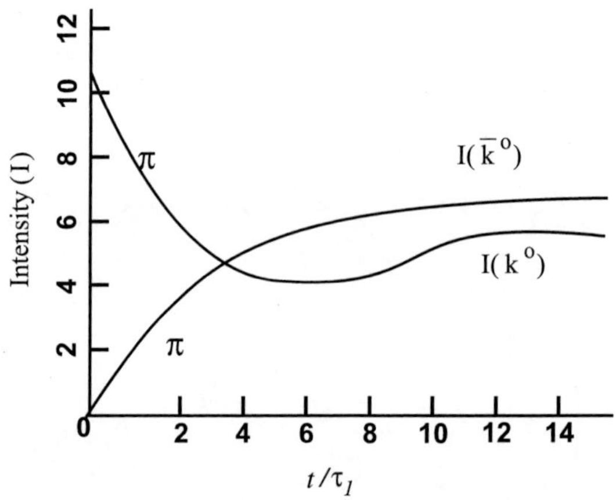

Fig. 12.7

The magnitude of the self energy of K_1 and K_2 can be estimated crudely by inspection of the figure (12.8) below. From these diagrams, it follows that

$$\Delta m \text{ (for } \Delta S = 1 \text{ coupling)} \sim G_F^2 m_K^5 \sin^2\theta_c \sim 10^{-4} \text{ eV} \quad (12.121)$$

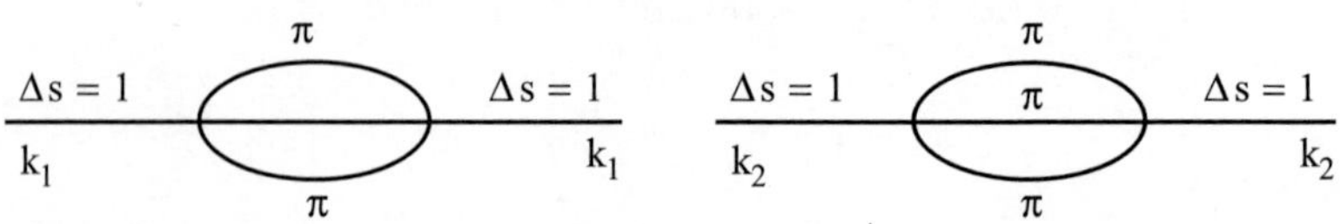

Fig. 12.8

within an order of magnitude of the observed value (12.120). By a similar argument one can set stringent limits on the strength of a possible direct

$\Delta S = 2$ transition between K^0 and $\overline{K}^0$ states. If fG_F denotes the strength of $\Delta S = 2$ coupling then

$$\Delta m \text{ (for } \Delta S = 2 \text{ coupling)} \sim fG_F m_K^3 \sim 10^3 f \text{ eV} \tag{12.122}$$

giving $f < 10^{-3}$

The Regeneration Phenomena

Suppose we start with a pure K^0 beam and let it coast in vacuo for $t \sim 10^2\tau_1$, so that all the K_1 component has decayed and we are left with the K_2 component only. Now let the K_2 beam traverse a slab of material of thickness L and interact with the nuclei of the material. The strong interactions will pick out the strangeness $S = \pm 1$ component of the beam

$$|K^0(t)\rangle = \frac{1}{\sqrt{2}}\{|K_1(t)\rangle + |K_1(t)\rangle\} \to \frac{1}{\sqrt{2}}|K_2(t)\rangle, \quad t > 10^2\tau_1$$

$$= \frac{1}{\sqrt{2}}\frac{1}{\sqrt{2}}\{|K^0(t)\rangle - |\overline{K}^0(t)\rangle\} \tag{12.123}$$

Thus of the original K^0 beam intensity 50% has disappeared by K_1-decay mode. The remainder K_2 component upon traversing a slab where its nuclear interactions can be observed should consist of 50% K^0 and 50% $\overline{K}^0$. The existence of $\overline{K}^0$ $(S = -1)$ a long way from the source of an originally K^0 $(S = +1)$ beam was confirmed by the observation of production of hyperons. $(\overline{K}^0 N \to \Lambda\pi, \sum\pi, \equiv K$ etc)

The K^0 and $\overline{K}^0$ component in K_2 [eqn. (12.123)] must be absorbed differently because strangeness is conserved in strong interactions. K^0 can only undergo elastic and charge exchange scattering ($K^0 p \to K^0 p$, $K^+ n$; $K^0 n \to K_0 n$) while $\overline{K}^0$ can also undergo strangeness exchange processes producing hyperons in addition to elastic and charge exchange scattering.

($\overline{K}^0 p \to \overline{K}^0 p$, $\Lambda\pi^+$, $\sum\pi$, $\equiv^0 K^+$; $\overline{K}^0 n \to \overline{K}^0 n$, $K^- p$, $\Lambda\pi^0$, $\sum^\pm \pi^\mp$, $\equiv^- K^+$, $\equiv^0 K^0$). With more final states accessible, the $\overline{K}^0$ is therefore absorbed more strongly than K^0. After emerging from the slab the K^0 and $\overline{K}^0$ amplitudes will respectively be $f|K^0\rangle$ and $\overline{f}|\overline{K}^0\rangle$ with $\overline{f} < f < 1$. Thus after traversing a length L

$$\frac{1}{\sqrt{2}}|K_2(t)\rangle - \frac{1}{\sqrt{2}}\frac{1}{\sqrt{2}}\{f|K^0\rangle - \overline{f}|\overline{K}^0\rangle\}$$

$$= \frac{1}{\sqrt{2}} \cdot \frac{1}{\sqrt{2}} \frac{1}{\sqrt{2}} \{f(|K_1\rangle + |K_2\rangle) - \overline{f}(|K_1\rangle - |K_2\rangle)\}$$

$$= \frac{1}{2\sqrt{2}} \{(f - \overline{f})|K_1(t)\rangle + (f + \overline{f})|K_2(t)\rangle\} \quad (12.124)$$

Since $f \neq \overline{f}$, it follows that a part of the K_1 state has been regenerated. The regeneration of short-lived K_1 in a long-lived K_2 beam was confirmed by experiment. In fact, the regeneration phenomenon was predicted even before the K_2 decay made of neutral kaons was observed.

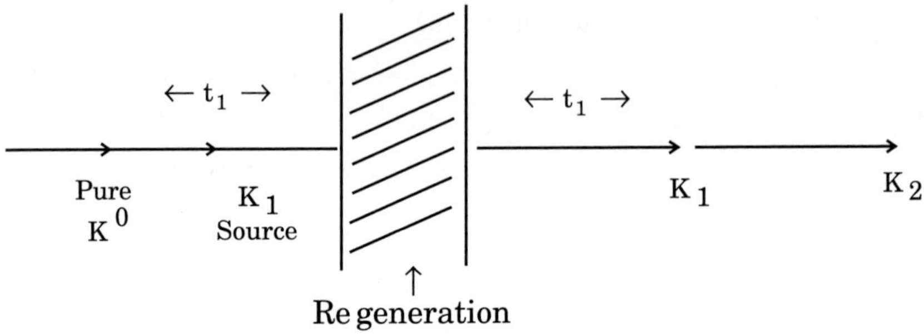

Fig. 12.9

The regeneration phenomenon observed in the neutral kaon system is simply a consequence of the superposition principle and non-commutativity of observables in quantum mechanics. Since no two components of the angular momentum operator $\vec{J}$ commute, only one component (J_z) can be diagonalized at a time. In a similar fashion the CP and S operators do not commute, so that one can have states which are eigenstates of either CP or S, but not both.

12.12 CP Violation in the Neutral kaon Decay

Following the discovery of parity violation in weak decay process in 1957, it was generally believed that the weak interaction was at least invariant under the CP-operation. The short-lived component K_S and the long-lived component K_L were therefore identified with the CP eigenstates K_1 and K_2, respectively. In 1964, however, an experiment by Christenson, Cronin, Fitch and Turlay first demonstrated that the long-lived state K_L could also decay to $\pi^+\pi^-$ with a branching ratio of order 10^{-3}. In fact, they found

$$\frac{\Gamma(K_L \to \pi^+\pi^-)}{\Gamma(K_L \to \text{ all})} = (1.57 \pm 0.05)10^{-3} \quad (12.125)$$

The argument quoted above on regeneration and mass difference which follow from the superposition principle of quantum mechanics, remain essentially unchanged although the formula for K_S and K_L will have to be modified slightly.

Thus $|K_L\rangle = (|K_2\rangle + \in |K_2\rangle) = \frac{1}{\sqrt{2}}[(1+\in)|K^0\rangle - (1-\in)|\overline{K}^0\rangle]$

$$|K_S\rangle = (|K_1\rangle + \in |K_2\rangle) = \frac{1}{\sqrt{2}}[(1+\in)|K^0\rangle + (1-\in)|\overline{K}^0\rangle] \tag{12.126}$$

where $\in$ = small quantity, $|\in|^2 \to o$

The measure of the degree of CP violation is usually quoted as the amplitude ratio

$$|\eta_{+-}| = \frac{Amp(K_L \to \pi^+\pi^-)}{Amp(K_S \to \pi^+\pi^-)} = (2.27 \pm 0.02) \times 10^{-3} \tag{12.127}$$

Since both K_S and K_L can decay into 2 pions, interference effects in the $\pi^+\pi^-$ signal are expected as a function of the time development of an initially pure K^o beam. From (12.118) we expect the intensity to vary as

$$I_{2\pi}(t) = I'_{2\pi}(o)[e^{-t/\tau_s} + |\eta_{+-}|^2 e^{-t/\tau_L} 2|\eta_{+-}|e^{-\frac{t}{2}\left(\frac{1}{\tau_s}+\frac{1}{\tau_2}\right)}\cos(t\Delta m + \phi_{+-})]$$

$$I'_{2\pi}(o) = I_{2n}(o)/|1 + \eta_{+-}|^2 \tag{12.128}$$

where t measures proper time and ϕ_{+-} is an appropriate phase angle between the $K_S \to 2\pi$ and $K_L \to 2\pi$ amplitudes

$$\eta_{+-} = |\eta_{+-}|e^{i\phi_{+-}} \tag{12.129}$$

Such interference effects have been observed.

Similar CP-violation effects have been found for the decay $K_L \to 2\pi^o$ with η_{+-} replaced by $\eta_{oo} = |\eta_{oo}|e^{i\phi_{oo}}$. If $\Delta I = \frac{1}{2}$ rule holds one experts $|\eta_{+-}| = |\eta_{oo}|$ and $\phi_{+-} = \phi_{oo}$. The experimental values are

$$|\eta_{+-}| = (2.274 \pm 0.022) \times 10^{-3}, \quad \phi_{+-} = (44.6 \pm 1.2)^\circ$$

$$|\eta_{oo}| = (2.33 \pm 0.08) \times 10^{-3}, \qquad \phi_{oo} = (54 \pm 5)^\circ$$

and are compatible with the $\Delta I = \frac{1}{2}$ rule.

CP violation is also observed in the leptonic decay modes of K_L,

$$K_L \to \pi^+ l\overline{\nu}_L, \quad \pi^- \overline{l}\nu_l, \quad l = e^-, \quad \mu^- \tag{12.130}$$

The 2 decay modes get interchanged under the CP transformation. If CP is violated, a small charge asymmetry is expected. The asymmetry is given by

$$\Delta = \frac{\Gamma(K_L \to \pi^- e^+ \nu_e) - \Gamma(K_L \to \pi^+ e^- \overline{\nu}_e)}{\Gamma(K_L \to \pi^- e^+ \nu_e) + \Gamma(K_L \to \pi^+ e^- \overline{\nu}_e)}$$

$$= (0.330 \pm 0.012) \times 10^{-2} \quad (12.131)$$

All CP-violation data in neutral kaon decay are consistent with the "superweak model" of Wolfenstein in which the effects are those due to a specific new type of interaction and observable only in the neutral kaon system. Among the predictions of the model are $|\eta_{+-}| = |\eta_{oo}| = \eta$ and $|\phi_{+-}| = |\phi_{oo}| = \phi$ and one obtains the relation

$$\Delta = 2 \text{ Re } \eta = 2|\eta| \cos\phi = 0.322 \times 10^{-3} \quad (12.132)$$

in agreement with (12.132). There are, however, other models of CP-violation which are for all practical purposes almost indistinguishable from the superweak theory. Models involving six or more quark flavors and generalized Cabibbo mixing do include a possible finite CP-violating phase angle. In contrast to the superweak theory in which CP violation is detectable only in the neutral kaon system, other theories predict CP violating effects also in bottom meson ($B^o - \overline{B}^o$ system) decays, nonzero value for E1 moment of the neutron, and baryon-anti baryon asymmetry.

[The quantities η_{+-} and η_{oo} depend on the amplitude A_o and A_2 for finding the final state pions in a state of i-spin $I = 0$ or $I = 2$ ($I = 1$ is forbidden because of Bose symmetry and angular momentum conservation. The former implies $L+I =$ even and the latter requires. $L = 0$). One obtains the expressions $\eta_{+-} = \in + \in'$, $\eta_{oo} = \in -2 \in'$ where $\in' = \frac{i}{\sqrt{2}} \cdot \frac{\text{Im}A_2}{A_o} e^{i(\delta_2 - \delta_1)}$]

12.13 Generalized Cabibbo Mixing. The CKM Matrix

It was first pointed out by Kobayashi and Maskawa in 1972 that one can have possible CP-violation in weak processes if one has at least 3 quark doublets (six quark flavors) just as absence of strangeness-changing weak neutral currents can be explained if one had a fourth quark (GIM mechanism). The weak charged current with 3 quark doublets

$$\begin{pmatrix} u \\ d \end{pmatrix}, \quad \begin{pmatrix} c \\ s \end{pmatrix}, \quad \begin{pmatrix} t \\ b \end{pmatrix}$$

can be expressed in a matrix form by

$$J^W = (\overline{u}\,\overline{c}\,\overline{t})(M)\begin{pmatrix} d \\ s \\ b \end{pmatrix} \tag{12.133}$$

where M is a 3×3 unitary matrix and the space-time structure operator has been omitted.

An $n \times n$ unitary matrix is described by n^2 real parameters of which $\frac{1}{2}n(n-1)$ are Euler angles (mixing angles) and $\frac{1}{2}n(n+1)$ are phases. Of the latter $(2n-1)$ are arbitrary and unobservable, leaving $\frac{1}{2}(n-1)(n-2)$ nontrivial phases. Thus a 2×2 unitary matrix corresponding to 2 quark doublets is described by only 1 mixing angle and no phase (Cabibbo mixing), whereas a 3×3 unitary matrix (corresponding to 3 quark doublets) is described by 3 mixing angles $(\theta_1, \theta_2, \theta_3)$ and 1 non-trivial phase δ. Thus to ensure possible CP violation one has to have atleast 3 quark doublets. The 3×3 CKM matrix (Cabibbo-Kobayashi-Maskawa)is given by

$$V = \begin{pmatrix} V_{ud} & V_{us} & V_{ub} \\ V_{cd} & V_{cs} & V_{cb} \\ V_{td} & V_{ts} & V_{tb} \end{pmatrix}$$

$$= \begin{pmatrix} c_1 & c_3 s_1 & s_1 s_3 \\ -c_2 s_1 & c_1 c_2 c_3 - s_2 s_3 e^{i\vec{\delta}} & c_1 c_2 s_3 + c_3 s_2 e^{i\vec{\delta}} \\ s_1 s_2 & -c_1 c_3 s_2 - c_2 s_3 e^{i\vec{\delta}} & -c_1 s_2 s_3 + c_2 c_3 e^{i\vec{\delta}} \end{pmatrix} \tag{12.134}$$

with $c_i = \cos\theta_i$ and $s_i = \sin\theta_i$ $(i = 1, 2, 3)$, since under time reversal $c^{i\delta} \to e^{-i\delta}$, the phase angle δ introduces the possibility of a T-or CP-violating amplitude. In terms of the CKM matrix parameters, the CP-violating parameter $\in$ in neutral Kaon decay is given by $\in\sim s_1 s_3 \sin\delta$. A finite value for $\in' / \in\sim 10^{-2}$ is also expected.

Comparison shows that the $u \to d$ mixing, specified by $\cos\theta_c$ in Cabibbo theory is given by $\cos\theta_1$ in the more general CKM model. The $u \to s$ mixing as determined from the semileptonic decays of hyperons and kaons is given by $\sin\theta_c$ in Cabibbo theory and $\cos\theta_3 \sin\theta_1$ in the CKM model. This implies that $\cos\theta_3 \sim 1$ and thus $\sin\theta_3$ must be small. In fact, it turns out that s_1, s_2 and s_3 are all small.

There is an alternative parameterization of the CKM matrix due to Wolfenstein.

$$V = \begin{pmatrix} V_{ud} & V_{us} & V_{ub} \\ V_{cd} & V_{cs} & V_{cb} \\ V_{td} & V_{ts} & V_{tb} \end{pmatrix}$$

$$= \begin{pmatrix} 1 - \frac{\lambda^2}{2} & \lambda & A\lambda^3(\rho - i\eta) \\ -\lambda(1 + iA^2\lambda^4\eta) & 1 - \frac{\lambda^2}{2} & A\lambda^2 \\ A\lambda^3(1 - \rho - i\eta) - A\lambda^2 & (1 + i\lambda^2\eta) & 1 \end{pmatrix} \quad (12.135)$$

In this parametrization the CKM matrix is almost diagonal, with the small off-diagonal terms described by small powers of λ multiplying simple functions of A, ρ, η.

Current Knowledge of the CKM Parameters

Six of the nine absolute values of the CKM matrix elements are measured correctly from tree level decay processes.

(1) From a comparison of the rates for superallowed nuclear β-decay with muon decay gives

$$|V_{ud}| = 0.9740 \pm 0.0010$$

(2) An analysis of $K \to \pi e \nu$ and semileptonic hyperon decays gives

$$|V_{us}| = 0.2196 \pm 0.0023$$

(3) Neutrino and antineutrino production of charm quarks off valence d quarks gives

$$|V_{cd}| = 0.224 \pm 0.016$$

(4) An analysis of semileptonic D meson decays $(D^+ \to \overline{K}^o e^+ \nu_e)$ similar to $K \to \pi e \nu$ gives

$$|V_{cs}| = 1.04 \pm 0.016$$

(5) From an analysis of exclusive B meson decays, $B \to \pi l \nu$ one $B \to \rho l \nu$ gets

$$|V_{ub}| = 0.0034 \pm 0.0006$$

(6) From the lifetime of the B-meson, $\tau_B \approx 10^{-12}$ sec and $|V_{ub}|$ one obtains

$$|V_{cb}| = 0.0395 \pm 0.0017$$

Finally $|V_{tq}|$'s $(q = d, s, b)$ are determined from the unitary bound $\sum_{q=d,s,b} |V_{tq}|^2 = 1$ and one obtains

$$\begin{aligned} &|V_{td}| = 0.0084 \pm 0.0018, \quad |V_{ts}| = 0.044 \pm 0.017 \\ \text{and} \quad &|V_{tb}| = 0.99 \pm 0.29 \end{aligned} \tag{12.136}$$

Lastly, with 3 generations of quarks one expects CP violation in the neutral B-meson system [eg $B_d(b\bar{d}) - \overline{B}_d(d\bar{b})$]. Moreever, while the CP-violating amplitude in the kaon system is expected to be small compared with the CP-conserving amplitude, in the B-system CP-violating and CP-conserving amplitudes are expected to be of similar magnitude.

12.14 Two-Component Neutrino Theory

Experimentally, it has been observed that only right-handed antineutrinos $(\overline{\nu}_R)$ are emitted in electron-decay. Correspondingly, only left-handed neutrinos (ν_L) are emitted in positron β-decay, which is the process inverse to the former. This is consistent with the (V-A) theory, because of the $(1-\gamma_5)$ factor in H^W_{int}. The same is true for all other known weak decays. Since right-handed neutrinos (ν_R) and their left-handed antiparticles $(\overline{\nu}_L)$ are absent from all weak processes and the neutrinos have only weak interaction, they represent an unnecessary extra degree of freedom in the Dirac equation for a massless particle which one may attempt to remove. The Dirac eqn. for a massless particle may be written as

$$i\left(\frac{\partial}{\partial t} + \vec{\alpha}\cdot\vec{\nabla}_x\right)\psi(\vec{x}, t) = o \tag{12.137}$$

where $\vec{\alpha} = (\alpha_1, \alpha_2, \alpha_3) \cdot \alpha_j'$'s are hermitian operators independent of $(\overline{x}, t)$. Any soln. of (12.137) also a solution-of massless KG eqn.

$$\left(\frac{\partial^2}{\partial t^2} - \vec{\nabla}_x^2\right)\psi(\vec{x}, t) = o \tag{12.138}$$

(although the converse need not be true). This means that the α_j's must satisfy the anti commutation relations

$$\alpha_j\alpha_k + \alpha_k\alpha_j = 2\delta_{jk}\cdot 1 \quad (j, k = 1, 2, 3) \tag{12.139}$$

Furthermore, from (12.139) one gets

$$\text{Tr}(\alpha_j) = o \quad (j = 1, 2, 3) \tag{12.140}$$

In the matrix representation, the above condition for 3 operators may be satisfied by 2 × 2 Pauli matrices

$$\alpha_j = \lambda\delta_j \tag{12.141}$$

$$\text{where } \lambda = \lambda^* \text{ (i.e. } \lambda \text{ is real) and } \lambda^2 = 1 \quad (\lambda = \pm 1) \tag{12.142}$$

(It was the necessity for constructing β as the fourth anticommuting matrix that necessiated α_i's and β to be 4 × 4 matrices in case of the Dirac eqn. for a massive particle.)

Substituting (12.141) in (12.137) one gets

$$i\left(\frac{\partial}{\partial t} + \lambda\vec{\sigma}\cdot\vec{\nabla}_x\right)\psi(\vec{x},t) = o \tag{12.143}$$

Eqn. (12.143) is known as the Weyl eqn. and its solution (a 2-component spinor) is known as a Weyl spinor.

A plane wave soln. of eqn. (12.143) is of the form

$$\psi(\vec{x},t) = \chi(p)e^{-i(Et-\vec{p}\cdot\vec{x})} \tag{12.144}$$

Substituting (12.144) in (12.143) and (12.138) we get

$$(E - \lambda\vec{\sigma}\cdot\vec{p})\chi(p) = o \tag{12.144}$$

$$\text{and} \quad (E^2 - \vec{p}^{\,2})\chi(p) = o \tag{12.145}$$

From (12.145) one gets $E = \pm E_p$ where $E_p = |\vec{p}|$

Thus there are four possible soln. of the Weyl eqn. [eqn. (12.143)] corresponding to $E = \pm E_p$ and $\lambda = \pm 1$. These may be written as

$$\left.\begin{aligned}
\nu_R &: \psi_{++}(\overline{x},t) = \chi_R(p)e^{-i(E_pt-\vec{p}\cdot\vec{x})}\\
\overline{\nu}_L &: \psi_{-+}(\overline{x},t) = \chi_L(p)e^{+i(E_pt+\vec{p}\cdot\vec{x})}\\
\nu_L &: \psi_{+-}(\overline{x},t) = \chi_L(p)e^{-i(E_pt-\vec{p}\cdot\vec{x})}\\
\overline{\nu}_R &: \psi_{--}(\overline{x},t) = \chi_R(p)e^{+i(E_pt+\vec{p}\cdot\vec{x})}
\end{aligned}\right\} \tag{12.146}$$

where $\chi_R(p)$ and $\chi_L(p)$ are solns. of the eqn.

$$\frac{\vec{\sigma}\cdot\vec{p}}{|\vec{p}|}\chi(p) = \omega\chi(p) \tag{12.147}$$

with $\omega = \pm 1$, i.e. they are eigenstates of the helicity operator

$$h = \vec{\sigma}\cdot\vec{p}/|\vec{p}| \tag{12.148}$$

If $\vec{p} = |\vec{p}|\hat{z}$, then

$$\chi_R = \begin{pmatrix} 1 \\ 0 \end{pmatrix} \quad \text{and} \quad \chi_L = \begin{pmatrix} 0 \\ 1 \end{pmatrix} \tag{12.149}$$

Only the last 2 solutions in (12.146) are realised in nature.

To understand better the relation of the 2 component solution of the Weyl eqn. (Weyl spinor) with the 4-component solns. of the Dirac eqn. (Dirac spinors) we return to the Dirac eqn. for a particle with mass m and choose a representation of Dirac matrices (Weyl representation) as follows:

$$\vec{\alpha} = \left(\begin{array}{c|c} \vec{\sigma} & o \\ \hline o & -\vec{\sigma} \end{array}\right), \quad \beta = \left(\begin{array}{c|c} o & -1 \\ \hline -1 & o \end{array}\right), \quad \gamma_5 = \left(\begin{array}{c|c} 1 & o \\ \hline o & -1 \end{array}\right) \tag{12.150}$$

which differ form the commonly used Dirac representation

$$\vec{\alpha} = \left(\begin{array}{c|c} o & \vec{\sigma} \\ \hline \vec{\sigma} & o \end{array}\right), \quad \beta = \left(\begin{array}{c|c} 1 & o \\ \hline o & -1 \end{array}\right), \quad \gamma_5 = \left(\begin{array}{c|c} o & 1 \\ \hline 1 & o \end{array}\right) \tag{12.151}$$

by a unitary transformation,

$$U = \frac{1}{\sqrt{2}}(1 + \gamma_5\gamma^o) \tag{12.152}$$

where $\gamma_5 = i\gamma^o\gamma'\gamma^2\gamma^3$, $\gamma^o = \beta$, $\gamma^j = \beta\alpha_j$ $(j = 1, 2, 3)$ (12.153)

In the Weyl representation the Dirac eqn.

$$i\frac{\partial}{\partial t}\psi(\overline{x}, t) = \frac{1}{i}\vec{\alpha} \cdot \vec{\nabla}_x\psi(\vec{x}, t) + \beta m\psi(\vec{x}, t) \tag{12.154}$$

may be expressed in a split 2-component notation

$$\psi = \begin{pmatrix} \xi \\ \eta \end{pmatrix} \tag{12.155}$$

$$\begin{aligned} i\left(\frac{\partial}{\partial t} + \vec{\sigma} \cdot \vec{\nabla}_x\right)\xi(\vec{x}, t) &= -m\eta(\vec{x}, t) \\ i\left(\frac{\partial}{\partial t} - \vec{\sigma} \cdot \vec{\nabla}_x\right)\eta(\vec{\alpha}, t) &= -m\xi(\vec{x}, t) \end{aligned} \tag{12.156}$$

The upper and lower component of ψ are mixed only by the mass terms in eqn. (12.156). In the limit $m \to o$, there results two uncoupled 2-component eqns. corresponding to eqn. (12.143) with $\lambda = \pm 1$

If one defines $\psi_R = \begin{pmatrix} \xi \\ 0 \end{pmatrix}$ and $\psi_L = \begin{pmatrix} o \\ \eta \end{pmatrix}$ $[\psi = \psi_R + \psi_L]$ then in the Weyl representation

$$\gamma_5 \psi_R = +\psi_R \quad \text{and} \quad \gamma_5 \psi_L = -\psi_L$$
$$\text{and} \quad \psi_R = \frac{1}{2}(1+\gamma_5)\psi, \quad \psi_L = \frac{1}{2}(1-\gamma_5)\psi \tag{12.157}$$

ψ_R and ψ_L are known as the right and left chiral components of the Dirac spinor ψ, since γ_5 is called the chirality operator.

Although the possibility of describing a massless Dirac particle by a 2 component eqn. was first discussed by Weyl in 1929, it was not taken seriously for a long time, since the Weyl eqn. is not invariant under parity (P) transformation. After parity violation was discovered in weak interaction in 1957, the 2 component Weyl eqn. was resurreceted by Landau, Lee and Yang and Salam who observed that charge conjugation (C) invariance is also lost but that a combined invariance under CP remains. Under charge conjugation (C) operation the Dirac spinor transforms as

$$\psi(\vec{i}, t) \xrightarrow{C} \psi_c(\vec{x}, t) = c\psi^*(x, t) \tag{12.158}$$

where $c = 4 \times 4$ matrix $= i\gamma^2$ ($c^+ = c^{-1}$, i.e. c is unitary)

In the Weyl representation,

$$c = i\gamma^2 = i\beta\alpha_2 = i\left(\begin{array}{c|c} 0 & -1 \\ \hline -1 & 0 \end{array}\right)\left(\begin{array}{c|c} \sigma_2 & 0 \\ \hline 0 & -\sigma_2 \end{array}\right) = \left(\begin{array}{c|c} 0 & i\sigma_2 \\ \hline -i\sigma_2 & 0 \end{array}\right)$$

$$\text{Thus } \psi_c = \begin{pmatrix} \xi_c \\ \eta_c \end{pmatrix} = c\psi^* = \left(\begin{array}{c|c} 0 & i\sigma_2 \\ \hline -i\sigma_2 & 0 \end{array}\right)\begin{pmatrix} \xi^* \\ \eta^* \end{pmatrix} = \begin{pmatrix} i\sigma_2\eta^* \\ -i\sigma_2\xi^* \end{pmatrix}$$

$$\text{Thus} \quad \begin{array}{l} \xi \xrightarrow{C} \xi_c = i\sigma_2\eta^* \\ \eta \xrightarrow{C} \eta_c = -i\sigma_2\xi^* \end{array} \tag{12.159}$$

Identifying $\xi = \nu_R$, $\xi^* = \overline{\nu}_L$, $\eta = \nu_L$, $\eta^* = \overline{\nu}_R$ we get

$$\nu_L \xrightarrow{C} \overline{\nu}_L, \quad \nu_R \xrightarrow{C} \overline{\nu}_R \tag{12.160}$$

Under parity (P) operator the Dirac spinor transform as

$$\psi(\vec{x}, t) \xrightarrow{P} \psi_o(\vec{x}, t) = \gamma^o\psi(-\vec{x}, t) \tag{12.161}$$

In the Weyl representation

$$\begin{pmatrix} \xi_P(\vec{x},t) \\ \eta_P(\vec{x},t) \end{pmatrix} = \left(\begin{array}{c|c} 0 & -1 \\ \hline -1 & 0 \end{array}\right) \begin{pmatrix} \xi(-\vec{x}_1,t) \\ \eta(-\vec{x},t) \end{pmatrix} = \begin{pmatrix} -\eta(-\vec{x},t) \\ -\xi(-\vec{x},t) \end{pmatrix} \tag{12.162}$$

Thus
$$\begin{aligned} \xi(\vec{x},t) &\xrightarrow{P} \xi_P = (\vec{x},t) = -\eta(-\vec{x},t) \\ \eta(\vec{x},t) &\xrightarrow{P} \eta_P = (\vec{x},t) = -\xi(-\vec{x},t) \end{aligned} \tag{12.163}$$

This means $\nu_R \xrightarrow{P} \nu_L$, $\nu_L \xrightarrow{P} \nu_R$

Under CP operator the Dirac spinor transforms as

$$\psi(\vec{x},t) \xrightarrow{CP} \vec{\psi}_{CP}(\vec{x},t) = (c\gamma^0)\psi^*(-\vec{x},t)$$

$$c\gamma^0 = i\gamma^2\gamma^0 = i\beta\alpha_2\beta = -i\alpha_2 = \begin{pmatrix} -i\sigma_2 & 0 \\ 0 & i\sigma_2 \end{pmatrix}$$

This means $\begin{pmatrix} \xi_{CP}(\vec{x},t) \\ \eta_{CP}(\vec{x},t) \end{pmatrix} = \begin{pmatrix} -i\sigma_2 & 0 \\ 0 & i\sigma_2 \end{pmatrix} \begin{pmatrix} \xi^*(-\vec{x},t) \\ \eta^*(-\vec{x},t) \end{pmatrix}$

Therefore,
$$\begin{aligned} \xi(\vec{x},t) &\xrightarrow{CP} \xi_{CP} = (\vec{x},t) = -i\sigma_2\xi^*(\vec{x},t) \\ \eta(\vec{x},t) &\xrightarrow{CP} \eta_{CP} = (\vec{x},t) = +i\sigma_2\eta^*(\vec{x},t) \end{aligned} \tag{12.164}$$

For a massless Dirac particle, η satisfy the eqn.

$$i\left(\frac{\partial}{\partial t} - \vec{\sigma}\cdot\vec{\nabla}_x\right)\eta(\vec{x},t) = 0 \tag{12.165}$$

Taking complex conjugate we get

$$-i\left(\frac{\partial}{\partial t} - \vec{\sigma}^*\cdot\vec{\nabla}_x\right)\eta^*(\vec{x},t) = 0 \tag{12.166}$$

multiplying from the left by σ_2 and using $\sigma_2\vec{\sigma}^*\sigma_2 = -\vec{\sigma}$ we get

$$-i\left(\frac{\partial}{\partial t} + \vec{\sigma}\cdot\vec{\nabla}_x\right)\sigma_2\eta^*(\vec{x},t) = 0$$

$$\text{or,} \quad -i\left(\frac{\partial}{\partial t} - \vec{\sigma}\cdot\vec{\nabla}'_x\right)\sigma_2\eta^*(-\vec{x}',t) = 0 \tag{12.167}$$

This means $\eta_{CP}(\vec{x},t)$ satisfies the same eqn. as $\eta(\vec{x},t)$. A similar result is obtained for $\xi(\vec{x},t)$. Thus the Weyl eqn. (12.143) is invariant under CP transformation.

Now $\nu_R \xrightarrow{CP} \overline{\nu}_L$, $\nu_L \xrightarrow{CP} \overline{\nu}_R$

Thus the fact that ν_L and $\overline{\nu}_R$ are realized in nature is a consequence of CP invariance. Weak interactions are known to be consistent with CP invariance except for neutral kaon (and possibly neutral B-meson) decays.

Problems

1. The muon lifetime is 2.2×10^{-6} s. Calculate the lifetime of the τ-lepton, given that the branching ratio for $\tau^- \to e^- + \overline{\nu}_e + \nu_\tau$ is 18% ($m_\tau = 1.78$ GeV, $m_\mu = 105.7$ MeV).

2. Use the V-A theory and the CVC hypothesis to write down the lowest order matrix element for the weak decay $\pi^- \to \pi^o + e^- + \overline{\nu}_e$. Evaluate the decay rate and obtain an estimate of the branching ratio

$$\Gamma(\Sigma^- \to \Lambda + e^- + \overline{\nu}_e)/\Gamma(\pi^- \to \mu^- + \overline{\nu}_\mu).$$

3. Obtain an estimate of the branching ratio

$$\Gamma(\Sigma^- \to \Lambda + e^- + \overline{\nu}_e)/\Gamma(\Sigma^- \to n + \pi^-)$$

assuming that the matrix element for the electron decay mode of Σ^- is the same as the matrix element for the neutron β-decay and that the baryon recoil can be neglected. ($\tau_\Sigma = 1.5 \times 10^{-10}$ s, $m_\Sigma = 1197$ MeV, $m_\Lambda = 1116$ MeV)

4. Assuming that the quark level process is $d \to u + e^- + \overline{\nu}_e$ show that

$$\Gamma(\pi^- \to \pi^o + e^- + \overline{\nu}_e) = \frac{G_F^2}{30\pi^3}(\Delta m)^5$$

where $|\Delta m = m(\pi^-) - m(\pi^o) = 4.6$ MeV. Evaluate the decay rate and compare with $\Gamma(\pi^- \to \mu^- + \overline{\nu}_\mu)$.

5. Nonleptonic decays of strange particles do not conserve i-spin but appear to obey the $\Delta I = \frac{1}{2}$ rule. Use this rule to show that

(a) $\dfrac{\Gamma(\Lambda \to \pi^- + p)}{\Gamma(\Lambda \to \pi^o + n)} = 2$

(b) $\dfrac{\Gamma(K_S \to \pi^+ + p^-)}{\Gamma(K_S \to \pi^o + \pi^o)} = 2$

(c) $\dfrac{\Gamma(K_L \to \pi^o + \pi^o + \pi^o)}{\Gamma(K_L \to \pi^+ + \pi^- + \pi^o)} = \dfrac{3}{2}$

6. Apply the $\Delta Q = \Delta S$ rule for semileptonic decays of strange particles to determine which of the following decays are suppressed and which are favored

$$\Sigma^- \rightarrow n + e^- + \overline{\nu}_e, \Sigma^+ \rightarrow n + e^+ + \nu_e, \Sigma^o \rightarrow p + e^- + \overline{\nu}_e$$

7. Consider the decay modes

$$\Sigma^+ \rightarrow n + \pi^+, \Sigma^+ \rightarrow p + \pi^o, \Sigma^- \rightarrow n + \pi^-$$

Denoting the decay amplitudes by A^+, A^o and A^-, respectively, show that

$$A^+ + \sqrt{2}A^o = A^-$$

8. Using the helicity argument or otherwise show that the V-A theory of weak interactions predicts that at high energies

$$R = \frac{\sigma(\overline{\nu}_e + e^- \rightarrow \overline{\nu}_e + e^-)}{\sigma(\nu_e + e^- \rightarrow \nu_e + e^-)} = \frac{1}{3}.$$

What would be the value of R if the charged weak current had a $V + A$ structure?

Chapter 13

Electroweak Unification and the Glashow-Salam-Weinberg (GSW) Model

13.1 Introduction

It is generally believed that all fundamental interactions of nature are describable by gauge theories, that is, field theories which are invariant under local gauge transformations in some abstract space. Part of the belief stems from the fact that experience with QED has shown that gauge symmetry ensures renormalizability. However, attempts to construct a gauge invariant, hence renormalizable, theory of weak interaction encountered many difficulties and they were overcome in a number of steps. This finally led to unification of weak interaction with electromagnetic interaction. The unified theory is called the electroweak theory and its first and minimal version is known as the Glashow-Salam-Weinberg (GSW) model.

The V-A theory which evolved out of Fermi's theory of nuclear β decay, was generally successful in explaining all known weak decays. However, it could not be applied to high energy weak processes (neutrino scattering and production processes). Experimental study of such processes became possible after high energy neutrino beams were made available by secondary means. In the calculation of the matrix elements of these processes one has to include the contributions of higher order diagrams and the V-A theory results were highly divergent. This is because the theory is not renormalizable as the coupling constant G_F has the dimension of $1/M^2$ (in natural units). One could take the view that the V-A theory is an effective theory, being the low-energy approximation of a more correct theory valid at all energies. The correct theory must incorporate all the known properties the V-A theory, eg, maximum parity violation and γ_5 invariance (chiral invariance). As a first step one introduced the idea of intermediate vector bosons. The weak interactions between the 2 pairs of fermions in β decay (or μ decay) are mediated by vector bosons. They must be charged ($W^{\pm}$) since the fermion currents involved are charge (flavor) changing. Assuming

$|q^2| \ll M_W^2$ the (effective) coupling constant in V-A theory is related to the coupling constant in the IVB theory, the relation being $G_F/\sqrt{2} \approx g_W^2/M_W^2$. Assuming $g_W \sim e$, the electromagnetic coupling constant, $M_W \sim 100 M_p$ ie the weak vector bosons must be very massive and therefore the range of the weak interaction $\sim 1/M_W \sim 10^{-16}$ cm.

That the weak interaction has an extremely short range also follows from the success of V-A theory. However it is known that field theories with massive vector bosons are not renormalizable and so the problem remained. On the other hand theories with massless vector bosons are renormalizable if one imposes local gauge invariance. Thus one could explore the possibility of starting with a gauge invariant theory and hence massless vector bosons and devise some mechanism of giving masses to the vector bosons in the theory without disturbing the manifest gauge invariance of the Lagrangian. It was found that this indeed could be done by incorporating complex scalar fields in the theory and spontaneously breaking the gauge symmetry so that the scalar fields have nonzero vacuum expectation values and then invoke the Higgs mechanism which makes use of the gauge freedom to gauge away some of the scalar degrees of freedom (would-be Goldstone bosons) so that they reappear as the longitudinal degrees of freedom of equal number of vector bosons, which thereby become massive. The two theories, the starting theory with massless vector bosons and the endup theory with massive vector bosons are unitarily related and hence equivalent. As a consequence of the symmetry-breaking mechanism used, some of the scalar degrees of freedom, known as Higgs bosons, are left over in the endup theory. Finally it was proved by t'Hooft etal that spontaneously broken gauge theories are renormalizable like exact gauge theories.

It was apparent from the beginning that in order to describe weak interaction the underlying symmetry group of the gauge theory must be non-Abelian. This is because the vector bosons involved in weak decays must be charged ($W^{\pm}$). Since the gauge bosons transform like the adjoint representation of the symmetry group, the number of generators of the group must be ≥ 2. Now the smallest non-Abelian group is $SU(2)$ with 3 generators. The third generator must correspond to a neutral vector boson which interacts with neutral currents. Thus there must exist weak neutral currents. Moreover, in order to cancel the divergences in higher order diagrams corresponding to the processes like $e^+e^- \rightarrow W^+W^-$ and also to satisfy the unitarity constraints order by order in perturbation theory there must exist a neutral vector boson with coupling of the same order as electromagnetic coupling and mass of the same order as the mass of the weak charged borons. Thus the gauge group must incorporate the gauge group of electromagnetism. That is to say, an acceptable (renormalizable) theory of the weak interaction must be unified with electromagnetism. Thus the minimal gauge group is $SU(2) \times U(1)_Y$ with 4 gauge bosons (2 charged and 1 neutral corresponding to $SU(2)$ and 1 neutral corresponding to $U(1)$).

$SU(2) \times U(1)_Y$ is then spontaneously broken to $U(1)_Q$, the gauge group of QED. The physical spectrum of the theory consists of 3 (2 charged +1 neutral) massive weak gauge borons ($W^\pm, Z^0$) and 1 massless gauge boson (corresponding to the photon γ). One has the Gell-Mann-Nishijima type relation.

$$Q = I_3 + \frac{Y}{2} \tag{13.1}$$

where Q is the electric charge and Y is the weak hypercharge and I_3 is the third component of 'weak' i-spin (one of the 3 generators of $SU(2)$).

13.2 Spontaneous Symmetry Breaking (SSB)

A symmetry is said to be spontaneously broken if the Lagrangian is invariant under the symmetry transformation but the physical vacuum is not. This leads to exact conservation laws but hides the symmetry of the theory (hidden symmetry).

The simplest example of SSB is the case of a real scalar field described by the Lagrangian.

$$\pounds = \frac{1}{2}(\partial_\mu \phi)(\partial^\mu \phi) - V(\phi) \tag{13.2}$$

where $V(\phi) = \frac{\mu^2}{2}\phi^2 + \frac{\lambda}{4}\phi^4, \quad \lambda > 0$

(13.2) is invariant under the discrete transformation $\phi \rightarrow -\phi$

The minimum of the potential is given by

$$\frac{\partial V}{\partial \phi} = 0 \quad \text{or} \quad \mu^2\phi + \lambda\phi^3 = 0 \tag{13.3}$$

Case 1. $\mu^2 > 0$. In this case $V(\phi)$ has a unique minimum at $\phi = 0$. The minimum of the potential in the classical theory corresponds to the ground state in QM and the vacuum in QFT. This is called "ordinary" symmetry. The physical spectrum corresponds to a particle of $(\text{mass})^2 = \mu^2$, $(\mu^2 > 0)$. (Fig. 13.1a).

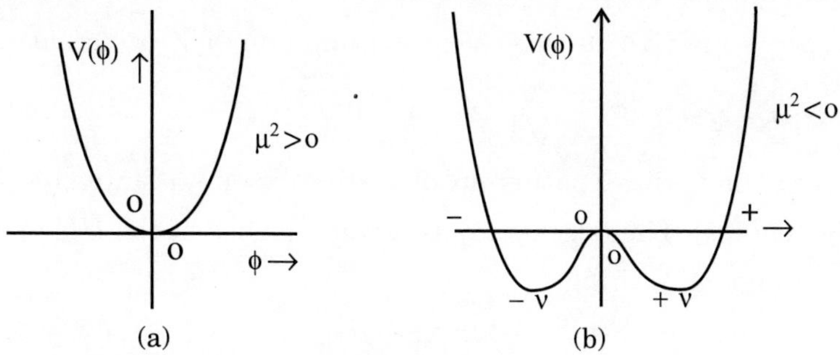

Fig. 13.1

Case 2. $\mu^2 < 0$. In this case $V(\phi)$ has minima at

$$\phi = \pm v, \; v = \sqrt{-\mu^2/\lambda} \text{ (v is real)} \tag{13.4}$$

which corresponds to two degenerate ground states or vacua. We may choose either of (say $\phi = +v$) the vacua. (Fig. 13.1b).

The transformation $\phi \to -\phi$ is then a symmetry of the Lagrangian, but not of the vacuum state. This situation is referred to as SSB. One defines a shifted field

$$\eta = \phi - v \tag{13.5}$$

so that $\eta_{\text{min}} = 0$, ie the vacuum corresponds to $\eta = 0$. In terms of the shifted field the effective potential is

$$V(\eta) = \frac{\mu^2}{2}(\eta + v)^2 + \frac{\lambda}{4}(\eta + v)^4 = (\lambda v^2)\eta^2 + (\lambda v)\eta^3 + \frac{\lambda}{4}\eta^4 + \frac{\lambda v^4}{4} \tag{13.6}$$

The physical spectrum corresponds to a particle with

$$(\text{mass})^2 = 2\lambda v^2 = -2\mu^2 \; (\mu^2 < 0). \tag{13.7}$$

13.3 Spontaneous Breaking of Continuous Symmetries

Let us consider the Lagrangian of a complex scalar field (equivalent to 2 real scalar fields), given by

$$\mathcal{L} = (\partial_\mu \phi)^*(\partial^\mu \phi) - V(|\phi|), \; |\phi|^2 = \phi^* \phi \tag{13.8}$$

$$V(|\phi|) = \mu^2 |\phi|^2 + \lambda |\phi|^4 \tag{13.9}$$

The Lagrangian is invariant under the transf. $\phi \to e^{i\alpha}\phi$. (α = real variable).

Case 1. $\mu^2 > 0$ (ordinary symmetry). The minimum occurs at $|\phi| = 0$. The ground state (vacuum) is unique and the physical spectrum corresponds to two degenerate real scalar fields of (mass^2) $= \mu^2 (> 0)$.

Case 2. $\mu^2 < 0$ (SSB). In this case the minimum of V occurs at

$$|\phi| = \frac{v}{\sqrt{2}}, \quad v = \sqrt{-\mu^2/\lambda} \tag{13.10}$$

which corresponds to a continuum of distinct vacuum states ($\phi = \frac{v}{\sqrt{2}}e^{i\alpha}$) with same energy. The degeneracy is a consequence of the $U(1)$ symmetry of the Lagrangian. Writing

$$\phi = \phi_1 + i\phi_2$$

we choose as the physical vacuum the configuration

$$(\phi_1)_0 = +\frac{v}{\sqrt{2}}, \ (\phi_2)_0 = 0, \ \alpha = 0,$$

(the subscript refers to minimum or vacuum. Expanding about the vacuum configuration we write

$$\phi = \frac{1}{\sqrt{2}}(\eta + v + i\xi) \tag{13.11}$$

such that $\eta_0 = \xi_0 = 0$

We get for the effective potential,

$$V(\eta, \xi) = (\lambda v^2)\eta^2 + (\lambda v)\eta^3 + \frac{\lambda}{2}\eta^2\xi^2 + (\lambda v)\eta\xi^2 + \frac{\lambda}{4}\eta^4 + \frac{\lambda}{4}\xi^4 - \frac{\lambda v^4}{4} \tag{13.12}$$

and $$L = \frac{1}{2}(\partial_\mu \eta)(\partial^\mu \eta) + \frac{1}{2}(\partial_\mu \xi)(\partial^\mu \xi) - V(\eta, \xi) \tag{13.13}$$

The physical spectrum consists of 2 real scalar particles, one described by η with (mass)$^2 = 2\lambda v^2 = -2\mu^2$ and the other described by ξ with (mass)$^2 =$ 0. There are also terms describing both self and mutual interaction of the particles.

This splitting of the spectrum is an example of Goldstone's theorem according to which if the Lagrangian of a theory has an exact continuous symmetry which is not a symmetry of the vacuum, a massless scalar particle must occur.

13.4 Spontaneous Breaking of Gauge Symmetry. The Higgs Mechanism (Abelian Case)

We next consider a Lagrangian with local gauge invariance which is spontaneously broken. The simplest example is the Abelian Higgs model described by the Lagrangian

$$L = (D_\mu\phi)^*(D^\mu\phi) - V(|\phi|) - \frac{1}{4}F_{\mu\nu}F^{\mu\nu} \tag{13.14}$$

where $D_\mu = \partial_\mu - igA_\mu,\ F_{\mu\nu} = \partial_\mu A_\nu - \partial_\nu A_\mu$ (13.15)

and $V(|\phi|) = \mu^2|\phi|^2 + \lambda|\phi|^4,\ |\phi|^2 = \phi^*\phi$

The Lagrangian is invariant under $U(1)$ local gauge transformation

$$\phi(x) \to \phi'(x) = e^{i\alpha(x)}\phi(x)$$

$$A_\mu(x) \to A'_\mu(x) = A_\mu(x) + \frac{1}{g}\partial_\mu\alpha \tag{13.16}$$

We consider the case $\mu^2 < 0$ (SSB). As before, the minimum of the potential occurs at $|\phi| = v/\sqrt{2},\ v = \sqrt{-\mu^2/\lambda}$ and one has a continuum of distinct but degenerate vacuum states. We choose a particular one and write,

$$\phi = \frac{1}{\sqrt{2}}(v + \eta + i\xi) \approx e^{i\xi/v}\frac{1}{\sqrt{2}}(v+\eta) \tag{13.17}$$

Substituting (13.17) for ϕ we get

$$V = (\lambda v^2)\eta^2 + (\lambda v)\eta^3 + \frac{\lambda}{4}\eta^4 + \frac{\lambda}{2}\eta^2\xi^2 + (\lambda v)\eta\xi^2 + \frac{\lambda}{4}\xi^4 - \frac{\lambda v^4}{4}$$

$$D_\mu\phi = \frac{1}{\sqrt{2}}(\partial_\mu - igA_\mu)(\eta + v + i\xi) \tag{13.12}$$

$$= \frac{1}{\sqrt{2}}\left[(\partial_\mu - igA_\mu)\eta - igv\left(A_\mu - \frac{1}{gv}\partial_\mu\xi\right) + gA_\mu\xi\right] \tag{13.18}$$

The η field has $(\text{mass})^2 = 2\lambda v^2 = -2$ and the ξ field zero mass (Goldstone boson). The gauge field appears to have acquired a mass (gv) but it is coupled with the ξ field. Eq. (13.18) suggests that we make a gauge transformation,

$$A_\mu \to A'_\mu = A_\mu - \frac{1}{gv}\partial_\mu\xi$$

and $\phi \to \phi' = e^{-i\xi/v}\phi = \frac{1}{\sqrt{2}}(v+\eta)$ (13.19)

Since $\pounds$ is invariant under local gauge transformation, we can write it in terms of the primed fields (ϕ', A'_μ)

$$\begin{aligned} L &= [(\partial_\mu + igA'_\mu)\phi'](\partial^\mu_\nu - igA'^\mu)\phi'] - V(\phi') - \frac{1}{4}F'_{\mu\nu}F'^{\mu\nu} \\ &= \frac{1}{2}[(\partial_\mu + igA'_\mu)(\eta + v)](\partial^\mu_\mu - igA'^\mu)(\eta + v)] - V(\eta) - \frac{1}{4}F'_{\mu\nu}F'^{\mu\nu} \\ &= \frac{1}{2}(\partial_\mu\eta)(\partial^\mu\eta) - V(\eta) - \frac{1}{4}F'_{\mu\nu}F'^{\mu\nu} + \frac{g^2v^2}{2}A'_\mu A'^\mu \\ &\qquad + (g^2v)\eta A'_\mu A'^\mu + \frac{g^2}{2}\eta^2 A'_\mu A'^\mu \end{aligned} \tag{13.20}$$

where $V(\eta) = (\lambda v^2)\eta^2 + (\lambda v)\eta^3 + \frac{\lambda}{4}\eta^4 - \frac{\lambda v^4}{4}$ (13.21)

and $F'_{\mu\nu} = \partial_\mu A'_\nu - \partial_\nu A'_\mu$

The particle spectrum is now manifest :

a real scalar field (η) with $(\text{mass})^2 = 2\lambda v^2$

a real massive vector field (A'_μ) with $(\text{mass})^2 = g^2v^2$

no Goldstone field (ξ) (massless scalar field)

There are interaction terms ηAA, $\eta^2 AA$ and self interaction terms η^3 and η^4 of the scalar field η.

Because of our choice of gauge, the ξ field has disappeared. The gauge transformation

$$A_\mu \to A'_\mu = A_\mu - \frac{1}{gv}\partial_\mu\xi$$

shows that what formerly was the ξ field has now become the longitudinal component of the A'_μ field. Before SSB we had 2 scalars ($\phi^\pm$) and 2 helicity states of A_μ, giving a total of 4 real degrees of freedom. After SSB, we are left with 1 scalar (η) and 3 helicity states of the A'_μ field, again giving a total of 4 real degrees of freedom.

It is commonly said that the massless gauge boson A_μ "ate up" the massless Goldstone boson (ξ) to become a massive vector boson (A'_μ). The remaining massive scalar field (η) is known as the Higgs boson. The gauge in which the particle spectrum is transparent is known as the unitary gauge-(U-gauge) because only physical fields appear in the Lagrangian. The gauge in which the Lagragian is mainfestly local gauge invariant and hence renormalizable is known as the renormalization gauge (R-gauge).

13.5 Spontaneous Breaking of Non-Abelian Gauge Symmetry

We consider a scalar field theory which is invariant under local gauge transformations corresponding to a nonabelian group G with N generators. The scalar field belongs to a representation of dimension M of the group G if it is real or a representation of dimension $M/2$, where M is even if it is complex ($M < N$). The symmetry is said to be spontaneously broken if the minimum of the effective scalar potential occurs for $\phi_0 \neq 0$. We can always choose

$$\phi_0 = \begin{pmatrix} 0 \\ 0 \\ : \\ 0 \\ \phi_M = v \end{pmatrix}, \text{ (if } \phi \text{ real); } \phi_0 = \begin{pmatrix} 0 \\ 0 \\ : \\ 0 \\ \phi_{M/2} = \dfrac{v}{\sqrt{2}} \end{pmatrix} \text{ (if } \phi \text{ complex)} \quad (13.23)$$

ϕ has M real components or $\frac{M}{2}$ complex components. In that case the symmetry is said to be spontaneously broken to a smaller symmetry group G' with N' generators where $N' = N - (M - 1) = N - M + 1$. Thus there will be ($M$-1) Goldstone bosons. Now the gauge bosons belong to the adjoint representation of the gauge symmetry group G so that the no. of gauge bosons equals the no. of generators. As in the abelian case we can expand ϕ about ϕ_0 and then make a suitable local gauge transformation such that the $(M - 1)$ Goldstone bosons disappear and equal no. of gauge bosons acquire longitudinal degrees of freedom so that they become massive vector bosons. Thus in the new gauge (U gauge) $(N - N')$ vector bosons are massive and the remaining N' vector bosons continue to remain massless so that the gauge symmetry corresponding to the gauge group G' is exact. Since $3(N - N') + 2N' + 1 = 2N + M$, the total no. of degrees of freedom remain the same.

13.6 Glashow-Salam-Weinberg (GSW) Model

It is the simplest model that provides a renormalizable theory of weak interaction by incorporating unification of weak and electromagnetic interactions. It is a gauge theory with the gauge group chosen to be $G =$

$SU(2)_L \times U(1)_Y$, direct product of weak i-spin group and weak hypercharge group. The subscript L refers to the fact that fermions appear in left handed doublets and right handed singlets. The gauge symmetry is spontaneously broken to $U(1)_Q$, the gauge symmetry of electromagnetic interaction which remains exact in the model. Q, I_3 and Y satisfy the Gell-Mann Nishijima relation for the group.

$$Q = I_3 + \frac{Y}{2} \tag{13.24}$$

The commutation relations are,

$$[I_a, I_b] = i \in_{abc} I_c, [I_a, Y] = 0, \quad [I_3, Q] = 0$$

Here Q is the electric charge (in unit of e), Y is the weak hypercharge and I_3 is the third component of the weak i-spin $\vec{I} = (I_1, I_2, I_3)$

Fermion sector

As stated above, Fermions appear in left-handed doublets and right-handed singlets, except for the neutrino which appears only in a left-handed doublet. Thus for the first generation leptons we have

$$\psi_L = \begin{pmatrix} \nu_e \\ e \end{pmatrix}_L, \quad \psi_R = e_R \tag{13.25}$$

with $Y_L = -1$, $Y_R = -2$ so that (13.24) is satisfied. Here

$$\psi_L = \frac{1}{2}(1 - \gamma_5)\psi \quad \text{and} \quad \psi_R = \frac{1}{2}(1 + \gamma_5)\psi \tag{13.26}$$

All the three lepton generations have the same multiplet structure. For the first generation quarks we have

$$\psi_L = \begin{pmatrix} u \\ d \end{pmatrix}_L, \quad \psi_R = u_R, d_R \tag{13.27}$$

with $Y_L = \frac{1}{3}$, $Y_R = +\frac{4}{3}$ for u_R and $Y_R = -\frac{2}{3}$ for d_R

All three quark generations have the same multiplet structure. The absence of FCNC is accounted for by incorporating GIM mechanism which amounts to replacing (d, s, b) by (d', s', b') where

$$\begin{pmatrix} d' \\ s' \\ b' \end{pmatrix} = V \begin{pmatrix} d \\ s \\ b \end{pmatrix}, \ V \text{ is the } 3 \times 3 \text{ CKM matrix. } V = \begin{pmatrix} V_{ud} & V_{us} & V_{ub} \\ V_{cd} & V_{cs} & V_{cb} \\ V_{td} & V_{ts} & V_{tb} \end{pmatrix} \tag{13.28}$$

Gauge Sector

Since the number of gauge bosons is equal the number of generators of the symmetry group, there are four gauge bosons in the GSW model.

They are $W_\mu^\alpha (\alpha = 1, 2, 3)$ for $SU(2)$ and B_μ for $U(1)$.

The gauge invariant Lagrangian involving the fermion fields and the gauge fields is given by

$$\pounds_1 = \pounds(\psi, D_\mu\psi, G_{\mu\nu}) \tag{13.29}$$

where D_μ is the covariant derivative

$$D_\mu = \partial_\mu - igA_\mu \tag{13.30}$$

and $G_{\mu\nu}$ is the covariant field lensor

$$G_{\mu\nu} = \frac{1}{ig}[D_\mu, D_\nu] \tag{13.31}$$

Here A_μ is a single gauge field in the abelian case and A_μ is a matrix valued field in the nonabelian case, i.e.

$$A_\mu = \sum_{\alpha=1}^{r} A_\mu^a t_a \tag{13.32}$$

$t_a = n \times n$ matrix $n =$ dimension of the representation to which the field ϕ belongs. For $SU(2)$, $n = 2$, $r = 3$ and $t_a = \frac{\tau_a}{2}$

In the GSW model,

$$\pounds_1 = \pounds_{\text{fermion}} + \pounds_{\text{gauge}} \tag{13.33}$$

$$\text{where} \quad \pounds_{\text{fermion}} = \overline{\psi}_L \gamma^\mu \left(\partial_\mu - igW_\mu^a \frac{\tau_a}{2} - ig'B_\mu \frac{Y_L}{2}\right)\psi_L + \overline{\psi}_R i\gamma^\mu \left(\partial_\mu - ig'B_\mu \frac{Y_R}{2}\right)\psi_R \tag{13.34}$$

$$\begin{aligned} L_{\text{gauge}} &= -\frac{1}{4}W_{\mu\nu}^a W^{\mu\nu,a} - \frac{1}{4}B_{\mu\nu}B^{\mu\nu} \\ W_{\mu\nu}^a &= \partial_\nu A_\mu^a - \partial_\mu A_\nu^a - g \in_{\text{abc}} W_\mu^b W_\mu^c \\ B_{\mu\nu} &= \partial_\nu B_\mu - \partial_\mu B_\nu \end{aligned} \tag{13.35}$$

The $U(1)$ coupling constant is denoted by $\frac{g'}{2}$ (the factor $\frac{1}{2}$ being chosen for simplifying later expressions) and the $SU(2)$ coupling constant is denoted by g.

The Lagrangian is invariant under infinitesimal $SU(2)$ and $U(1)$ gauge transformations separately.

$$SU(2)_L\colon \psi_L \to \left[1 + i\vec{\alpha}(x)\cdot\frac{\vec{\tau}}{2}\right]\psi_L$$

$$\psi_R \to \psi_R$$

$$W_\mu^a \to W_\mu^a + \frac{1}{g}\partial_\mu\alpha_a(x) + \in_{abc}\alpha_b W_\mu^c$$

$$B_\mu \to B_\mu$$

$$U(1)_Y\colon \psi_L \to \left[1 + i\beta(x)\frac{Y_L}{2}\right]\psi_L$$

$$\psi_R \to \left[1 + i\beta(x)\frac{Y_R}{2}\right]\psi_R$$

$$W_\mu^a \to W_\mu^a$$

$$B_\mu \to B_\mu + \frac{1}{g'}\partial_\mu\beta$$

Local gauge invariance forbids mass terms for the gauge fields and chiral invariance forbids mass terms for the fermion fields. In a chiral gauge theory left-handed fermions and right-handed fermions appear in different multiplets and hence there can not be any coupling between the two and hence no term of the form $m\overline{\psi}\psi(\overline{\psi}\psi = \overline{\psi}_L\psi_R + \overline{\psi}_R\psi_L)$

Higgs Sector

In order to give masses to the weak gauge bosons the gauge symmetry is spontaneously broken to $U(1)_Q$ so that the photon remains massless even after SSB. To accomplish this one has to introduce a scalar field with 4 real degrees of freedom. In GSW model this is done by including a complex scalar doublet field (Higgs field) in the Lagrangian,

$$\phi = \begin{pmatrix} \phi^+ \\ \phi^0 \end{pmatrix} = \begin{pmatrix} \dfrac{\phi_1 + i\phi_2}{\sqrt{2}} \\ \dfrac{\phi_0 + i\phi_3}{\sqrt{2}} \end{pmatrix} \quad . \tag{13.38}$$

which transforms like an $SU(2)$ doublet and must therefore have $Y_\phi = +1$. Its self interaction provides the mechanism for the SSB giving masses to gauge bosons (and also to fermions). The gauge invariant Lagrangian for the scalar field is given by

$$\pounds_H = (D_\mu\phi)^+(D^\mu\phi) - V(\phi^+\phi) \tag{13.39}$$

where $D_\mu\phi = \left(\partial_\mu + igW_\mu^a\frac{\tau_a}{2} + ig'B_\mu\frac{Y_\phi}{2}\right)\phi$ (13.40)

and $V(\phi^+\phi) = \mu^2(\phi^+\phi) + \lambda(\phi^+\phi)^2,$ (13.41)

$$\left[\phi^+\phi = \frac{1}{2}(\phi_0^2 + \phi_1^2 + \phi_2^2 + \phi_3^2)\right]$$

We are also free to add interaction terms which involve Yukawa coupling of the scalars to the fermions. Such terms are of the form

$$\pounds_{\text{Yukawa}} = -G_f[\overline{\psi}_L\phi\psi_R + \overline{\psi}_R\phi^+\psi_L] \tag{13.42}$$

The full Lagrangian in the GSW model is then given by

$$\pounds = \pounds_{\text{fermion}} + \pounds_{\text{gauge}} + \pounds_{\text{scalar}} + \pounds_{\text{Yukawa}}$$

For $\mu^2 < 0$ the minimum of the effective potential occurs at

$$|\phi| = \frac{v}{\sqrt{2}}, \quad v = \sqrt{\frac{-\mu^2}{\lambda}} \tag{13.43}$$

Of the infinitely many choices we choose the configuration, $(\phi_0, \phi_1, \phi_2, \phi_3) = (v, 0, 0, 0)$. This choice corresponds to the following choice for the vacuum expectation value (VEV) of the scalar (Higgs) field.

$$\langle\phi\rangle_0 = \begin{pmatrix}\phi^+ \\ \phi^0\end{pmatrix}_0 = \begin{pmatrix}0 \\ \dfrac{v}{\sqrt{2}}\end{pmatrix} \tag{13.44}$$

One finds that $\tau_a\langle\phi\rangle_0 \neq 0$, $Y\langle\phi\rangle_0 \neq 0$. However

$$(\tau_3 + Y)\langle\phi\rangle_0 = 0 \tag{13.45}$$

This means $Q\langle\phi\rangle_0 = 0$ or $e^{i\alpha Q}\langle\phi\rangle_0 = \langle\phi\rangle_0$ (13.46)
i.e. the vacuum is left invariant by $U(1)_Q$ transformation. One says that the $SU(2)_L \times U(1)_Y$ gauge symmetry is spontaneously broken to $U(1)_Q$, the (abelian) gauge symmetry of electromagnetism.

Next we expand the Lagrangian about the minimum of V by writing

$$\phi(x) = e^{i\vec{\xi}(x)\cdot\vec{\tau}/2v}\begin{pmatrix}0 \\ \dfrac{v + H(x)}{\sqrt{2}}\end{pmatrix} = U^{-1}(\xi)\begin{pmatrix}0 \\ \dfrac{v + H(x)}{\sqrt{2}}\end{pmatrix} \tag{13.47}$$

where $U(\xi) = e^{i\vec{\xi}(x)\cdot\vec{\tau}/2v}$ and transforming to U gauge

$$\begin{aligned}
&\phi \to \phi' = U(\xi)\phi = \begin{pmatrix} 0 \\ \dfrac{v + H(x)}{\sqrt{2}} \end{pmatrix} \\
&W_\mu \to W'_\mu = UW_\mu U^{-1} + \frac{1}{ig}(\partial_\mu U)U^{-1}, \\
&B_\mu \to B_\mu \qquad\qquad\qquad\qquad\qquad\qquad\qquad\qquad (13.48) \\
&R \to R \\
&\psi_L \to \psi'_L = U(\xi)\psi_L
\end{aligned}$$

$$\text{Here,} \begin{cases} W_\mu = \displaystyle\sum_{\alpha=1}^{3} W_\mu^a \cdot \frac{\tau^a}{2} \\ W'_\mu = \displaystyle\sum_{\alpha=1}^{2} W_\mu^{\prime a} \cdot \frac{\tau^a}{2} \end{cases} \qquad (13.49)$$

Henceforth we drop the primes on ψ'_L and $W_\mu^{\prime a}$. The new fields are as good as the old ones since the theory is gauge invariant. The new fields are said to be fields in the U-gauge, while the old fields are said to be fields in the R-gauge. In the U gauge the physical particle spectrum is manifest, while in the R gauge the theory is mainfestly gauge invariant. The form of $\pounds_{\text{fermion}}$ and $\pounds_{\text{gauge}}$ remain unchanged, while the form of $\pounds_{\text{Higgs}}$ and $\pounds_{\text{Yukawa}}$ get changed. In terms of the new fields the Higgs Lagrangian becomes

$$\begin{aligned} L_{\text{Higgs}} = \frac{1}{2}(\partial_\mu H)(\partial^\mu H) - V(H) + \frac{1}{8}(H+v)^2\{g^2(W_\mu^1 W^{\mu 1} + W_\mu^2 W^{\mu,2}) \\ + (g' B_\mu - g W_\mu^3)(g' B^\mu - g W^{\mu,3})\} \end{aligned} \qquad (13.50)$$

where $V(H) = \dfrac{\mu^2}{2}(H+v)^2 + \dfrac{\lambda}{4}(H+v)^4$

$$= (\lambda v^2)H^2 + (\lambda v)H^3 + \frac{\lambda}{4}H^4 - \frac{\lambda v^4}{4}, \left[v = \sqrt{\frac{-\mu^2}{\lambda}}\right] \qquad (13.51)$$

Defining $W_\mu^\pm = \dfrac{1}{\sqrt{2}}(W_\mu^1 \pm iW_\mu^2)$ (13.52)

and the orthogonal combinations

$$Z_\mu = \frac{g' B_\mu - g W_\mu^3}{\sqrt{g^2 + g'^2}}, \; A_\mu = \frac{g B_\mu + g' W_\mu^3}{\sqrt{g^2 + g'^2}} \qquad (13.53)$$

the last term involving the Higgs field and the gauge fields in Eq. (13.50) can be written as

$$\mathcal{L}_{\text{Higgs}}^{\text{gauge}} = \frac{1}{8}(H+v)^2[2g^2 W_\mu^+ W^{\mu-} + (g^2 + g'^2)Z_\mu Z^\mu] \tag{13.54}$$

Apart from interaction terms involving the Higgs fields and weak gauge bosons fields, $H_{\text{Higgs}}^{\text{gauge}}$ contain terms quadratic in gauge fields alone which may be interpreted as mass terms for charged vector fields $W_\mu^\pm$ and the neutral vector boson field Z_μ, ie terms of the form

$$m_W^2 W_\mu^+ W^{\mu-} \quad \text{and} \quad m_Z^2 Z_\mu Z^\mu$$

$$\text{where } m_W^2 = \frac{1}{4} g^2 v^2 \quad \text{and} \quad m_Z^2 = \frac{1}{4}(g^2 + g'^2)v^2 \tag{13.55}$$

one can write,

$$m_Z^2 = \frac{m_W^2}{\cos^2\theta_W} \tag{13.56}$$

$$\text{where } \tan\theta_W = \frac{g'}{g} \tag{13.57}$$

θ_W is known as the Weinberg angle. In terms of θ_W one can write

$$\begin{pmatrix} A_\mu \\ Z_\mu \end{pmatrix} = \begin{pmatrix} \cos\theta_W & \sin\theta_W \\ +\sin\theta_W & -\cos\theta_W \end{pmatrix} \begin{pmatrix} B_\mu \\ W_\mu^3 \end{pmatrix} \tag{13.58}$$

and the inverse relation

$$\begin{pmatrix} B_\mu \\ W_\mu^3 \end{pmatrix} = \begin{pmatrix} \cos\theta_W & +\sin\theta_W \\ \sin\theta_W & -\cos\theta_W \end{pmatrix} \begin{pmatrix} A_\mu \\ Z_\mu \end{pmatrix} \tag{13.59}$$

A_μ and Z_μ are eigen states of the mass matrix involving B_μ and W_μ^3 with eigen values

$$m_A^2 = 0 \quad \text{and} \quad m_Z^2 = (g^2 + g'^2)v^2/4.$$

There is no mass term for the A_μ field which may therefore be identified with the photon field, the gauge field corresponding to the surviving $U(1)_Q$ symmetry.

Interaction of Fermions with the Gauge Fields

One can write $\mathcal{L}_{\text{fermion}}$ given by eq. (13.34) as

$$\mathcal{L}_{\text{fermion}} = \mathcal{L}_{\text{free}} + \mathcal{L}_{\text{int}} \tag{13.60}$$

where $\mathcal{L}_{\text{free}}$ is the Lagrangian for the free fermion fields, and is given by

$$\mathcal{L}_{\text{free}} = \overline{\psi}_L i\gamma^\mu \partial_\mu \psi_L + \overline{\psi}_R i\gamma^\mu \partial_\mu \phi_R \tag{13.61}$$

and $\mathcal{L}_{\text{int}}$ is the part of the $\mathcal{L}_{\text{fermion}}$ which describes the interactions between the fermion fields and the gauge fields,

$$\mathcal{L}_{\text{int}} = g\overline{\psi}_L \gamma^\mu \frac{\tau_a}{2} W_\mu^a \psi_L + g'\overline{\psi}_L \frac{Y_L}{2} B_\mu \psi_L + g'\overline{\psi}_R \gamma^\mu \frac{Y_R}{2} B_\mu \psi_R \tag{13.62}$$

In terms of the physical gauge fields defined by (13.52) and (13.53), eq. (13.62) becomes

$$L_{\text{int}} = \frac{g}{\sqrt{2}}(J_+^\mu W_\mu^- + J_-^\mu W_\mu^+) + \frac{gg'}{\sqrt{g^2+g'^2}} J^{\mu,em} A_\mu + \frac{1}{2}\sqrt{g^2+g'^2}\, J^{\mu,N} Z_\mu \tag{13.63}$$

$$\text{where } J_\pm^\mu = \overline{\psi}_L \gamma^\mu \tau_\pm \psi_L, \quad \tau_\pm = \frac{1}{2}(\tau_1 \pm i\tau_2) \tag{13.64}$$

$$\begin{aligned} J^{\mu,em} &= \overline{\psi}_L \gamma^\mu \frac{1}{2}(\tau_3 + Y_L)\psi_L + \overline{\psi}_R \gamma^\mu \frac{Y_R}{2} \psi_R \\ &= \overline{\psi}_L \gamma^\mu Q_L \psi_L + \overline{\psi}_R \gamma^\mu Q_R \psi_R \end{aligned} \tag{13.65}$$

$$Q_L = \frac{1}{2}(\tau_3 + Y_L), \quad Q_R = \frac{Y_R}{2}$$

and

$$\begin{aligned} \frac{1}{2} J^{\mu,N} &= -\overline{\psi}_L \gamma^\mu \frac{\tau_3}{2} \psi_L + \frac{g'^2}{g^2+g'^2} J^{\mu,em} \\ &= -J^{\mu,3} + \frac{g'^2}{g^2+g'^2} J^{\mu,em} \end{aligned} \tag{13.66}$$

In terms of the Weinberg angle θ_w defined by Eq. (13.57) we can write

$$\frac{gg'}{\sqrt{g^2+g'^2}} = g\sin\theta_w, \quad \sqrt{g^2+g'^2} = \frac{g}{\cos\theta_w} \quad \text{and} \quad \frac{g'^2}{g^2+g'^2} = \sin^2\theta_w \tag{13.67}$$

Using (13.67), Eqns, (13.63) and (13.66) can be written in simpler forms

$$\mathcal{L}_{\text{int}} = \frac{g}{\sqrt{2}}(J_+^\mu W_\mu^- + J_-^\mu W_\mu^+) + g\sin\theta_W J^{\mu,em} A_\mu + \frac{g}{2\cos\theta_W} J^{\mu,N} Z_\mu \tag{13.68}$$

and

$$\frac{1}{2} J^{\mu,N} = -J^{\mu,3} + \sin^2\theta_W J^{\mu,em} \tag{13.69}$$

Restricting only to the first generation leptons, we get

$$J_+^\mu = \overline{\psi}_L \gamma^\mu \tau_+ \psi_L = \overline{\nu}_{eL} \gamma^\mu e_L = \frac{1}{2}\overline{\nu}_e \gamma^\mu (1-\gamma_5) e = \frac{1}{2} J^{\mu,W} \tag{13.70}$$

where $J^{\mu,W} = \overline{\nu}_e \gamma^\mu (1-\gamma_5) e$ (13.71)

[$J^{\mu,W}$ is the charged leptonic current involved in weak decays (neutron and muon decays) as described by the V-A theory]

$$J^{\mu,em} = -\overline{e}_L \gamma^\mu e_L - \overline{e}_R \gamma^\mu e_R = -\overline{e}\gamma^\mu e \tag{13.72}$$

and
$$\begin{aligned} J^{\mu,N} &= -\overline{\nu}_{eL}\gamma^\mu \nu_{eL} + \overline{e}_L \gamma^\mu e_L - 2\sin^2\theta_W \overline{e}\gamma^\mu e \\ &= -\overline{\nu}_{eL}\gamma^\mu \nu_{e_L} + (1-2\sin\theta_W)\overline{e}_L\gamma^\mu e_L - 2\sin\theta_W \overline{e}_R \gamma^\mu e_R \\ &= -\frac{1}{2}\overline{\nu}_e \gamma^\mu (1-\gamma_5)\nu_e + \overline{e}\gamma^\mu (g_V^e - g_A^e \gamma 5) e \end{aligned} \tag{13.73}$$

$$\text{where } g_V^e = \frac{1}{2} - 2\sin^2\theta_W, \quad g_A^e = \frac{1}{2} \tag{13.74}$$

The interactions of the fermions with charged weak vector bosons are consistent with the low-energy phenomenology described by the V-A theory provided we identify

$$\frac{G_F}{\sqrt{2}} = \left(\frac{g}{2\sqrt{2}}\right)^2 \cdot \frac{1}{M_W^2} = \frac{g^2}{8M_W^2} = \frac{1}{2v^2} \tag{13.75}$$

where v is the v.e.v. of the Higgs field.

The interactions of the fermions with the neutral weak vector boson Z_μ is given by the last term in (13.68).

The interaction of the fermions with the massless vector boson A_μ is given by the third term in (13.68), we can identify A_μ with the electromagnetic field (photon) and the electromagnetic coupling parameter e with $g\sin\theta_W$

$$g\sin\theta_W = e \tag{13.76}$$

Using (13.57) we can write (13.76) in the form

$$\frac{1}{g^2} + \frac{1}{g'^2} = \frac{1}{e^2} \tag{13.77}$$

Eq. (13.77) implies that both g and g' are $\geq e$

From (13.75) and (13.76) one gets

$$m_W^2 = \frac{\pi\alpha}{\sqrt{2}G_F} \cdot \frac{1}{\sin^2\theta_W} = \frac{(37.4\ \text{Gev/c})^2}{\sin^2\theta_W} \tag{13.78}$$

(13.78) is consistent with the experimental values of θ_W and m_W. The weak neutral current corresponding to the neutrino is pure V-A. However, for other fermions it is an unequal mixture of V and A. The effective weak neutral current coupling at low energies is given by

$$G_N = \left(\frac{g}{2\cos\theta_W}\cdot\frac{1}{2}\right)^2\cdot\frac{1}{M_Z^2} = \frac{g^2}{16M_Z^2\cos^2\theta_W} = \frac{g^2}{16M_W^2} = \frac{G_F}{2\sqrt{2}} \tag{13.79}$$

Fermion Masses

In the GSW model fermion masses are generated by introducing a Yukawa type interaction between the fermion field and the scalar field followed by SSB. For leptons, this is simple since the neutrinos are assumed to be massless in the SM. Thus considering the first generation leptons, we have

$$\pounds_{\text{Yukawa}} = -G_e[\overline{\psi}_L\phi\psi_R + h.c], \tag{13.80}$$

In the U-gauge,

$$\phi = \begin{pmatrix} 0 \\ \dfrac{v+H}{\sqrt{2}} \end{pmatrix} \tag{13.81}$$

and (13.80) becomes

$$\begin{aligned}\pounds_{\text{Yukawa}} &= -\frac{G_e}{\sqrt{2}}[\overline{e}_L(v+H)e_R + h\cdot c] \\ &= -\frac{G_e}{\sqrt{2}}[v(\overline{e}_Le_R + \overline{e}_Re_L) + \overline{e}_LHe_R + \overline{e}_RHe_L] \\ &= -m_e(\overline{e}e) - \frac{G_e}{\sqrt{2}}\overline{e}He \end{aligned} \tag{13.82}$$

$$\text{where } m_e = \frac{G_e v}{\sqrt{2}} = \frac{G_e\sqrt{2}}{g}m_W \tag{13.83}$$

The first term in (13.82) may be interpreted as the mass term for the electron. Quark masses can also be generated by Yukawa coupling to scalars followed by SSB. However since both members of a left-handed quark doublet are to be massive and hence have right-handed components (as singlets), we need two separate Yukawa couplings, one for generating mass for the u type quark and the other for the d-type quark. Considering only the first quark generation, we have

$$\pounds_{\text{Yukawa}} = -G_d[\overline{\psi}_L\phi d_R + h\cdot c] + G_u[\overrightarrow{\phi}_L\phi_c u_R + h\cdot c] \tag{13.84}$$

where $\psi_L = \begin{pmatrix} u \\ d \end{pmatrix}_L$ (neglecting generation mixing)

$$\phi = \begin{pmatrix} \phi^+ \\ \phi^0 \end{pmatrix} \quad \phi_c = i\tau_2\phi^* = \begin{pmatrix} \overline{\phi}^0 \\ -\phi^- \end{pmatrix}$$

In the U-gauge,

$$\phi \to \phi' = \begin{pmatrix} 0 \\ \dfrac{v+H}{\sqrt{2}} \end{pmatrix}, \quad \phi_c = \begin{pmatrix} \dfrac{v+H}{\sqrt{2}} \\ 0 \end{pmatrix}$$

$$\pounds_{\text{Yukawa}} = -\frac{G_d v}{\sqrt{2}}\overline{d}d - \frac{G_u v}{\sqrt{2}}\overline{u}u - \frac{G_d}{\sqrt{2}}\overline{d}Hd - \frac{G_u}{\sqrt{2}}\overline{u}Hu \tag{13.85}$$

The first 2 terms in (13.85) may be interpreted as the mass terms for the d-quark and u-quark respectively with

$$m_d = \frac{G_d v}{\sqrt{2}}, \quad m_u = \frac{G_u v}{\sqrt{2}}$$

The Higgs Boson

One of the consequences of the symmetry breaking mechanism (Higgs mechanism) used in the GSW model to generate masses for the 3 weak gauge bosons with the photon remaining massless is that one real scalar degree of freedom is left over which should occur as a physical particle. This neutral scalar boson is known as the Higgs boson. Its mass depends on both λ and $v, (m_H^2 = 2\lambda v^2)$. The gauge boson mass determines v but λ is undetermined without any experimental information on the Higgs boson itself. Therefore the mass of the Higgs boson is undetermined in the GSW model. The Higgs boson has 3 types of coupling—self-couplings (H^3, H^4), coupling with the weak gauge bosons (HVV and HHVV) and coupling with the fermions $(\overline{f}fH)$. The interaction terms determine the production mechanisms and decay modes of the Higgs boson. Of these, the HW^+W^-, HZ^0Z^0 and $H\overline{f}f$ couplings are the most important for phenomenology. No Higgs boson has been seen at LEP II. This implies $m_H > m_Z$. The dominant process in LEP II ($\sqrt{s} \leq 200$ GeV) is $e^-e^+ \to Z^* \to Z + H$. LEP data suggests $m_H > 114.4$ GeV (C.L. 95%). From various considerations one expects $m_H \leq 1$ TeV.

13.7 Experimental Tests of the GSW Model

There are three important predictions of the GSW model that can be experimentally tested. These are—(1) neutral weak currents, (2) massive vector bosons that mediate the weak interactions and (3) the existence of a neutral scalar boson (Higgs boson). The first two have been experimentally observed and the direct experimental evidence of the third one is eagerly awaited. There are indirect evidences of the existence of the Higgs boson from the precision tests of the SM with some reasonable bounds on its mass. Hopefully it will be discovered when the next generation accelerators, particularly the LHC, become operational.

Neutral Weak Currents

In addition to charged weak currents which are responsible for weak decays of particles ($\mu^- \to e^-\overline{\nu}_e\nu_\mu, n \to pe^-\overline{\nu}_e$ etc.) the GSW model predicts the necessary existence of neutral weak currents in which there is no transfer of electric charge. Even when generation mixing is allowed to account for flavor-changing charge currents, GIM mechanism ensures there is no flavor-changing neutral currents (FCNC). The neutral weak current events were first seen experimentally in neutrino-induced high energy processes ("muon-less" events) by the Gargamelle group in 1973 using secondary ν_μ and $\overline{\nu}_\mu$ beams from the CERN SPS.

Charged weak current process: $\nu_\mu + e^- \to \mu^- + \nu_e,\ \nu_\mu + N \to \mu^- + X$

Neutral weak current process: $\nu_\mu + e^- \to \nu_\mu + e^-,\ \nu_\mu + N \to \nu_\mu + X$

Form of the weak neutral current: While the charged current is of the form $f\gamma^\mu(1-\gamma_5)f'$, where f and f' are members of a weak i-spin doublet, the weak neutral current is of the form

$$J^{\mu,N} = \overline{f}\gamma^\mu\,(g_V^f - g_A^f\gamma_5)f \tag{13.87}$$

where $g_V^f = -I_3^f + 2Q_f x_W,\ g_A^f = -I_3^f$ (13.88)

Here I_3 is the third component of the weak i-spin, Q_f is the electric charge (in unit of e) and $x_W = \sin^2\theta_W$, θ_W being the weak mixing angle (Weinberg angle).

$$f = \nu_l \ (l = e, \mu, \tau)\text{: } g_V^f = g_A^f = -\frac{1}{2}$$

$$f = l\text{: } g_V^f = \frac{1}{2} - 2x_W, \ g_A^f = +\frac{1}{2}$$

$$f = u, c, t\text{: } g_V^f = -\frac{1}{2} + \frac{4}{3}x_W, \ g_A^f = -\frac{1}{2}$$

$$f = d, s, b\text{: } g_V^f = \frac{1}{2} - \frac{2}{3}x_W, \ g_A^f = \frac{1}{2} \tag{13.89}$$

Neutrino-electron Elastic Scattering ($\nu_\mu e^- \to \nu_\mu e^-$)

It is a basic process free from the complications of strong interaction effects and can be used to determine the weak mixing angle θ_W. Moreover, while in the V-A theory it is a higher order process compared to the corresponding charged current process ($\nu_\mu + e^- \to \mu^- + \nu_e$), in the GSW model it is of the same order as the charged current process and occurs at the tree level itself. We have

$$A = +i\left(\frac{g}{2\cos\theta_W}\right)^2 \frac{1}{2}\overline{u}_\nu(k')\gamma^\mu(1-\gamma_5)u_\nu^{(k)}\frac{1}{(k-k')^2 - m_z^2}$$

$$\times\, \overline{u}_e(p')\gamma_\mu(g_\nu^e - g_A^e\gamma_5)u_e(p)$$

$$\approx -i\left(\frac{g}{2\cos\theta_W}\right)^2 \frac{1}{2m_z^2}[\overline{u}_\nu(k')\gamma^\mu(1-\gamma_5)u_\nu(k)][\overline{u}_e(p')\gamma_\mu(g_V^e - g_A^e\gamma_5)u_e(p)]$$

$$= -\frac{iG_F}{\sqrt{2}}[\overline{u}_\nu(k')\gamma^\mu(1-\gamma_5)u_\nu(k)][\overline{u}_e(p')\gamma_\mu(g_V^e - g_A^e\gamma_5)u_e(p)]$$

$$|\overline{A}|^2 = \frac{1}{2}\sum_{spins}|A|^2 = 4G_F^2\left[(g_V^e + g_A^e)^2 s^2 + (g_V^2 - g_A^e)^2 u^2\right]$$

$$= 4G_F^2[(1-2x_W)^2 s^2 + 4x_W^2 u^2]$$

$$\left(\frac{d\sigma}{d\Omega}\right)_{\text{CM}}^{el} = \frac{1}{64\pi^2 s}|\overline{A}|^2$$

$$s = (k+p)^2$$

$$t = (k-k') \approx -\frac{s}{2}(1-\cos\theta)$$

$$u = (k'-p)^2 \approx -\frac{s}{2}(1+\cos\theta)\} \Rightarrow d\Omega_{\text{CM}} \approx 4\pi dy$$

$$y = \frac{E - E'}{E} \approx \frac{s+u}{s}, 0 \le y \le 1$$

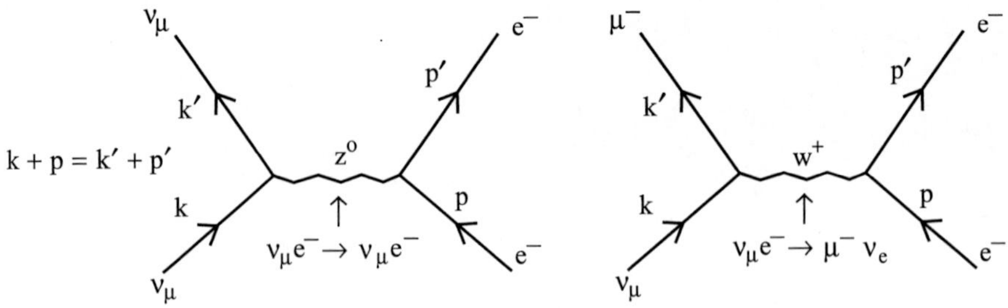

Fig. 13.2

In the CM frame, at high energies, we have

$$k = \frac{\sqrt{s}}{2}(1, \hat{z}), \; p = \frac{\sqrt{s}}{2}(1, -\hat{z})$$

$$k' = \frac{\sqrt{s}}{2}(1, \hat{n}), \; p' = \frac{\sqrt{s}}{2}(1, -\hat{n})$$

$$\hat{z} \cdot \hat{n} = \cos\theta$$

Thus we get

$$\left(\frac{d\sigma}{dy}\right)_{\nu_\mu e^- \to \nu_\mu e^-} = \frac{G_F^2 s}{4\pi}[(1 - 2x_W)^2 + 4x_W^2(1-y)^2] \tag{13.90}$$

Integrating we get

$$\sigma_{\text{tot}}(\nu_\mu e^- \to \nu_\mu e^-) = \frac{G_F^2 s}{4\pi}[(1 - 2x_W)^2 + \frac{4}{3}x_W^2] \tag{13.91}$$

For the charged current process $\nu_\mu + e^- \to \mu^- + \nu_e$, we get

$$\left(\frac{d\sigma}{dy}\right)_{\nu_\mu e^- \to \mu^- \nu_e} = \frac{G_F^2 s}{\pi} \tag{13.92}$$

This gives

$$R = \frac{\sigma_{NC}(s)}{\sigma_{CC}(s)} = \frac{\sigma_{\text{tot}}(\nu_\mu e^- \to \nu_\mu e^-)}{\sigma_{\text{tot}}(\nu_\mu e^- \to \mu^- \nu_e)} \tag{13.92}$$

$$= \frac{1}{4}\left(1 - 4x_W + \frac{16}{3}x_W^2\right) \tag{13.93}$$

From the observed value of R one can compute x_W. which agrees with the values obtained by other means. Present accepted value of x_W is 0.23.

Weak Gauge Bosons

As seen from eqns. (13.75), (13.76), (13.78) and (13.57), with v, g and $\sin\theta_W$ determined from G_F, $\alpha = e^2/4\pi$ and neutral current data, the W

and Z boson masses are fixed in the GSW model. The weak vector bosons were first discovered in $\overline{p}p$ collider experiments. at CERN in 1983 and their masses and decay widths measured. These agreed with the values predicted by the GSW model. Later on, in the 1990's the Z-boson and W-boson properties were studied with much better statistics at the e^-e^+ colliders LEP I and LEP II at CERN. The processes studied were:

$\overline{p}p$ collider: $\overline{p}p \to W^\pm X, W^- \to e^-\overline{\nu}_e, W^+ \to e^+\nu_e$

$$\overline{p}p \to Z^0 X, Z^0 \to e^-e^+$$

LEP I: $e^-e^+ \to Z^0$

LEP II: $e^-e^+ \to W^+W^-$ (13.94)

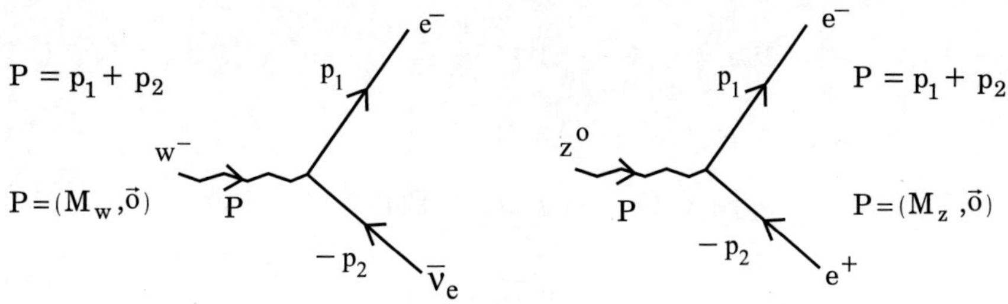

Fig. 13.3

***W*-decay:** We first consider the decay $W^- \to e^-\overline{\nu}_e$

$$P = p_1 + p_2$$

In the rest frame of $W, P = (M_W, \vec{0})$

The decay amplitude is

$$A = -\frac{ig}{2\sqrt{2}}\overline{u}_e(p_1)\gamma^\mu(1-\gamma_5)v_{\nu_e}(p_2) \in_\mu \tag{13.95}$$

$$\Gamma = \frac{1}{2M_W}\int |\overline{A}|^2 \cdot (2\pi)^4\delta_4(p_1+p_2-P)\frac{d^3p_1}{(2\pi)^3(2E_1)} \cdot \frac{d^3p_2}{(2\pi)^3(2E_2)} \tag{13.96}$$

$$|\overline{A}|^2 = \frac{1}{3}\sum_{spins}|A|^2 \tag{13.97}$$

Substituting (13.95) in (13.97) we get

$$|\overline{A}|^2 = \frac{1}{3}\cdot\left(\frac{g}{2\sqrt{2}}\right)^2 \sum_{\text{spins}}[\overline{u}_e(p_1)\gamma^\mu(1-\gamma_5)v_{\nu_e}(p_2) \in_\mu]$$

$$\cdot\,[\overline{u}_e(p_1)\gamma^\nu(1-\gamma_5)v_{\nu_e}(p_2) \in_\nu]^+$$

$$= \frac{1}{3}\frac{G_F M_W^2}{\sqrt{2}} \sum_{\text{spins}} \in_\mu \in_\nu^* \sum_{\text{spins}} [\overline{u}_e(p_1)\gamma^\mu 1 - \gamma_5)v_{\nu_e}(p_2)\overline{v}_{\nu_e}(p_2)\gamma^\nu (1-\gamma_5)u_e(p_1)]$$

After carrying out the spin sums we get

$$|\overline{A}|^2 = \frac{1}{3}\frac{G_F M_W^2}{\sqrt{2}} \left(-g_{\mu\nu} + \frac{P_\mu P\nu}{M_W^2}\right) \cdot 8[p_1^\mu p_2^\nu - g^{\mu\nu}(p_1 \cdot p_2) + p_2^\mu p_1^\nu + i \in^{\mu\nu\alpha\beta} p_{1\alpha}p_{2\beta}]$$

$$= \frac{1}{3}\frac{G_F M_W^2}{\sqrt{2}} 8M_W^2 = \frac{8}{3\sqrt{2}} G_F M_W^4 \tag{13.98}$$

After carrying out the phase space integration in the rest frame of W boson we get from (13.96) and (13.98)

$$\Gamma_W^e = \Gamma(W^- \to e^-\overline{\nu}_e) = \frac{1}{2M_W} \cdot |\overline{A}|^2 \cdot \frac{1}{8\pi} = \frac{G_F M_W^3}{6\pi\sqrt{2}} \tag{13.99}$$

$$\Gamma_W^e = 0.228 \text{ GeV} \quad \text{for} \quad M_W = 80.4 \text{ GeV}$$

$$\Gamma_W^l = \Gamma(W^- \to e^-\overline{\nu}_e) + \Gamma(W^- \to \mu^-\overline{\nu}_\mu) + \Gamma(W^- \to \tau^-\overline{\nu}_\tau) = 3\Gamma_W^e \tag{13.100}$$

$$\Gamma_W^h = N_c \sum_{\substack{q'=d,s,b \\ q=u,c}} |V_{qq'}|^2 \Gamma(W^- \to q'\overline{q}) = 6\Gamma_W^e \left[1 + \frac{\alpha_s(M_W)}{\pi}\right] \tag{13.101}$$

The second term on rhs. of eq. (13.101) gives the QCD correction. We get

$$\Gamma_W^{\text{tot}} = \Gamma_W^l + \Gamma_W^h = 9\Gamma_W^e \left[1 + \frac{2}{3}\frac{\alpha_s(M_W)}{\pi}\right] \approx 2.10 \text{ GeV}$$

$$B_e = \frac{\Gamma(W^- \to e^-\overline{\nu}_e)}{\Gamma(W^- \to \text{all})} = 10.86\% \tag{13.102}$$

the predicted values agree reasonably with experimental values:

$$\Gamma_W^{\text{tot}} = (2.124 \pm 0.041) \text{ GeV}$$

$$B_e = (10.72 \pm 0.16)\%$$

$$M_W = (80.425 \pm 0.038) \text{ GeV} \tag{13.103}$$

***Z* decay:** The Z-boson was first detected in $\overline{p}p$ collider experiment at CERN in 1983 through its leptonic decay, $Z^0 \to e^-e^+$. In the GSW model the amplitude for this decay (in lowest order) is given by

$$A = -\frac{ig}{2\cos\theta_W}\overline{u}_e(p_1)\gamma^\mu(g_\nu^e - g_A^e\gamma_5)v_e(p_2) \in_\mu$$

$$|\overline{A}|^2 = \frac{1}{3}\sum_{\text{spins}} |A|^2 \tag{13.104}$$

$$= \frac{1}{3}\cdot\left(\frac{g}{2\cos\theta_W}\right)^2 \sum_{\text{spins}} \in_\mu \in_\nu^* \sum_{\text{spins}} [\overline{u}_e(p_1)\gamma^\mu(g_\nu^e - g_A^e\gamma_5)v_e(p_2)]$$
$$[\overline{v}_e(p_2)\gamma^\nu(g_\nu^e - g_A^e\gamma_5)u_e(p_1)]$$

$$= \frac{1}{3}\cdot\frac{2G_F M_z^2}{\sqrt{2}}\left(-g_{\mu\nu} + \frac{P_\mu P_\nu}{M_z^2}\right)$$
$$\times 4[\{(g_\nu^e)^2 + (g_A^e)^2\}\{p_1^\mu p_2^\nu - g^{\mu\nu}(p_1 p_2) + p_2^\mu p_1^\nu\}$$
$$+ i(2g_\nu^e g_A^e) \in^{\mu\nu\alpha\beta} p_{1\alpha} p_{2\beta}]$$

$$= \frac{1}{3}\cdot\frac{2G_F M_z^2}{\sqrt{2}} 4\left[(g_\nu^e)^2 + (g_A^e)^2\right][2p_1\cdot p_2]$$

$$= \frac{1}{3}\cdot\frac{2G_F M_z^2}{\sqrt{2}} 4\left[(g_\nu^e)^2 + (g_A^e)^2\right]\cdot M_z^2 \tag{13.105}$$

$$\Gamma_z^e = \Gamma(Z^0 \to e^- e^+) = \frac{1}{2M_z}|\overline{A}|^2\cdot\frac{1}{8\pi} = \frac{G_F M_z^3}{6\pi\sqrt{2}}[(g_\nu^e)^2 + (g_A^e)^2]$$

$$= 2\left[(g_\nu^e)^2 + (g_A^e)^2\right]\cdot\Gamma_z^0$$

$$= [1 - 4x_W + 8x_W^2]\Gamma_z^0 \tag{13.106}$$

where $$\Gamma_z^0 = \frac{G_F M_z^3}{12\pi\sqrt{2}} = \Gamma(Z^0 \to \nu_e\overline{\nu}_e) \tag{13.107}$$

$$\Gamma_z^q = \Gamma(Z^0 \to q\overline{q}) = 2N_c\left[(g_\nu^q)^2 + (g_A^q)^2\right]\Gamma_z^0\left[1 + \frac{\alpha_s(M_z)}{\pi}\right] \tag{13.108}$$

$(q = u, d, c, s, b)$

$$\Gamma_z^{\text{tot}} = 3[\Gamma(Z^0 \to e^- e^+) + \Gamma(Z^0 \to \nu_e\overline{\nu}_e)]$$
$$+ [2\Gamma(Z^0 \to u\overline{u}) + 3\Gamma(Z^0 \to d\overline{d})]$$

$$= \Gamma_z^0\left[21 - 40x_W + \frac{160}{3}x_W^2\right]\left[1 + \lambda\frac{\alpha_s(M_z)}{\pi}\right]$$

where $$\lambda = \frac{15 - 28x_W + \frac{88}{3}x_W^2}{21 - 40x_W + \frac{160}{3}x_W^2} \tag{13.109}$$

with $M_z = 91.18$ GeV and $\sin^2\theta_W = 0.23$ we get

$$\Gamma(Z^0 \to \nu_e\overline{\nu}_e) = \Gamma_z^0 = 0.166 \text{ GeV} = 166 \text{ MeV}$$

$$\Gamma(Z^0 \to e^- e^+) = 0.0835 \text{ GeV} = 83.5 \text{ MeV}$$

$$\Gamma_z^{\text{tot}} = 2.50 \text{ GeV} \tag{13.110}$$

These agree well with the experimental values

$$\Gamma_z^{\text{tot}} = (2.4952 \pm 0.0023)\ \text{GeV}$$
$$M_z = (91.1876 \pm 0.0021)\ \text{GeV}$$
$$\Gamma(z^0 \to e^- e^+) = (83.984 \pm 0.086)\ \text{MeV}$$
$$\Gamma(\text{invisible}) = (499.0 \pm 1.5)\ \text{MeV}$$
$$N_\nu = \text{No. of massless neutrinos} = 3.0 \pm 0.2 \tag{13.111}$$

Problems

1. (a) Explain why $\nu_\mu + e^- \to \nu_\mu + e^-$ is a better test of weak neutral current than $\nu_e + e^- \to \nu_e + e^-$

   (b) Obtain the total cross section for the process $\nu_e + e^- \to \nu_e + e^-$ in the GSW model and also that for the process $\nu_\mu + e^- \to \nu_\mu + e^-$ and compare the two results.

2. (a) Show that if one includes the contributions of both γ and Z^0, then the GSW model predicts that in the CM frame

$$\frac{d\sigma}{d\pi}(e^- e^+ \to \mu^- \mu^+) = \frac{\alpha^2}{4s}[A_0(1 + \cos^2\theta) + A_1 \cos\theta]$$

   where $A_0 = 1 + 2\text{Re}(r)c_V^2 + |r|^2(c_V^2 + c_A^2)^2$

   and $\quad A_1 = 4\text{Re}(r)c_A^2 + 8|r|^2 c_V^2 c_A^2$

   with $\quad r = \dfrac{\sqrt{2}G_F M_z^2}{(s - M_z^2) + iM_z\Gamma_z}\left(\dfrac{s}{4\pi\alpha}\right)$

   assuming $\sqrt{s} \sim M_z$ and the weak neutral current of the form $\overline{f}\gamma^\mu(c_V - c_A\gamma^5)f$

   (b) Calculate the total cross section $\sigma_{\text{tot}}(e^- e^+ \to \mu^- \mu^+)$ and the forward-backward asymmetry

$$A_{FB} = \frac{F - B}{F + B}$$

   where $F = \displaystyle\int_{\theta=0}^{\pi/2} \frac{d\sigma}{d\Omega} d\Omega,\ B = \int_{\pi/2}^{\pi} \frac{d\sigma}{d\Omega} d\Omega$

3. (a) Give the Feynman rules for the interactions of the Higgs boson in the standard model (SM).

(b) Calculate the decay width $\Gamma(H \rightarrow f\overline{f})$ for the decay of the Higgs boson into a fermion-antifermion pair (quark/lepton) with $2m_f < m_H$ in the SM. What is the decay width of a 150 GeV Higgs boson decaying to a $\tau^-\tau^+$ pair.

4. Show that the cross section for the Higgs boson production process $e^- + e^+ \rightarrow Z^0 + H$ for $\sqrt{s} > M_Z + M_H$ (for example, at LEP II) is given by (in the SM)

$$\sigma(e^- + e^+ \rightarrow Z^0 + H) = \frac{(G_F M_z^2)^2}{3\pi}(g_v^{e^2} + g_A^e)^2 \cdot \frac{|\vec{p}|}{\sqrt{s}} \cdot \frac{3M_z^2 + |\vec{p}|^2}{(s - M_Z^2)^2 + M_Z^2\Gamma_Z^2}$$

where $|\vec{p}|$ = CM momentum of the final state and the Higgs boson mass is given by the relation $M_H^2 = S + M_Z^2 - 2\sqrt{s}E_Z$.

Chapter 14

Beyond The Standard Model (BSM) Theories & New Physics

14.1 Imperfections and Inadequacies of The Standard Model

The standard model (SM) of particle physics is one of the best tested and most well-established theories in physics. It has been successful in accounting for almost all the known results of experiments in particle physics that have been conducted so far upto accelerator energies of a few hundred GeV. In the early days, the agreement between the SM and experiment, although persuasive was often crude. But the more recent precision tests of the SM have attained accuracies at one part per-thousand level. In spite of this remarkable success, the SM is generally believed to be neither a complete nor a perfect theory. There are several reasons, both theoretical and observational behind this belief. According to many the SM is at best an effective theory, being the low energy version of a more exact theory valid at all energies. The physical cut-off scale upto which the SM is valid and beyond which new physics is expected to appear depends on the choice of the more exact theory.

We now briefly discuss the issues that have provided a strong motivation for considering theories that involve physics beyond the standard model (BSM). These may be classified into 3 broad categories-aesthetic questions, theoretical problems and observational shortcomings.

Aesthetic questions: The standard model is not an elegant theory since it contains a large number of arbitrary parameters, 19 in all, that are to be fixed by experimental data. These are—3 coupling constants, 2 parameters of the effective scalar field potential, 1 QCD scale parameter, 9 fermion masses (6 quarks and 3 charged leptons) and 4 quark mixing parameters (3 angles and 1 phase). Secondly, the electroweak sector of SM is not really unified, strictly speaking, because there are still 2 independent coupling constants, one for $SU(2)$ and another for the $U(1)$ part of the EW gauge group $SU(2)_L \times U(1)_Y$. Thirdly, the fermion masses are not only arbitrary but also vary over a wide range. The ratio of the top quark mass and the electron mass is 3×10^5. More over,there is no explanation for the existence of three fermion generations and their mixing. Fourthly, Gravitation, which is one of the four fundamental interactions of nature and is the dominant one in the macroscopic world, is completely left out of consideration in the SM. Moreover, a quantum theory of gravity cannot be worked out in the standard framework of field theory without leading to serious divergence difficulties. Finally, the choice of the EW symmetry-breaking mechanism in the SM is quite arbitrary. The Higgs boson, one of the major consequences of SSB mechanism in the SM, has not yet been seen in experiments and the nature of the symmetry-breaking mechanism is, therefore, still an open question.

Theoretical problems: The major theoretical problem faced by the SM is concerned with the mass of the Higgs boson. It goes by various names—the hierarchy problem, the fine-tuning problem and the naturalness problem. The presence of fundamental scalar fields in a quantum field theory (QFT) is always problematical. In the SM, the weak gauge boson masses are protected from getting large quantum loop corrections by gauge symmetry. Similarly the fermions are protected from being very massive by chiral symmetry. But there is no such symmetry to protect the Higgs boson mass from becoming very large due to quantum corrections. One-loop quantum correction to the scalar field mass (self-energy) is known to be (quadratically) divergent, being proportional $\wedge^2$, where $\wedge$ is the cut-off energy upto which the SM is valid and beyond which new physics is expected to appear. This could be the Planck mass M_P, (10^{19} GeV), the energy scale at which gravitational interaction becomes as strong as the electromagnetic interaction. Alternatively, it could be the grand unification mass M_X (10^{16} GeV), the energy scale at which the strong interaction becomes as weak as the electromagnetic interaction. However, the Higgs boson mass in the SM is expected to be below or near the EW scale (300 GeV), the energy at which the weak interaction strength equals that of the electromagnetic interaction. One has, therefore, to perform delicate cancellations by manipulating (fine tuning) the parameters of the scalar sector in the SM in order to keep the physical Higgs boson mass in the desired (natural) range. One often says that the relative lightness of the Higgs boson seems to arise from a bizarre conspiracy of different effects.

Another theoretical problem the proponents of the SM have to find a solution to is the "Strong CP problem". This is the problem of understanding why CP violation is so small in the strong interaction sector. Since CP is not a symmetry of the SM, there is nothing to prevent the appearance of a CP-violating term in the QCD Lagrangian, i.e., a term of the form

$$G_\theta = \theta F\tilde{F} \tag{14.1}$$

where F is the field tensor and $\tilde{F}$ is the corresponding pseudotensor,

$$\tilde{F}_{\mu\nu} = \frac{1}{2} \in_{\mu\nu\rho\sigma} F^{\rho\sigma} \tag{14.2}$$

We usually ignore such terms because in classical field theory they are inconsequential. They are total derivatives and hence do not modify the equation of motion. In a $U(1)$ theory, for example,

$$F\tilde{F} = 2 \in^{\mu\nu\rho\sigma} \partial_\mu A_\nu \partial_\rho A_\sigma = 2\partial_\mu(\in^{\mu\nu\rho\sigma} A_\nu \partial_\rho A_\sigma) \tag{14.3}$$

This has a non-Abelian generalization. However, despite being a total divergence, these terms have physical effects at the quantum level. In QCD, they turn out to be highly constrained. From the limits on the neutron electric dipole ($E1$) moment, one can show that $\theta < 10^{-9}$. Perhaps there is some explanation of the smallness of the θ-parameter, and perhaps this is a clue to new physics beyond the SM. The most compelling explanation of the smallness of θ so far, involves a hypothetical particle, called the "axion", first proposed by Peceei and Quinn.

The third theoretical problem of the SM is known as "the Flavor problem". This corresponds to understanding the origin of fermion masses, the quark mixing angles and the CP violation parameter, In the SM, this is accomplished by introducing Yukawa couplings involving the fermion fields and the scalar field and then making use of the Higgs mechanism. However, this is quite arbitary and involves a large number of parameters.

Observational shortcomings: The so called observational shortcomings of the SM do not concern the strict particle physics domain (high-energy accelerator-based experiments) but to observations in cosmology and astrophysics. In particular, there are some conflicts of the SM of particle physics with the standard model of cosmology (The Hot Big Bang Model) and the standard model of the sun, provided one can apply particle physics concepts to cosmological and astronomical problems. Incidentally, this emerging area of physics is known as Astroparticle Physics.

There are at present three compelling and firmly established observational facts that the SM fails to account for-neutrino oscillations, the existence of dark matter and dark energy in the universe.

The observed electron neutrino (ν_e) flux from the sun in all underground solar neutrino experiments has been found to be lower by a factor of about 2–3 compared to that predicted theoretically by the otherwise successful standard solar model (SSM)). This is known as the solar neutrino problem. A similar deficit in the more recently observed atmospheric muon-neutrino flux ($\nu_\mu, \overline{\nu}_\mu$) compared to that expected theoretically has been noticed. This is known as the atmospheric neutrino problem. Both the solar neutrino problem and the atmospheric neutrino problem can be best explained if one assumes that there are oscillations between two or more neutrino flavors so that one is converted into another by the time the neutrinos travel from the point of production to the point of observation. This is possible if the flavor eigenstates and the mass eigenstates of the neutrinos are not the same so that a given flavor eigenstate is a linear combination of mass eigenstates (just like quark generation mixing). As a result of this mixing there will be oscillations between different neutrino flavors. It turns out that the frequencies of oscillations are functions of mixing angles and mass differences. This means that at least one or some of the neutrinos should have non-zero mass. However, in the SM one has only left-handed neutrinos and right-handed antineutrinos. As a result, there cannot be any neutrino mass term. Thus one can say that the observation of a massive neutrino or its expectation is a definite signal for new physics beyond the SM.

By now there exists an impressive evidence that not only most of the matter in the universe is "dark" (non-luminous), i.e., it does not emit any radiation, but what is more crucial is that such dark matter is non-baryonic in nature. The SM does not provide any viable non-baryonic dark matter candidate. (It has been proved that massive neutrinos cannot account for such large amount of non-baryonic, non-luminous matter).

More recently, there is a persuasive evidence that the energy density of the universe is largely in some unfamiliar mysterious form. It could be the energy of the vacuum, a "cosmological constant" (first introduced by Einstein to ensure that the universe is static, i.e. non-expanding). If, so, it is even more puzzling than the hierarchy problem. In QFT the vacuum energy is "quartically" divergent. Thus the cosmological constant, if identified with the vacuum energy is proportional to $\wedge^4$, where $\wedge$ is again some cutoff upto which the SM is valid. If $\wedge = M_X \sim 10^{14}$ GeV, then the vacuum energy $\sim 10^{56}$ GeV^4. The measured value of the dark-energy density, by contrast is 10^{-47} GeV^4. This wide discrepancy is probably one of the most troubling problem facing physicists to-day.

Given that we have a strong evidence that the present universe is matter-antimatter asymmetric (i.e. no sizeable amount of primordial antimatter has survived), it is appealing to have a dynamical mechanism which gives rise to the observed baryon-antibaryon asymmetry starting from a symmetric situation. One signature of this asymmetry is the observed value of the baryon-to-photon ratio n_B/n_γ. The number density of photons n_γ as measured from the observed microwave background radiation (2.7 K) is ~ 400 cm^{-3}. The observed baryon density in the universe to-day is $\sim 10^{-31}$ gm cm^{-3}, yielding a baryon number density $\sim 10^{-8}$ cm^{-3}. If baryon number is conserved this implies that $n_B/n_\gamma \geq 10^{-10}$. On the other hand if we start from a Big Bang which is symmetric with respect to matter and antimatter and require baryon number conservation then we expect that at present $n_B/n_\gamma \leq 10^{-19}$.

Attempts to explain this asymmetry have so far been unsuccessful. Sakharov in 1967 first stated precisely the conditions under which the laws of physics could lead to a prediction of this asymmetry. These are baryon number violation, CP violation and departure from thermal equilibrium during some period of time in the history of the universe. It turns out that an efficient mechanism to generate the observed baryon-antibaryon asymmetry (baryogenesis) requires the introduction of new particles and new interactions beyond the SM. Grand unifications theories (GUTS) offer a possible solution, at least qualitatively, to the matter-antimatter asymmetry problem but so far they have failed to attain quantitative agreement.

Finally, we briefly discuss what is known as "Inflation" hypothesis in cosmology. The standard model of cosmology (The Big Bang Model) has been highly successful in accounting for some of the most striking characteristics of the universe—the expansion of the universe (Hubble law) the cosmic microwave background radiation (CMBR) and the primordial light element abundance. It does however suffer from a number of serious general difficulties the solution of which demands an important addition or modification to the basic model. The various problems are conventionally known as the horizon problem, the flatness problem, the cosmological constant problem, the monopole problem and the small scale inhomogeneity problem. The nature of these problems suggests that there must have been a period in the early history of the universe when the rather smooth evolution of the universe as postulated in the standard model was interrupted by a period of much more rapid and dramatic change. This idea is the basis of the "Inflation" hypothesis first proposed by Guth in 1980 and subsequently developed by a number of authors. Inflation requires a period in the early history of the universe when the expansion was exponential. It turns out that inflation can offer possible solutions to all the problems mentioned above except the cosmological constant problem. The SM with its single Higgs doublet fails to generate such an inflationary stage. Again some extensions of the SM, where in particular new scalar fields are introduced, are able to produce a temporary inflation of the early universe.

The Standard Model as an Effective Low-energy Theory

From the above discussions, it is apparent that there are sufficient theoretical and observational reasons to go beyond the standard model (SM). On the other hand, the well-established and elegant success of the SM in accounting for all the known particle physics phenomenology upto energies of the order of the electroweak scale ($\sim$ 300 GeV) requires that the SM has to be recovered at least as the low-energy limit of such more complete theories. Indeed, it may well be the case that one has a tower of underlying theories which show up at different energy scales.

14.2 Beyond the Standard Model (BSM)

There have been several approaches to construct theories that go beyond the standard model (SM) of particle physics. Most of them have the SM as an effective low-energy version. BSM theories involve either higher symmetrics or extra space dimensions or both. They can be generally classified into five categories. These are Grand unified theories(GUTS), Models with low-energy supersymmetry (SUSY), Models with dynamical symmetry breaking (DSB), String theories and Models with large extra dimensions or warped extra dimensions (EDM).

(a) Grand Unified Theories (GUTS)

The major motivation for grand unified theories (GUTS) is to overcome the problems of the large number of parameters and the imperfect unification of the fundamental interactions in the SM. In GUTS, the strong, electromagnetic and weak interactions become part of a single interaction at a very high energy scale. This is accomplished by the use of a gauge group which is simple (not semisimple as in the SM) and has rank ≥ 4 (the rank of the SM gauge group). $SU(5)$, $SO(10)$ and E_6 are some of the possible choices for the gauge group. GUTs predict the strength of the strong interaction in terms of the weak and electromagnetic interaction strengths and also the tau-lepton mass in terms of the b-quark mass. GUTS have made two additional predictions—(1) the proton has a finite lifetime and (2) magnetic monopoles exist. The simplest GUTS, however, have failed experimental tests. They predict a proton lifetime $\leq 10^{28}$ yrs. and, in the light of precision tests of the SM, don't get the strong coupling constant right. But the proton lifetime and monopole prediction have stimulated new experimental activities and theoretical studies. Simple GUTS have also answered other long-standing questions such as the origin of electric charge quantization and provided the first concrete realization of Sakharov conditions required to understand the matter-antimatter asymmetry of the universe.

(b) Models with Dynamical Symmetry Breaking (DSB)

Such models provide an alternative to the EW gauge symmetry breaking mechanism of the SM. The motivations are—(1) so far the Higgs boson has not been seen in experiments and the present experimental bound on its mass lies above the Z-boson mass which agrees with the theoretical bound obtained from the analysis of precision tests of the SM and (2) the presence of elementary scalar fields creates problems in any QFT. In models with DSB there are new fundamental fermions with new gauge interactions. Although fundamental scalar bosons are absent in such models, pseudoscalar (and scalar) bosons which are composites of these new fundamental fermions (and their antiparticles) can arise (analogons to the way pions appear in QCD) and may provide an experimental signature of new physics.

The most popular and the most studied of models with DSB is the Technicolor model. In this model, a new gauge interaction, (similar to QCD) called technicolor acts between new massless fermions named technifermions. The technifermions carry the quantum numbers of the SM gauge group plus their own technicolor charges. The technicolor interaction is asymptotically free (like QCD) and at an energy scale of order 1 TeV it becomes strong. At this energy scale, it is assumed that condensates of technifermions and antitechnifermions form·giving rise to a set of pseudoscalar Goldstone bosons (similar to pions in QCD). The technifermions interact with SM gauge bosons through techinifermion loops. By the Higgs mechanism three of the technipions are "gauged away" to appear as longitudinal components of the weak gauge bosons (W^+, W^-, Z^0) of the SM and as a result they acquire mass. After three technipions are removed to give masses to the weak gauge bosons, some "uneaten" technipions remain in the physical mass spectrum of the model and are, in principle, experimentally detectable. The number of such physical and massive technipions depends on the choice of the technicolor gauge group and the technifermion representations. However, SM fermions (quarks and leptons) still remain massless and further new interactions (called extended technicolor, etc.) must be postulated to generate masses of the SM fermions. While such a method can be made to work, it is not a compelling one, and no version has been able to naturally give quark masses and mixing angles of the right magnitudes without also giving rise to possible difficulties with the flavor-changing neutral currents (FCNC). No simple equivalent of the GIM mechanism has been found in these models. Moreover, the predicted masses of the lightest physical technipions are low enough to have been detected at LEP and Tevatron experiments. Hence the technicolor models are presently ruled out.

(c) Models with low-energy Supersymmetry (SUSY)

Supersymmetry relates fermions to bosons and vice versa. In a supersymmetric theory half-integral-spin particles (fermions) and integral-spin particle (bosons) appear in the same multiplet (called the supermultiplet) and hence have the same mass and coupling with other particles. At present, low-energy SUSY is the most popular approach to BSM physics. There are several motivations for considering theories with SUSY-(1) It is the most general symmetry consistent with a local RQFT. (2) It softens the degree of divergences in a theory because of a possible cancellation between fermionic and bosonic loop corrections due to their opposite sign (a $f\overline{f}$ loop contribution always has an extra negative sign). It can thus provide a solution to the hierarchy problem without artificial fine turning. (3) Grand unification of the gauge couplings (strong, electromagnetic and weak) of the SM can be achieved by using SUSY. In the minimal supersymmetric extension of the standard model (MSSM) the gauge couplings unify nicely if the scale of SUSY breaking is about 1 TeV. (4) Assumption of R-parity invariance in low-energy SUSY almost automatically yields a candidate, the "neutralino", for the "dark matter" of the universe with a matter density in the required range. (5) low-energy SUSY is strongly suggested by superstring theories (string theories with local SUSY) which look promising for a consistent quantum description of Gravitation (Quantum Gravity) together with strong, electromagnetic and weak interactions, that is, for unification of all four fundamental interactions observed in nature.

Clearly, SUSY cannot be exact. If it were, then all SM fermions (quarks and leptons) would be accompanied by bosonic superpariners (squarks and sleptons) of the same mass and electric charge. Similarly all SM bosons (gauge bosons and Higgs bosons) would be accompanied by fermionic superpartners (gauginos and Higgsinos) also of the same mass and electric charge. However, this is not the case as they have not so far been seen in experiments at LEP and Tevatron. But the gauge hierarchy problem and to some extent unification of gauge couplings require low-energy SUSY. This requires that SUSY should remain unbroken down to an energy scale

($\sim$ 1 TeV) not very far from the electroweak scale ($\sim$ 300 GeV). Thus superparticles are predicted to have masses in the range 100 GeV-1 TeV. They are, therefore, very much expected to be seen at the next higher energy accelerator (LHC and ILC) experiments.

(d) String theories

The major motivation for studying string theories is their possible realization as a quantum theory of gravity and its unification with the other three fundamental interactions (strong, electromagnetic and weak) of nature. Conventional field theoretical methods fail for obtaining a quantum theory of gravity because of the appearance of severe divergences which, unlike in QED cannot be removed by an appropriate renormalization procedure as the gravitational coupling constant is not dimensionless (it has negative mass dimension 2 in natural units). Most of the divergences in a QFT are known to arise from the behavior of the theory at short distances. These can be avoided to a large extent if the fundamental particles in the theory are not dimensionless point-like objects but are extended objects. In string theory the "particles" are regarded as 1-dimensional string-like objects. The string size which is usually taken to be of order Planck length (10^{-33} cm) provides a natural short-distance scale so that one cannot probe distances shorter than the string size and the divergence problem is thereby avoided. String theories were first proposed in the 1960's as theories of hadron physics (strong interaction). But it was soon realised that although hadrons have a number of string-like features, string theories were not suitable for a detailed description of the physics of hadrons. (In particular string theory failed to explain the asymptotic freedom of strong interaction and the parton-like behavior of hadrons at high energies or equivalently at short distances). In the 1970's a small group of theorists realised that the string idea could be fruitfully applied in constructing a theory of quantum gravity. This is because in string theories different "particles" can be thought of as different "excitations" (different modes of vibration) of the same string and there is one mode that behaves like a massless spin-2 particle which can be identified with the "graviton", the quantum of quantum gravity. Furthermore, the presence of such a particle implied that the string theory possessed general coordinate covariance and thus possibly can describe quantum gravity.

Moreover, the theory contains only one dimensional parameter, the string scale.

In the 1980's Green, Schwarz and a few other theorists realised that when supersymmetry (SUSY) is included in string theory, the resulting theory, named "superstring" theory, rather successfully incorporates quantum mechanics without the troublesome infinities provided one works in 10 dimensional ($D = 10$) space-time (9 + 1 dimensional). Thus one has a candidate theory of quantum gravity as long as one could accept that our conventional 4-dimensional ($D = 4$) world has an extra 6 spatial dimensions that are compactified to string size as in the old Kaluza-Klein theory (first proposed in the 1920's for unifying Einstein's gravity and Maxwell's electromagnetism but later abandoned because of many difficulties).

Shortly after this breakthrough a certain type of superstring theory, known as the "Heterotic String" theory was proposed which possessed a gauge symmetry with the gauge group large enough to include the SM gauge group as a subgroup and thus could describe all the known fundamental interactions in a natural unified way. There are 2 possible choices of the gauge group, $SO(32)$ and $E_8 \times E_8$, in order to make the theory both divergence-free and anomaly-free (both the groups have rank 16 with total number of generators equal to 496).

The problem of string theory has been its lack of uniqueness. There are a large number of solutions to the equations of string theory. There is no dynamical mechanism that prefers one solution to another. The SM would hopefully correspond to a particular internal space or vacuum configuration chosen by nature by some yet unknown mechanism Unfortunately, however, no unique string theory having the SM as its effective low-energy limit has so far emerged.

Great progress has been made in the 1990's in understanding the deeper structure of string theory. All the known string theories (type I, type IIA, type IIB, $E_8 \times E_8$ heterotic and $SO(32)$ heterotic) have now been found to be actually related by various duality transformations to one another and to a larger mysterious 11-dimensional ($D = 11$) theory, the so-called "M-theory".

There is another discouraging feature of string theory. The typical size of a string, the so-called string length, is the Planck length $(G_N\hbar/c^3)^{\frac{1}{2}} =$

10^{-33} cm. The energy needed to probe such small distances is the energy equivalent to what is known as the Planck mass and is of order 10^{19} GeV. Such high energies are far beyond the accessible limits of the present day accelerators and most probably cannot be achieved in any forseeable future accelerator. From the phenomenological point of view it is very important to have some possible low-energy signatures of string theory.

(e) Models with large extra dimensions and models with warped extra dimensions (EDM)

The idea of extra space dimensions originated in the search for a unified theory of fundamental interactions observed in nature. In the 1920's Kaluza and Klein proposed to unify Einstein's gravity and Maxwell's electromagnetism in a geometrical scheme containing an extra spatial dimension ($D = 5$, i.e. $4 + 1$ dimensional space-time).However, the Kaluza-Klein theory was abandoned since it stumbled on many insoluble difficulties. With the success of electroweak unification attention was shifted to the use of larger symmetry (larger gauge group) in implementing "grand unification" i.e. unification of all known fundamental interaction of nature.

Interest in extra dimensions was revived with the development of string theory as a possible candidate for quantum gravity. In string theory it is natural to imagine that the compactification scale is not very much different from the Planck scale However, if one is willing to give up this prejudice, an extraordinary possibility opens up. Perhaps the extra dimensions are not of Planck size but much larger, even macroscopic. Arkami-Hamed, Dimopoulos and Dvali (ADD) first pointed out that from an experimental point of view, the limits on the size of extra dimensions are surprisingly weak. Such a viewpoint places the hierarchy problem in a new light, perhaps allowing solutions which are different from those in technicolor or supersymmetric model.

In the ADD model, the only force one can use to probe the extra dimensions is gravity. Unfortunately, we have no knowledge of gravity at distances less than about a mm. This is because direct tests of gravitational forces i.e. Newtonian gravity which is the low velocity limit $\left(\frac{v}{c} \ll 1\right)$, of Einstein's gravity are based on torsion balance experiments that measure the attraction between oscillating spheres and the smallest scale on which this type of experiment has so far been performed is 0.2 mm. (These limits have recently been improved). However, we know from precision tests of the SM that strong and electroweak forces cannot be modified at distance larger than about 10^{-16} cm. This prompted ADD to suggest that these forces might be trapped in a submanifold (subspace) called "brane" that has the usual three spatial dimensions of our everyday world, yet a "thickness" of about 10^{-16} cm in the extra dimensions (The full multidimensional space-time is called "bulk").

In the ADD picture, the Newton's gravitational constant G_N is a derived quantity that depends on the volume of the extra dimensions and the force of gravity is expected to be $\sim 1/r^{D-2}$, where D = total number of dimensions ($D - 1$ spatial and 1 time dimension). When viewed from the full multidimensional space, known as the "bulk", there might be only one fundamental scale. The most radical suggestion lowers the fundamental scale (M_f) of the theory from the Planck scale (M_p) to almost the EW scale ($M_f \sim 1$ TeV). This assumption leads to the estimate of the size of the extra dimensions in terms of their number. Higher number of extra dimensions implies smaller compactification scale (size of extra dimensions). If there are d large extra dimensions, any other being comparable in size to the fundamental scale (M_f^{-1}), then one has the relation,

$$M_p^2 = M_f^{2+d} R^d \tag{14.4}$$

where R = size of the large extra dimensions. If $d = 2$, $R \approx 0.01$ cm. For $d = 3$, $R \approx 10^{-7}$ cm (1 nm). But $d = 1$ would make R larger than earth-to-sun distance (1 A.U.) and can be safely ruled out by observations. Thus two extra large dimensions ($d = 2$) would give a compactification scale which is close to the present experimental limit for gravity effect detection.

"Branes" that correspond to the manifold (subspace) where SM particles and interactions reside, are crucial to the ADD picture. The theoretical constructs necessary for the brane world picture came from independent developments in string theory. Extended objects of various dimensionalities are integral to string theory. These were generally called p-branes, where p = number of spatial dimensions of the fundamental objects. Thus $p = 0$ corresponds to a point-like particle, $p = 1$ is a string-like object and $p = 2$ corresponds to a membrane. Branes in string theory possess gauge symmetries on their "surfaces" (submanifolds) and automatically incorporate quantum gravity. The gauge symmetry arises from "open" strings (i.e. strings that have their end points stuck on the brane). Two of these open strings can collide to from a "closed" string that can travel onto the higher dimensional "bulk". The simplest excitation modes of these closed strings correspond to gravitons.

One important experimental signature of the brane-world picture is the possible appearance of new states called the Kaluza-Klein (KK) excitations at future high-energy colliders (LHC, ILC, ...). These excited states are a feature of models with compactified dimensions. As a result of the compactification, fields propagating in the "bulk" expand into a series of states known as the Kaluza-Klein (KK) tower, with the individual KK excitations labelled by mode numbers. Similar to a particle in a box, the momentum of the bulk field is quantized in the compactified dimensions. For an observer trapped in the "brane" each quantum of momentum in the compactified volume appears as a KK excited state with $m^2 = p_d^2$, where m = mass of the state and d = no. of compactified dimensions. This builds a tower of states each of which carries identical spin and gauge quantum numbers. If the size R of the compactified dimensions is large, the KK modes are light. However, if the extra dimensions are of infinite extent ($R \to \infty$) instead of being compactified, the d-dimensional momentum and the resulting KK spectrum is continuous. The detailed properties of the KK states are determined by the geometry of the compactified dimensions and their measurement would reveal the underlying geometry of the bulk. The signature of KK excitations is the production of high energy photons accompanied by missing energy, with a cross section showing a dramatic rise with energy.

Although the "brane world" idea is appealing as an explanation of the weakness of gravity, there are a number of constraints and problems that have to do with cosmology, astrophysics and aesthetics. First, the naive statement that 2 extra dimensions ($d = 2$), 0.1 mm in size, is allowed may not be quite correct. Such large extra dimensions would quite significantly affect the behavior of astrophysical objects such as supernova, because they would cause these objects to lose energy by emitting gravitons in the bulk. Such graviton emission would show up as an anomalous cooling of their interiors. Calculations show that the extra dimensions must be smaller than submm size ($R \ll 0.1$ mm).

Secondly, the 'brane world' idea has dramatic implications for the early evolution of the universe. With 2 extra dimensions one gets an unacceptable modification of the primordial light element abundance. One way to avoid this problem is to raise the new fundamental scale (M_f) of gravity above 1 TeV. The third problem associated with the "brane world" picture is more aesthetic and has to do with the unification of weak, electromagnetic and strong forces into a single force. In most of the GUTs, the unification is achieved in the conventional $D = 4$ space-time world and the grand unification scale (M_{GU}) is taken to be around 10^{16} GeV which is very much larger than 1 TeV, the fundamental scale (M_f) of gravity in the "brane-world" picture. To evade this problem, it has been suggested by some theorists that the gauge forces can feel "some" extra dimensions but not very large one (> 1 mm). In this way it is possible to achieve a different form of unification. The field content that is allowed to propagate in the "bulk" as well as the size and geometry of the "bulk" itself vary between different models.

Against the difficulties mentioned above, the "brane world' picture also offers new solution to old problems, for example, the dark matter problem, fermion mass hierarchy problem and the cosmological constant (dark energy) problem. To explain why the cosmological constant is so significantly small, there has been a recent novel approach motivated by a variation of the "brane world" idea. For this, one considers "branes" in which the "bulk" dimensions are curved or "warped" but not necessarily compactified. It is hoped that by "warping" the extra dimensions in the right way, it might be possible to explain why the cosmological constant appears to be so small.

Randall and Sundrum (RS) have made a radical proposal: they argue that "warping" might be enormous and might account for the large hierarchy between the Planck scale and the electroweak (EW) scale. In the simplest version of the RS model there is one extra dimension. It may be called angle ϕ with $0 < \phi < \pi$. The model contains 2 branes, one at $\phi = 0$ and another at $\phi = \pi$. The tensions in the 2 branes are taken to be equal and opposite and one imagines that the SM fields propagate in one brane, the "visible" sector brane, while some other "hidden" fields propagate in the other brane.

There are other variants of the RS proposal which have been put forward. Perhaps the most interesting is the one in which the extra dimensions are not compactified at all but are extremely "warped" with gravity localised on the "visible" brane. These ideas suggest a rich set of possibilities and a potentially important area for further research.

14.3 The $SU(5)$ Model

As discussed earlier, Grand unified theories (GUT) address themselves to 3 drawbacks of the SM-(1) presence of a large number of free parameters (2) lack of a rationale for charge quantization and (3) imperfect unification of weak and electromagnetic interactions. The third defect can easily be corrected if one chooses for the gauge symmetry a Lie group which is "simple" and has rank ≥ 4 (The SM gauge group $SU(3) \times SU(2) \times U(1)$ is semi-simple and has rank 4) so that it incorporates the SM. The smallest group that satisfies these 2 criteria is the $SU(5)$ group. A model of grand unification based on the $SU(5)$ gauge group was first formulated by Georgi and Glashow in 1974 and is also known as the Georgi-Glashow model.

The group $SU(5)$ has rank 4 and 24 generators. There are thus 4 fundamental representations and the adjoint representation has dimension 24. Since $SU(5)$ includes $SU(3) \times SU(2)$, it is often helpful to write the low-dimensional representations of $SU(5)$ as direct sum of $SU(3) \times SU(2)$ representations. This is particularly true of the fundamental representations and the adjoint representation of $SU(5)$

YT	$SU(5)$ repr.	$SU(3) \times SU(2)$ decomposition
	5	$(3, 1) + (1, 2)$
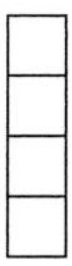	5^*	$(3^*, 1) + (1, 2)$
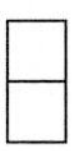	10	$(3, 2) + (3^*, 1) + (1, 1)$

$$10^* \qquad (3^*, 2) + (3, 1) + (1, 1)$$

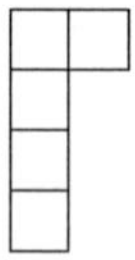

$$(8, 1) + (1, 3) + (1, 1)$$

$$24 = 24^* \quad +(3, 2) + (3^*, 2) \tag{14.5}$$

In the SM, each generation of fermions consists of 2 quarks and 2 leptons. All the left-handed particles appear in $SU(2)_I$ doublets and all the right-handed particles appear in $SU(2)_I$ singlets, except for the neutrino which can be only left-handed. Similarly, all the right-handed antiparticles appear in $SU(2)_I$ doublets and all the left-handed antiparticles appear in $SU(2)_I$ singlets, except for the antineutrino which can be only right-handed. Thus there are 15 left-handed fields and 15 right-handed fields. The 15 left-handed fields can be put in $5^* + 10$ representation of $SU(5)$. Similarly, 15 right-handed fields can be put in $5 + 10^*$ representations of $SU(5)$

$$5^* = (3^*, 1) + (1, 2), \qquad 5 = (3, 1) + (1, 2)$$

$$\overline{d}_L \quad \begin{pmatrix} \nu_e \\ e^- \end{pmatrix}_L \qquad\qquad d_R \quad \begin{pmatrix} e^+ \\ \nu_e \end{pmatrix}_R$$

$$10 = (3, 2) + (3^*, 1) + (1, 1), \quad 10^* = (3^*, 2) + (3, 1) + (1, 1)$$

$$\begin{pmatrix} u \\ d \end{pmatrix}_L \quad \overline{u}_L \quad e_L^+ \qquad\qquad \begin{pmatrix} \overline{d} \\ \overline{u} \end{pmatrix}_R \quad u_R \quad e_R^- \tag{14.6}$$

Similarly, the gauge bosons of the $SU(5)$ model belong to the adjoint representation 24 of $SU(5)$ whose $SU(3) \times SU(2)$ decomposition as given above is

$$\begin{aligned} 24 &= (8, 1) + (1, 3) + (1, 1) + (3^*, 2) + (3, 2) \\ &= \text{gluons} + \underbrace{(W^1, W^2, W^3) + B^0}_{(W^+, W^-, Z^0, \gamma)} + \text{Leptoquarks} \\ &\qquad\qquad\qquad\qquad\qquad\qquad (X, Y \text{ bosons}) \end{aligned} \tag{14.7}$$

The first 3 correspond to the SM gauge fields (gluons and the EW gauge bosons) while the last 2 correspond to new additional 12 gauge bosons which appear in both $SU(3)$ and $SU(2)$ nonsinglets. They are called "leptoquarks" since they connect leptons and quarks at a vertex involving their emission or absorption. Moreover, they induce processes that violate both B and L, for example $p \to \pi^0 e^+, \pi^+ \overline{\nu}_e$ which correspond to proton decay.(However, $(B\text{-}L)$ is conserved in such processes). X and Y bosons have fractional electric charges ($X_Q = \pm\frac{4}{3}, Y_Q = \pm\frac{1}{3}$ in unit of e).

Renormalization of couplings: If the $SU(5)$ GUT is to describe the SM, the $SU(5)$ gauge symmetry must break at some high energy scale to $SU(3) \times SU(2) \times U(1)$. Above this scale, the full $SU(5)$ symmetry holds to a good approximation and all gauge couplings renormalize in the same way. Below this scale the three couplings renormalize differently. We can write the expressions for the renormalized coupling constants

$$\alpha_i^{-1}(\mu) = \alpha_{\text{GUT}}^{-1}(M_{\text{GUT}}) + \frac{b_0^i}{4\pi} ln\left(\frac{\mu}{M_{\text{GUT}}}\right) \tag{14.8}$$

where $i = 1, 2, 3$ for $U(1)$, $SU(2)$ and $SU(3)$ couplings, respectively. We can calculate the β-function at 1-loop level starting with the usual formula. Here

$$b_0^i = \frac{11}{3} c_A - \frac{4}{3} c_f^{(i)} N_f^{(i)} - \frac{1}{3} c_\phi^{(i)} N_\phi^{(i)} \tag{14.9}$$

where $N_f^{(i)}$ = no. of fermions and $N_\phi^{(i)}$ = no. of scalars. For $SU(N)$, $c_A = N$, and for fermions and scalars in the fundamental representation, $c_f = c_\phi = 1/2$.

For $SU(3)$ and $SU(2)$ the β-function coefficient b_0^i is readily calculated. For the $U(1)$ one needs the relative normalization.

$$b_0^{(2)} = \frac{181}{6}, \quad b_0^{(3)} = 7, \quad b_0^{(1)} = \frac{61}{15} \tag{14.10}$$

We can run these eqns. backwards. The $SU(2)$ and $U(1)$ couplings are the best measured, so it makes sense to run them up to the unification scale. This determines α_{GUT} and M_{GUT} (α = fine structure constant). We can then predict the value of the $SU(3)$ coupling at say (M_Z). One finds that the grand unification scale is of order 10^{15} GeV and $\alpha_3(M_Z)$ is off by about 7 standard deviations from the observed value.

Proton Decay

The fermion-gauge boson couplings present in the $SU(5)$ Lagrangian include those which occur in the SM:

$$U(1) : g_1 B^\mu \sum_f \overline{f}\gamma^\mu \frac{Y}{2} f$$

$$SU(2) : g_2 W_\mu^i \left(\sum_Q \overline{Q}\gamma^\mu \tfrac{t_i}{2} Q + \sum_L \overline{L}\gamma^\mu \frac{t_i}{2} L\right)$$

$$SU(3) : g_3 G_\mu^a \sum_q \overline{q}\gamma^\mu \frac{\lambda_a}{2} q \tag{14.11}$$

Here $f = q, l$; Q and L are left-handed quark and lepton doublets; q and l refer to quark and lepton of all flavors (generations).

In addition, there are couplings of the fermions with the new "lepto-quark" bosons.

$$g_4 X_\mu^i (\overline{d}_{iL}^c \gamma^\mu e_L + \in_{ijk} \overline{u}_{jL}\gamma^\mu u_{kL}^c + \overline{e}_L^+ \gamma^\mu d_{iL} + h \cdot c)$$
$$+ g_5 Y_\mu^i (\overline{d}_{iL}^c \gamma^\mu \nu_{eL} + \in_{ijk} \overline{u}_{jL}\gamma^\mu d_{kL}^c + \overline{e}_L^+ \gamma^\mu u_{iL} + h \cdot c) \tag{14.12}$$

where (ijk) are color-triplet indices and q^c denotes a charge conjugate quark spinor. The coupling constants $g_i s (i = 1, \ldots, 5)$ are all related at the grand unification scale (M_{GUT}) by the group manifold. The new vertices in the $SU(5)$ model are summarized in Fig. 14.1 below.

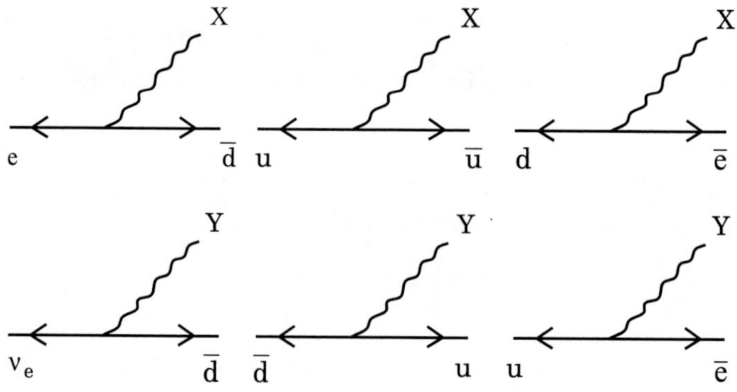

Fig. 14.1

They mediate transitions that lead to proton decay, which proceeds via 3 elementary processes that change baryon number (B) and lepton number (L) each by 1 unit. (However, B-L is conserved)

In analogy to calculation of muon lifetime, one can calculate the contribution of the above diagrams to proton decay and finds

$$\frac{1}{\tau_p} = \frac{g^2 A M_p^5}{M_X^4} \tag{14.13}$$

where g is the $SU(5)$ gauge coupling, M_p is proton mass, M_X is the lepto-quark boson mass, ($M_X \sim M_{\text{GUT}} \sim 10^{15}$ GeV) and A is a known constant. Eq. (14.13) gives $\tau_p \sim 10^{28}$ yrs, if $M_{\text{GUT}} \sim 10^{15}$ GeV. The current limit on proton lifetime is $\geq 5 \times 10^{33}$ yrs.

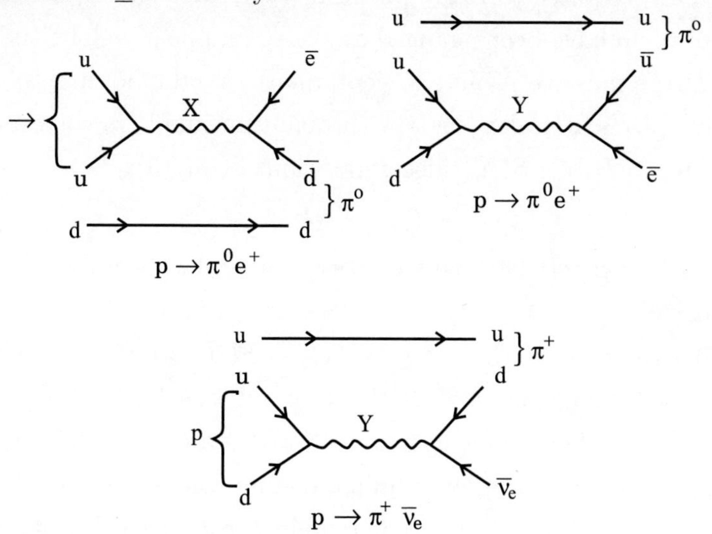

Fig. 14.2

The fact that B violation is likely in any GUT could provide a solution to the observed excess of matter over antimatter in the universe for which baryon number violation is a precondition.

Breaking of $SU(5)$ gauge symmetry

The weak gauge bosons in the SM have to acquire masses (of order EW scale). Moreover, the leptoquark gauge bosons must be endowed with enormous masses for the $SU(5)$ GUT to survive the existing bounds on the proton lifetime. Both can be achieved by the same spontaneous symmetry breaking (SSB) mechanism as used in the SM. The necessary symmetry breaking can be achieved in 2 steps. At the first step, $SU(5)$ gauge symmetry is broken to the SM gauge symmetry $SU(3) \times SU(2) \times U(1)$ as a result of which the lepto-quark gauge bosons acquire mass. At the next step, $SU(3) \times SU(2) \times U(1)$ symmetry is broken down to $SU(3) \times U(1)_{\text{em}}$ symmetry so that the 3 weak gauge bosons acquire masses but the gluons and the photon continue to remain massless.

$$SU(5) \xrightarrow{M_{\text{GUT}}} SU(3) \times SU(2) \times U(1) \xrightarrow{M_{\text{EW}}} SU(3) \times U(1)_{\text{em}} \tag{14.14}$$

To implement the first symmetry breaking step a real 24 (of scalar fields) is introduced. To implement the second symmetry breaking step a complex 5 (of scalar fields) is introduced (both are $SU(5)$ representations).

The SSB gives rise to many physical (real) Higgs scalars. From the 24 after 12 of them have been "gauged away" to reappear as longitudinal components of the massive X and Y (leptoquark) vector bosons, there remain 12 massive physical scalar fields with quantum numbers corresponding to the following $SU(3) \times SU(2)$ decompositions of $SU(5)$:

$$(8,1) + (3,1) + (1,1) \tag{14.15}$$

which may be expected to have masses comparable to masses of the X, Y vector bosons.

In order to break $SU(3)\times SU(2)\times U(1) \to SU(3)\times U(1)_{em}$ one introduces a complex scalar multiplet 5 of $SU(5)$. 5 complex scalar fields are equivalent to 10 real scalar fields of which only 3 are "eaten up" by Higgs-Kibble mechanism to become the longitudinal components of the 3 weak massive vector bosons (W^+, W^-, Z^0). There remain 7 real physical scalar fields with $SU(3) \times SU(2)$ quantum numbers (color and weak i-spin) specified by

$$(1,1) + (3,1) + (3^*,1) \tag{14.16}$$

The first one corresponds to the Higgs boson of the SM but still with unknown mass. The other 2 correspond to weak iso-singlet, color-triplet and color anti-triplet scalar bosons $h^{\pm/3}$ with electric change $\pm\frac{1}{3}e$. These "lepto-quark" Higgs bosons can mediate proton decay through transitions analogous to those mediated by $Y^{\pm 1/3}$ vector "leptoquark" bosons. It is, therefore, necessary to include locally gauge invariant interactions among the 5 and 24 scalar field multiplets in the $SU(5)$ Lagrangian which will give enormous masses to $h^{\pm 1/3}$ scalar fields.

The most general $SU(5)$ invariant potential involving the scalar fields is of the form

$$V = V_\phi + V_H + V_{\phi-H}$$

where $V_\phi = -m^2 \mathrm{Tr}\phi^2 + \lambda \mathrm{Tr}\phi^4 + \lambda'(\mathrm{Tr}\phi^2)^2$

$$V_\phi = -\mu^2|H|^2 + \lambda|H|^4$$

$$V_{\phi-H} = \Gamma H^*\phi H + \lambda' H^* H\, \mathrm{Tr}\phi^2 + \lambda' H^*\phi^2 H \tag{14.17}$$

with ϕ and H corresponding to 24 and 5, respectively.

5 of complex scalar fields can also be used to give masses to the SM fermions. The $SU(5)$ model of grand unification also predicts relations among fermion masses. For example, one has the tree level prediction

$$m_b = m_\tau \tag{14.18}$$

This "prediction" is off by a factor of about 3, but like the prediction of the gauge couplings of the SM, it is corrected by renormalization by roughly the same amount. For lighter fermions, the prediction fails. However, it is hoped that unlike the gauge couplings such predictions can be modified if there are additional Higgs fields in other $SU(5)$ representations.

The Weinberg Angle

Another important consequence of grand unification is the prediction of the Weinberg or weak-mixing angle. The couplings to both Z^0 and γ summed over all fermions in a $SU(5)$ multiplet have to vanish, giving us the relation

$$\sum_f Q(I_3 - Q^2 \sin^2\theta_W) = 0 \tag{14.19}$$

This gives $$\sin^2\theta_W = \frac{\sum_f QI_3}{\sum_f Q^2} = \frac{3}{8} \tag{14.20}$$

Here Q is electric charge and I_3 is the third component of the weak i-spin of a fermion and the sum is over all fermions in a $SU(5)$ multiplet. The value refers to the grand unification scale $M_{GUT} \sim 10^{15}$ GeV. Its value at the EW unification scale $M_{EW} \sim 10^3$ GeV can be obtained by applying RG eqns. and one gets,

$$\sin^2\theta_W = 0.21 \pm 0.01 \tag{14.21}$$

which is close to the experimental value.

Charge Quantization and Magnetic Monopoles

Two other successes of GUTs may be mentioned here. First, one has a possible explanation for electric charge quantization. In the $SU(5)$ model, it is automatic. Since the electric charge Q is an $SU(5)$ generator, it is quantized, just as color and i-spin are quantized. Thus Tr $Q = 0$ for each $SU(5)$ multiplet. Applying it to 5^* or 10 we get $e_d = -\left(\frac{1}{3}\right)$.

However, according to Dirac there is another explanation of electric charge quantization-existence of magnetic monopoles. Dirac showed that consistency of quantum mechanics demands that even if a single magnetic monopole exists anywhere in the universe, all electric charges must be integer multiples of a fundamental charge. Therefore, we must suspect that magnetic monopoles must be hidden somewhere in the GUTs.

Other groups

While $SU(5)$ is the minimal gauge group for grand unification, there are many other possibilities. Perhaps, the next simplest choice is unification with $SO(10)$ as the gauge group. Now $SO(2n)$ group has rank $l = n$ and $r = n(2n - 1)$ generators. Therefore for $SO(10)$, $l = 5$ and $r = 45$. Thus $SO(10)$ has 5 commuting generators. Presumably, this symmetry must be broken at some high energy scale. The most important fact about $SO(10)$ unification is that a single fermion generation of the SM fits nicely into the irreducible spinor representation 16 of $SO(10)$. The $SO(10)$ has an $SU(5)$ subgroup under which 16 decomposes as $10 + 5^* + 1$. The $SU(5)$ singlet has precisely the right SM quantum numbers (all zero) to play the role of the right-handed neutrino in the "see saw" mechanism to generate a small neutrino mass. The adjoint representation of $SO(10)$ to which the gauge bosons must belong has dimension $N = r = 45$ whose $SU(5)$ decomposition is $45 = 24 + 10 + 10^* + 1$. There is thus a larger proliferation of gauge bosons (and hence also Higgs bosons) as compared to the $SU(5)$ model of grand unification.

14.4 The Minimal Supersymmetric Standard Model (MSSM)

As discussed earlier, the major motivation for exploring a supersymmetric extension of the SM is the possibility of overcoming the hierarchy problem. However, as a price there appears a large number of new particles, the supersymmetric partners (super partners) of the SM particles, which have to be experimentally seen for the extended model to be acceptable. Therefore, one would like to keep the supersymmetric extension of the SM at a minimal level so that the predicted new physics is not very extensive.

The SUSY algebra

The supersymmetry (SUSY) algebra is an extension of the Poincare algebra and includes the SUSY generators which are fermionic "charges" Q_α^A ($A = 1, \dots N$) obeying anticommunication relation. Here N is the number of SUSY charges and α is a spinor index (for a Dirac Spinor, $\alpha = 1, 2, 3, 4$). Thus the SUSY algebra is said to be a "graded" Lie algebra since it involves both commutators and anticommutators. The possible value of N are 1, 2, 4, or 8, the symmetry is correspondingly said to be $N = 1, 2, 4$, or 8 SUSY.

If nature is supersymmetric, it is likely that the low-energy symmetry is $N = 1$ (corresponding to only one possible value for the index A above). It is very difficult to break $N > 1$ SUSY spontaneously; this is not the case for $N = 1$. Moreover, only $N = 1$ SUSY allows fermions in chiral representations (as in the case of the SM).

For $N = 1$ SUSY, the SUSY algebra contains only one self-conjugate spin-$\frac{1}{2}$ generator (Majorana spinor) Q_α with the property that it changes the angular momentum J by $1/2$ unit. It thus turns a boson into a fermion and vice-versa.

$$Q_\alpha|\text{boson} >\sim |\text{fermion} >, Q_\alpha|\text{fermion} >\sim |\text{boson} \tag{14.22}$$

The commutation and the anticommutation relations of the $N = 1$ SUSY algebra are as follows:

$$[P^\mu, P^\nu] = 0 \tag{14.23}$$

$$[P^\mu, J^{\rho\sigma}] = i(g^{\mu\rho}P^\sigma - g^{\mu\sigma}P^\rho) \tag{14.24}$$

$$[J^{\mu\nu}, J^{\rho\sigma}] = i(g^{\nu\rho}J^{\mu\sigma} - g^{\mu\rho}J^{\nu\sigma} - g^{\nu\sigma}J^{\mu\rho} + g^{\mu\sigma}J^{\nu\rho}) \tag{14.25}$$

$$[Q_\alpha, P^\mu] = 0 \tag{14.26}$$

$$[Q_\alpha, J^{\mu\nu}] = i(\sigma^{\mu\nu}Q)_\alpha \tag{14.27}$$

$$\{Q_\alpha, \overline{Q}_\beta\} = -2(\gamma^\mu C)_{\alpha\beta}P_\mu \tag{14.28}$$

$$\{Q_\alpha, Q_\beta\} = 0 = \{\overline{Q_\alpha}, \overline{Q_\beta}\} \tag{14.29}$$

where $\sigma^{\mu\nu} = \frac{i}{2}[\gamma^\mu, \gamma^\nu]$, $\overline{Q_\alpha} = Q_\alpha^+\gamma^0$, $[A, B] = AB - BA$

$$\{A, B\} = AB + BA, \quad C = \text{charge conjugation matrix} \tag{14.30}$$

(14.26) implies that the SUSY charge is conserved and (14.28) implies that two successive SUSY transformations amount to a space-time translation.

The smallest irreducible representation (supermultiplets) of $N = 1$ SUSY contain massless particles of different spins as follows:

$$\begin{array}{ccc} \text{Chiral} & \text{Vector} & \text{Tensor} \\ \begin{pmatrix} f, \frac{1}{2} \\ \tilde{f}, 0 \end{pmatrix} & \begin{pmatrix} g, 1 \\ \tilde{g}, \frac{1}{2} \end{pmatrix} & \begin{pmatrix} G, 2 \\ \tilde{G}, \frac{3}{2} \end{pmatrix} \end{array}$$

$$\begin{aligned} &f = \text{chiral fermion},\ g = \text{gauge boson},\ G = \text{graviton} \\ &\tilde{f} = s\text{-fermion (scalar)},\ \tilde{g} = \text{gaugino (chiral fermion)} \\ &\tilde{G} = \text{gravitino} \end{aligned} \tag{14.31}$$

Local gauge symmetry involves spin-1 gauge fields and local Poincare symmetry involves spin-2 gravitons. One can work in terms of these fields, writing SUSY transformation laws and constructing invariants. However, great 'simplification is achieved by enlarging the ordinary $(3 + 1)$ space-time to include commuting as well as anticommuting c-number variables (Grassman variables). The resulting space is called "superspace".

The MSSM is the $N = 1$ supersymmetric extension of the SM. The gauge group in MSSM is the same as in SM, namely, $G = SU(3)_C \times SU(2)_L \times U(1)_Y$. The spin-1 gauge fields of the SM are replaced by vector superfields in MSSM. The MSSM, therefore, contains some new spin-1/2 Majorana fermions, called "gauginos": the $SU(3)$ "gluino" ($\tilde{g}$), $SU(2)_L$ "Wino" ($\tilde{W}$) and the $U(1)_Y$ "Bino" ($\tilde{B}$). The supersymmetric partners (superpartners) of the photon (γ) and Z-boson are similarly called the "photino" ($\tilde{\gamma}$) and "Zino" ($\tilde{Z}$).

The spin-1/2 fermions of the SM (quarks and leptons) are replaced in MSSM by the corresponding chiral superfields including as new degrees of freedom, a complex spin-0 field for each quark and lepton chirality state: the "s-quarks" $\tilde{q}_L = (\tilde{u}_L, \tilde{d}_L)$, $\tilde{u}_R$, $\tilde{d}_R$ and the "sleptons" $\tilde{l}_L = (\tilde{\nu}_{eL}, \tilde{e}_L), e_R$ in 3 generations.

Finally, we must also replace the spin-0 fields (Higgs fields) of the SM by the corresponding chiral superfields with the superpartners called "Higgsinos" ($\tilde{H}$) being spin 1/2 Majorana spinors. However, there is one important change from the SM. In the MSSM, one must have at least 2 complex scalar doublets with $Y = \pm 1$ instead of just 1 complex scalar doublet of the SM. This is necessary in order to (1) give masses to all quarks and charged leptons, (2) cancel the gauge anomalies in the SUSY sector, and (3) avoid a massless fermion with electric charge ± 1. Moreover, unlike in the SM, one cannot have both $\tilde{\phi}$ and $\tilde{\phi}_c = i\tau_2\tilde{\phi}^*$ in the SUSY scalar protential which must be a 'holonomic' function of the fields (Here $\tilde{\phi}$ is the superpartner of the scalar field ϕ). Thus the Higgs sector in the MSSM consists of two weak i-spin complex doublets $\left(I = \frac{1}{2}\right)$ with $Y = \pm 1$,

$$H_1 = \begin{pmatrix} H_1^0 \\ H_1^- \end{pmatrix} \text{ with } Y = -1, \; H_2 = \begin{pmatrix} H_2^+ \\ H_2^0 \end{pmatrix} \text{ with } Y = +1 \tag{14.32}$$

R-Parity

In order to avoid the phenomenological embarrassment of having lepton number-violating or baryon number-violating terms in the most general gauge invariant and renomalizable super potential, one usually introduces in the MSSM a discrete symmetry called R-Parity invariance. (It can be a gauge symmetry and as such will not be broken in a consistent quantum theory). The corresponding multiplicative quantum number is called R-parity and is given by

$$R = (-)^{3B+L+2S} \tag{14.33}$$

where B, L and S are, respectively, the baryon number, lepton number and spin of the particle.

$R = +1$ (even), for all ordinary (SM) particles (quarks, leptons, gauge bosons and Higgs bosons)

$R = -1$ (odd), for all SUSY particles (squarks, sleptons, gauginos and Higgsiono) (14.34)

There are 2 immediate consequences of R invariance:

(1) existence of at least one stable superparticle called LSP (lightest supersymmetric particle)

(2) Superparticle production and decay must involve even number of superparticles.

The LSP is expected to be neutral under all gauge groups. It is, therefore, very weakly interacting. This in turn means the following:

(1) The generic signature of R-parity conserving SUSY theories is events with missing energy.

(2) SUSY is likely to produce an interesting dark-matter candidate.

The second point is one of the principal reasons that many physicists consider the possibility of low-energy SUSY so compelling. If one calculates the dark-matter mass density, one automatically finds a density in the right range, if the scale of SUSY breaking is ~ 1 TeV. (Another circumstantial evidence of low-energy SUSY is the unification of all the gauge couplings within the MSSM).

(Soft) SUSY breaking

If supersymmetry is exact, the ordinary (SM) particles and their superpartners (superparticles) must have the same mass. Phenomenologically, however, this is not the case. SUSY particles have not been seen at the LEP or the Tevatron. Thus SUSY must be broken. One usually assumes that SUSY breaking is "soft" so that mass splittings (between SM particles and superparticles) are of same order as the electroweak scale (~ 300 GeV–1 TeV). Thus the Lagrangian in the MSSM has 2 parts, one with exact SUSY and the other with moderately-broken SUSY.

$$L_{\text{MSSM}} = L_{\text{SUSY}} + L_{\text{Soft}} \tag{14.35}$$

The MSSM particle spectrum

To discuss $SU(2)_L \times U(1)_Y$ electroweak (EW) gauge symmetry breaking, it is generally assumed that all squark and slepton fields have vanishing vacuum expectation values (VEV) and the attention is restricted to the scalar (Higgs) potential.

$$\begin{aligned} V_H &= m_1^2|H_1|^2 + m_2^2|H|^2 + m_3^2(H_1H_2 + h\cdot c) \\ &\quad + \frac{g^2}{8}(H_2^+\vec{\sigma}H_2 + H_1^+\vec{\sigma}H_1)^2 + \frac{g'^2}{8}(|H_1|^2 - |H_2|^2)^2 \\ &= m_1^2|H_1|^2 + m_2^2|H_2|^2 + m_3^2(H_1H_2 + h\cdot c) \\ &\quad + \frac{1}{8}(g^2 + g'^2)(|H_1|^2 - |H_2|^2)^2 + \frac{g^2}{8}|H_1H_2|^2 \end{aligned} \tag{14.36}$$

$$\text{with } \langle H_1\rangle_0 = \frac{1}{\sqrt{2}}\begin{pmatrix} v_1 \\ 0 \end{pmatrix}, \langle H_2\rangle_0 = \frac{1}{\sqrt{2}}\begin{pmatrix} 0 \\ v_2 \end{pmatrix} \tag{14.37}$$

To minimize the Higgs potential it is convenient to use the auxiliary variables,

$$\tan\beta = v_2/v_1,\ \ v^2 = v_1^2 + v_2^2,\ \ v = (G_F\sqrt{2})^{-1/2} \tag{14.38}$$

To obtain a suitable vacuum, there are 2 constraints which the 'soft' breaking must satisfy:

(1) Without the soft breaking terms, $H_1 = H_2(v_1 = v_2)$ makes the $SU(2)$ and $U(1)$ H_2 terms vanish, i.e., there is no quartic coupling in this direction so the energy is unbounded from below unless

$$m_1^2 + m_2^2 - 2m_3^2 > 0 \tag{14.39}$$

(2) In order to obtain symmetry-breaking, the Higgs mass-matrix must have a negative eigen value. This gives the requirement

$$m_3^2 > m_1^2 m_2^2 \tag{14.40}$$

When these conditions are satisfied, it is straightforward to minimize the potential and determine the particle spectrum. In terms of the above variables, the minimization condition become

$$\sin 2\beta = \frac{2m_3^2}{m_2^1 + m_2^2},\ \ v^2 = \frac{4}{g^2 + g'^2}\cdot\frac{m_1^2 - m_2^2\tan\beta}{\tan^2\beta - 1} \tag{14.41}$$

one finds that

$$m_A^2 = \frac{m_3^2}{\sin\beta\cos\beta} = m_1^2 + m_2^2 \tag{14.42}$$

is a convenient parameter to work with.

R-even Sector

The R-even sector of the MSSM contains to begin with all the spin-1 and spin 1/2 particles of the SM. One finds

$$M_W^2 = \frac{g^2v^2}{4}, M_Z^2 = \frac{1}{4}(g^2 + g'^2)v^2 \tag{14.43}$$

Quarks of charge + 2/3 have tree-level masses proportional to ν_2, quarks of charge $-1/3$ and charged leptoms have trace-level masses proportional to ν_1. Neglecting generation mixing, one gets

$$m_u^2 = \frac{G_U^2}{2}v_2^2, m_d^2 = \frac{G_d^2}{2}v_1^2, m_e^2 = \frac{G_e^2}{2}v_1^2 \tag{14.44}$$

where $G_i^s(i = u, d, e)$ are dimensionless Yukawa couplings.

A nontrivial structure arises in the Higgs (spin-0) sector, where we have, to begin with, 2 complex scalar doublets H_1 and H_2 amounting to 8 real degree of freedom. After the 3 would-be Goldstone bosons have been "gauged away" to reappear as longitudinal components of massive weak vector bosons $(W^{\pm}, Z^0)$ we are left with 5 real physical degrees of freedom. 2 of them constitute a physical complex scalar field corresponding to charged Higgs particles $H^{\pm}$ and the other 3 become physical real scalar fields corresponding to 3 neutral Higgs particles A^0, H^0 and h^0. If one assumes that CP is a good symmetry, (The potential V_H by itself conserves CP, a simple field redefinition removing any phase in m_3^2) then A^0 is CP-odd and the other 2 (H^0 and h^0 are CP-even (By convention, $m_{h^0} < m_{H^0}$). One finds

$$m_H^2 \pm = m_W^2 + m_A^2 \tag{14.45}$$

$$m_{A^\circ}^2 = m_A^2 \tag{14.46}$$

and $$m_{H^0,h^0}^2 = \frac{1}{2}\left[(m_Z^2 + m_A^2) \pm \sqrt{(M_Z^2 + m_A^2)^2 - 4m_Z^2 m_A^2 \cos^2 2\beta}\right] \tag{14.47}$$

The following inequalities may be noted,

$$\begin{aligned} m_{H^\pm} &\geq m_W \\ m_{h^0} &\leq m_Z \\ m_{h^0} &\leq m_{A^0} \end{aligned} \tag{14.48}$$

Furthermore, at tree level all Higgs masses and couplings can be expressed in terms of 2 parameters only (m_A and $\tan\beta$) in contrast to 6 parameters needed for a general 2 Higgs-doublet model.

R-odd sector

The R-odd sector of the MSSM consists of those SUSY particles, which are supersymmetric partners (superpartners) of the SM particles.

The spin-0 superparticles of the MSSM are the squarks and the sleptons. The spin-1/2 superparticles of the MSSM are the gauginos and the Higgsinos. Of the gauginos the strongly-interacting gluinos do not mix with other spin-1/2 superparticles whose mass is an independent parameter of L_{soft}. However, the weakly interacting gauginos and the Higgsinos mix. As a consequence, the physical weakly-interacting spin-1/2 SUSY particles are the 2 pairs of charged and 4 neutral gaugino-Higgsino mixtures and are called "charginos" and "neutralinos", respectively.

The 2 "chargino" mass eigenstates ($\tilde{\chi}_1^\pm$ and $\tilde{\chi}_2^\pm$) are the superposition of the "winos" $\tilde{W}^\pm$ and the charged "Higgsinos" $\tilde{H}_2^+$ and $\tilde{H}_1^-$. More specifically, $(\tilde{\chi}_1^+, \tilde{\chi}_2^+)$ are linear superpositions of $(\tilde{W}^+, \tilde{H}_2^+)$ and $(\tilde{\chi}_1^-, \tilde{\chi}_2^-)$ are superpositions of $(\tilde{W}^-, \tilde{H}_1^-)$. Similarly the 4 "neutralino" mass eigenstates $\tilde{\chi}_j^0$ with $j = 1, 2, 3, 4$ are linear superpositions of the neutral weak 'gauginos' $(\tilde{W}_3, \tilde{B})$ and the neutral 'Higgsinos' $(\tilde{H}_1^0, \tilde{H}_2^0)$.

The masses and couplings of the 2 "chargino" pairs and 4 "neutralinos" are characterized by 4 parameters: the gaugino masses $\tilde{M}_1$ and $\tilde{M}_2$, the superpotential Higgs mass parameters m_A^2 and $\tan\beta$.

The lightest neutralino mass eigen state $\tilde{\chi}_1^0$ is the favorite candidate for the LSP (lightest supersymmetric particle in the MSSM). We may write

$$\tilde{\chi}_1^0 = N_{11}\tilde{B} + N_{12}\tilde{W}_3 + N_{13}\tilde{H}_1^0 + N_{14}\tilde{H}_2^0 \tag{14.49}$$

the case of pure 'photino' ($\tilde{\gamma}$) being the LSP ($\tilde{\chi}_1^0$) corresponds to

$$(N_{11}, N_{12}, N_{13}, N_{14}) = (\cos\theta_W, \sin\theta_W, 0, 0) \tag{14.50}$$

where θ_W is the weak mixing angle (Weinberg angle) in the SM.

14.5 Neutrino Oscillation, Mass and Mixing

(A) Solar neutrino production

It is by now well-established that energy production in stars takes place through thermonuclear fusion reactions in which lighter elements are fused into heavier elements with the mass excess being transformed into energy in accordance with the Einstein relation $E = (\Delta m)c^2$. The energy so produced is mostly emitted in the form of electromagnetic radiation from the stellar surface. In 'main sequence' stars this mostly involves the fusion of H-atoms to form He-atoms through a number of intermediate steps and accompanied by a number of secondary processes. The reactions can be broadly grouped into what are known as the proton-proton (pp) chain and the carbon-nitrogen-oxygen (CNO) cycle. In the case of the sun (a relatively young and not too hot main sequence star) the pp chain overwhelmingly dominates with the overall process as follows.

$$4\,^1\text{H} \rightarrow 4_{\text{He}} + 2e^+ + 2\nu_e + 25\text{ MeV} \tag{14.51}$$

The neutrinos produced in the solar core carry away only a small fraction of the energy released in the fusion process and being only weakly interacting pass mostly unabsorbed through the solar matter to reach the sun's surface. From there they propagate through space and some of them eventually reach the earth where they may be detected and their flux measured. The sun thus provides a rich source of neutrinos (mostly ν_e type) which can be utilised to derive information about the solar interior (core temperature etc.) and also to study neutrino properties (mass, magnetic moment etc.) and interactions with matter.

(B) Solar neutrino problem

Soon after the first experimental detection of antineutrinos (mostly $\overline{\nu}_e$ type) produced in a nuclear reactor by Reines etal in 1953, Davis etal started an experiment with detectors placed deep underground inside a mine (Homestake mine, south Dakota, USA) to detect and measure the flux of solar neutrinos. They used the chlorine radiochemical method for neutrino detection. Their experiment is still continuing to obtain a better statistics and has provided impetus to several other experiments designed for the same purpose. In all these experiments, it has been found that the observed solar neutrino flux after allowing for all possible corrections is only (30–60)% of the flux theoretically predicted on the basis of the standard solar model (SSM). since the SSM has been very successful in explaining many other experimental data about the sun, the above discrepancy known as the "solar neutrino problem" has led to intensive theoretical studies in order to find a reasonable solution to the problem.

The most plausible solution of the solar neutrino problem found so far is the phenomena of "neutrino oscillations" first suggested by Pontecorvo in the 1960's. The frequency of oscillations between 2 neutrino flavors is a functions of Δm^2 and θ,

$$\Omega_{\text{osc}} = f(\Delta m^2, \theta) \tag{14.52}$$

where $\Delta m^2 = |m_1^2 - m_2^2|$ and θ = neutrino mixing angle. Thus, for neutrino oscillations to occur, there have to be a number of non-degenerate neutrino mass eigen states (at least 2, with one having a non-zero eigen value). Since in the SM, the neutrino is taken to be massless (because of helicity requirement) the solar neutrino problem is often said to provide the strongest observational evidence so far for BSM physics.

(C) Neutrino oscillations

Like baryon number violation, there can be lepton number violating interactions in the SM regarded as an effective theory so that the leptonic $SU(2) \times U(1)$ eigenstates (flavor eigenstates) are not the same as the mass eigenstates leading to mixing between different flavor states and consequent neutrino flavor oscillations. The problem is similar to the $K^0 - \overline{K}^0$ mixing and "strangeness oscillations" (because of strangeness nonconservation in weak interactions). For simplicity, we consider the case of 2 neutrino flavors, which can be easily generalized to the case of 3 or more lepton flavors (generations). In the 2-flavor case, there is only one mixing angle θ. We first consider neutrino propagation through vacuum so that there is no modification due to interaction with matter. Let (ν_e, ν_μ) be the 2 flavor eigenstates and (ν_1, ν_2) the 2 mass (energy) eigenstates. From general postulates of quantum mechanics one set of eigenstates can be written as a linear superpoition of the other set of eigenstates.

$$\begin{pmatrix} \nu_e \\ \nu_\mu \end{pmatrix} = \begin{pmatrix} \cos\theta & \sin\theta \\ -\sin\theta & \cos\theta \end{pmatrix} \begin{pmatrix} \nu_1 \\ \nu_2 \end{pmatrix} \tag{14.53}$$

with the transformation matrix being orthogonal. The time evolution of the flavor eigenstates is given by

$$\begin{aligned} |\nu_e\rangle_t &= \cos\theta|\nu_1\rangle_t + \sin\theta|\nu_2\rangle_t \\ &= \cos\theta e^{-iE_1t}|\nu_1\rangle + \sin\theta e^{-iE_2t}|\nu_2\rangle \end{aligned}$$

and

$$\begin{aligned} |\nu_\mu\rangle_t &= -\sin\theta|\nu_1\rangle_t + \cos\theta|\nu_2\rangle_t \\ &= -\sin\theta e^{-iE_1t}|\nu_1\rangle + \cos\theta e^{-iE_2t}|\nu_2\rangle \end{aligned} \tag{14.54}$$

In the above,

$$\begin{aligned} &H|\nu_j\rangle = E_j|\nu_j\rangle \\ &\langle\nu_j|\nu_i\rangle = \delta_{ij}, \ (i, j = 1, 2) \end{aligned} \tag{14.55}$$

For an initially $(t = 0)$ pure ν_e beam the ν_μ amplitude after time t is given by

$$\begin{aligned} \langle\nu_\mu|\nu_e\rangle_t &= (-\sin\theta\langle\nu_1| + \cos\theta\langle\nu_2|)(\cos\theta e^{-iE_1t}|\nu_1\rangle + \sin\theta e^{-iE_2t}|\nu_2\rangle) \\ &= \sin\theta\cos\theta(e^{-iE_2t} - e^{-iE_1t}) \end{aligned} \tag{14.56}$$

The probability of obtaining a ν_μ at some later time $t(t > 0)$ in the initially $(t = 0)$ pure ν_e beam is then given by

$$P_{\nu_e \to \nu_\mu}(t) = |\langle \nu_\mu | \nu_e \rangle_t|^2 = \sin^2 2\theta \sin^2 \frac{1}{2}(E_2 - E_1)t \tag{14.57}$$

Similarly, the probability of the initially pure ν_e beam remaining pure ν_e after a time t is given by

$$\begin{aligned} P_{\nu_e \to \nu_e}(t) &= |\langle \nu_e | \nu_e \rangle_t|^2 = |(\cos^2 \theta e^{-iE_1 t} + \sin^2 \theta e^{-iE_2 t})|^2 \\ &= 1 - \sin^2 2\theta \sin^2 \frac{1}{2}(E_2 - E_1)t \\ &= 1 - P_{\nu_e \to \nu_\mu}(t) \end{aligned} \tag{14.58}$$

Now, $E_j^2 = m_j^2 + p^2$, $(j = 1, 2)$ and $p = |\vec{p}|$

Assuming $p \gg m_j$, we have

$$E_j \approx p + \frac{m_j^2}{2p} \quad \text{and} \quad E_2 - E_1 \approx \frac{m_2^2 - m_1^2}{2p} \tag{14.59}$$

Substituting (14.59) in (14.57) we get

$$P_{\nu_e \to \nu_\mu}(t) = \sin^2 2\theta \sin^2 \frac{(\Delta m^2)t}{4E} \tag{14.60}$$

$$\text{where } \Delta m^2 = |m_2^2 - m_1^2| \quad \text{and} \quad E \approx |\vec{p}| = p \tag{14.61}$$

The oscillatory term in (14.60) is often written in terms of the distance travelled L, and an oscillation length L_0, where

$$L_0 = \frac{4\pi E}{\Delta m^2} \tag{14.62}$$

L_0 is the distance over which the neutrino returns to its initial state. Eq. (14.60) then becomes

$$P_{\nu_e \to \nu_\mu}(t) = \sin^2 2\theta \sin^2 \left(\frac{\pi L}{L_0}\right) \tag{14.63}$$

In the above we have assumed that the neutrino travels with velocity $v \sim c = 1$ (natural units, $\hbar = c = 1$)

In ordinary units, (14.62) becomes

$$L_0 = \frac{4\pi E \hbar}{(\Delta m^2)c^3} = \frac{2.48E}{\Delta m^2} \text{ metre} \tag{14.64}$$

where E is in MeV and Δm^2 in $(\text{eV}/c^2)^2$. Alternatively if E is in GeV and Δm^2 in (eV/c^2), L_0 will be in Km. From the above we see that smaller is Δm^2, the larger is L_0, and larger must be L in order to have a finite probability to obtain flavor-transformed neutrinos.

From (14.63) we see that the $\nu_e \to \nu_\mu$ oscillation probability reaches maxima at $L = (2n+1)\frac{L_0}{2}$, while it vanishes at $L = nL_0$ (n = integer = 1,2,3,...). Thus for a large mixing angle ($\sin 2\theta \approx 1$), we can identify 3 distinct distance scales of oscillation,

$L \ll \frac{L_0}{2}$	$L \approx \frac{L_0}{2}$	$L \gg \frac{L_0}{2}$
$P_{\nu_e \to \nu_\mu}(t) = 0$	1	1/2

(14.65)

where the last value is obtained by time averaging $P_{\nu_e \to \nu_\mu}(t)$. From (14.65) it is evident that assuming large mixing angle ($\sin 2\theta \approx 1$), one can see the effect of neutrino oscillation for $L \geq \frac{L_0}{2} \cdot E \sim 1$ MeV, $L \sim 10^{11}$ m

For solar neutrinos, $E \sim 1$ MeV, $L \sim 10^{11} m$.

Therefore, to detect neutrino oscillations in solar neutrino experiments on earth,

$$\Delta m^2 \geq 10^{-11} (\text{eV}/c^2)^2 \left(\text{so that } L \geq \frac{L_0}{2} \right).$$

It may be mentioned that soon after the discovery of the neutrino, it was first pointed out by Pontecorvo that solar neutrino oscillation expts. can probe for neutrino mass down to a small fraction of 1 ev, if there is significant mixing between the neutrino species. In fact, this provided the major motivation for starting solar neutrino expts. in the 1960's.

(D) Neutrino oscillations in matter. MSW effect

In the case of propagation through matter, the phenomena of neutrino oscillations are modified compared to propagation through vacuum discussed above due to interaction of the neutrino with matter. The resulting modification was systematically worked out by Mikheyev and Smirnov following an earlier work by Wolfenstein and is known as the MSW effect. The MSW effect is analogous to the effect of a material medium on the propagation of light, which imparts an induced mass to the photon, resulting in an index of refraction. Similarly, due the MSW effect the neutrino propagating through the sun (from the core to the surface) acquires an induced mass.

The electrons in the solar medium have both charged current and neutral current interactions with ν_e but only neutral current interaction with neutrinos of other flavors. Since the neutral current interactions are identical for all neutrino flavors (ν_e, ν_μ, ν_τ) they have no effect on the propagation of solar neutrinos. So we have to consider only the effect of charged current interactions. The corresponding interaction energy is given by

$$H_{\text{int}} = \frac{G_F}{\sqrt{2}} n_e \tag{14.66}$$

where G_F is the Fermi coupling constant and n_e electron density in matter.

In order to study the MSW effect, it is convenient to use the formalism of the mass matrix. After including the effect of matter the phase differences arise from two sources-(1) neutrino mass difference as in the vacuum case and (2) difference in the interactions of the neutrinos of different flavors. As in the vacuum case, we restrict ourselves to the 2-flavor case and assume the neutrino to be ultrarelativistic so that $E \approx p + m^2/2p$ with $p = |\vec{p}| \approx E$. We start with propagation in vacuum and write the mass eigenstates as $|\nu_1\rangle$ and $|\nu_2\rangle$. Using (14.53) and the eqn. of motion of the mass eigenstates,

$$i\frac{\partial}{\partial t}\begin{pmatrix} |\nu_1\rangle \\ |\nu_2\rangle \end{pmatrix} = \begin{pmatrix} E_1 & 0 \\ 0 & E_2 \end{pmatrix}\begin{pmatrix} |\nu_1\rangle \\ |\nu_2\rangle \end{pmatrix}$$

the eqn. of motion of the flavor eigenstates in vacuum can be written in the form

$$i\frac{\partial}{\partial t}\begin{pmatrix} |\nu_e\rangle \\ |\nu_\mu\rangle \end{pmatrix} = \frac{\Delta_V}{2}\begin{pmatrix} -\cos 2\theta_V & \sin 2\theta_V \\ \sin 2\theta_V & \cos 2\theta_V \end{pmatrix}\begin{pmatrix} |\nu_e\rangle \\ |\nu_\mu\rangle \end{pmatrix} = H_V\begin{pmatrix} |\nu_e\rangle \\ |\nu_\mu\rangle \end{pmatrix} \tag{14.68}$$

where $\Delta_V = E_2 - E_1 = \Delta m^2/2E$ and θ_V is the mixing angle in the vacuum (θ_V is the same as θ used before). Including the effects of interaction, the eqn. of motion of the flavor eigenstates in matter can be written in a similar form,

$$\begin{aligned} i\frac{\partial}{\partial t}\begin{pmatrix} |\nu_e\rangle \\ |\nu_\mu\rangle \end{pmatrix} &= \frac{\Delta_M}{2}\begin{pmatrix} -\cos 2\theta_M & \sin 2\theta_M \\ \sin 2\theta_M & \cos 2\theta_M \end{pmatrix}\begin{pmatrix} |\nu_e\rangle \\ |\nu_\mu\rangle \end{pmatrix} \\ &= H_M\begin{pmatrix} |\nu_e\rangle \\ |\nu_\mu\rangle \end{pmatrix} \end{aligned} \tag{14.69}$$

$$\begin{aligned} \text{where} \quad & \Delta_M \cos 2\theta_M = \Delta_V \cos 2\theta_V - 2H_{\text{int}} \\ \text{and} \quad & \Delta_M \sin 2\theta_M = \Delta_V \sin 2\theta_V \end{aligned} \tag{14.70}$$

The matrices H_V and H_M on the rhs in the eqns. of motion (14.68) and (14.69) are known as mass matrices corresponding to neutrino propagation in vacuum and matter respectively. From (14.70) we get

$$\tan 2\theta_M = \frac{\tan 2\theta_V}{1 - a \sec 2\theta_V} \tag{14.71}$$

and $$\Delta_M = \Delta_V [1 - 2a \cos 2\theta_V + a^2]^{1/2} \tag{14.72}$$

where $$a = \frac{2H_{\text{int}}}{\Delta_V} = \frac{2\sqrt{2} G_F n_e E}{\Delta m^2} = \frac{L_0}{L_e} \tag{14.73}$$

where $$L_0 = \frac{4\pi E}{\Delta m^2} = \frac{2\pi}{\Delta_V} \quad \text{and} \quad L_e = \frac{\pi\sqrt{2}}{G_F n_e} \tag{14.74}$$

where L_0 is the oscillation length in vacuum (Eq. 14.62) and L_e is called the neutrino-electron scattering length. In ordinary units,

$$L_0 = \frac{4\pi E \hbar c}{(\Delta m^2) c^4} \quad \text{and} \quad L_e = \frac{\sqrt{2}\pi \hbar c}{G_F n_e} \tag{14.75}$$

Similar to the vacuum oscillation length L_0, one can define a matter oscillation length L_M, where

$$L_M = \frac{2\pi}{\Delta_M} \tag{14.76}$$

L_0 and L_M are related; from (14.72) and (14.73)

$$L_M = L_0 \left[1 - 2\left(\frac{L_0}{L_e}\right) \cos 2\theta_V + \left(\frac{L_0}{L_e}\right)^2 \right]^{-\frac{1}{2}} \tag{14.77}$$

Eq. (14.71) allows for a "resonance" in which the mixing angle θ_M in matter goes through a maximum. The "mixing" is maximum ($\theta_M = \pi/4$) if

$$a = \frac{L_0}{L_e} = \cos 2\theta_V \tag{14.78}$$

We can write the resonance condition (14.78) in terms of a "resonant" electron density" $(n_e)_{res}$, where

$$(n_e)_{res} = \frac{(\Delta m^2) \cos 2\theta_V}{2\sqrt{2} G_F E} \tag{14.79}$$

In ordinary units, (14.79) becomes

$$(n_e)_{res} = \frac{(\Delta m^2) c^3 \cos 2\theta_V}{2\sqrt{2} G_F E \hbar} \tag{14.80}$$

If its energy exceeds the resonance value for the electron density at the solar core, then a ν_e will always pass through resonance as it propagates progressively along regions of smaller electron densities on its way out of the sun. Mixing can become very large at resonance since even for a small θ_V, θ_M can reach the resonance value $\pi/4$. That is why it called matter-enhanced (resonant) oscillation.

It is instructive to examine the behavior of the flavor and mass eigenstates of the neutrino as a function of the electron density n_e. At the large densities near the solar core, (a is a large positive quantity and hence $\tan 2\theta_M$ is a small negative quantity) θ_M is close to $\pi/2(2\theta_M$ is close to $\pi)$ and the ν_e's are in the mass state $|\nu_2\rangle$. As the neutrinos move to regions of lower electron densities there is a very small probability of $|\nu_2\rangle \rightarrow |\nu_1\rangle$ transition, where the energy levels approach closer to each other. Thus the neutrinos are still in the state $|\nu_2\rangle$ when they reach the solar surface. Here, however, $\theta_M \approx \theta_V \approx$ small quantity, so that the neutrino state is mostly $|\nu_\mu\rangle$. Thus in going from the solar core to the solar surface through solar matter some ν_e's have transformed into ν_μ's. It appears from the above analysis that there can be regions in $\Delta m^2 - \theta_V$ space for which the neutrino oscillation modified by the MSW effect produces substantial reduction in the ν_e flux from the solar core.

(E) Neutrino Oscillation Experiments.

(1) Solar neutrinos: The first solar neutrino expt. was carried out by Davis etal. in the Homestake gold mines in South Dakota, USA. The expt. is still continuing to obtain a better statistics and uses the radiochemical method for neutrino detection. The detector consists of 615 tons of perchloethylene and is kept 1500 m undergound to minimize background effects. The reaction used is $\nu_e + {}^{37}_{17}\text{Cl} \rightarrow e^- + {}^{37}_{18}\text{Ar}$ with $E_{th} \approx 0.8$ MeV. Radioactive argon is collected and counted and from the data ν_e-flux coming from the sun is estimated. There have been 2 other expts. Using the radiochemical method. Both involve the reaction $\nu_e + {}^{71}_{31}Ga \rightarrow e^- + {}^{71}_{32}Ge$, with a lower threshold energy $E_{th} \approx 0.2$ MeV. The GALLEX expt. is being done in the Grand Sasso mountains in Italy and uses 30 tons of gallium dissolved as a chloride solution. The SAGE expt. is being done in the Cacausian mountains in Russia and uses 57 tons of metallic gallium. The activated 71_{Ga} is chemically treated before its decay into 71_{Ge} and is counted in a special low-activity proportional counter. The flux of 71_{Ge} so detected is used to compute the flux of incident solar neutrinos. The gallium expts. get the largest contributions from pp neutrinos (55%) followed by $7_{\text{Be}}(25\%)$ and $8_B(10\%)$ neutrinos. The chlorine expt. with a higher threshold gets the largest contributions from 8_B neutrinos (75%) followed by 7_{Be} neutrinos (15%).

Both the Kamiokande and the super Kamiokande expts. are being carried out in the Kamioka mountains in central Japan and is a collaboration between USA and Japan. It is based on the neutrino-electron scattering,

$$\nu_e + e^- \xrightarrow{W,Z} e^- + \nu_e$$

$$\text{and} \quad \nu_\mu + e^- \xrightarrow{Z} \nu_\mu + e^- \tag{14.81}$$

$\nu_e - e$ scattering can take place through Z as well as W exchange, whereas $\nu_\mu - e$ scattering can take place only through Z exchange. Thus one is able to distinguish between the 2 neutrino flavors (ν_e and ν_μ). The detector consists of 2000 tons of pure water in a tank 1 km below ground. The scattered charged particles (e^-) produce Cerenkov radiation which is detected. The Cerenkov radiation also indicates the direction of the incident neutrinos. $E_{th} = 7.5$ MeV for Kamiokande and $E_{th} = 6.5$ MeV for the super Kamiokande expt. Only 8_B neutrinos are detected. In their latest results, the authors claim to have obtained evidence for neutrino oscillations.

A similar experiment has been carried out recently at the Sadbury Neutrino observatory (SNO) located in a deep mine in Canada and is a collaboration among Canada, U.K and USA. The expt. uses 1000 tons of heavy water as the detector and the Cerenkov radiation produced by the emitted charged particles (e^-) is detected by using an array of photomultiplier tubes. Four different processes are studied

$$\text{(i)} \ \nu_e + d \xrightarrow{W} p + p + e^-, \qquad E_{th} = 2.5 \text{ MeV}$$

$$\text{(ii)} \ \nu_f + d \xrightarrow{Z} p + n + \nu_f, \qquad E_{th} = 2.2 \text{ MeV } (f = e, \mu)$$

$$\text{(iii)} \ \nu_e + e^- \xrightarrow{W,Z} e^- + V_e$$

$$\text{(iv)} \ \nu_\mu + e^- \xrightarrow{Z} \nu_\mu + e^- \ E_{th} = 7.0 \text{ MeV} \tag{14.82}$$

Reaction (i) is a charged-current process and is thus sensitive only to ν_e. Reaction (ii) is a neutral-current process and is thus "flavor -blind". Reaction (iii) is sensitive to both charged and neutral currents and reaction (iv) is sensitive only to neutral currents. The great depth of the mine results in a very low cosmic ray background. The relative ratio of the charged and neutral current reactions produces incontrovertible evidence that the neutrinos have undergone flavor oscillations.

(2) Atmospheric neutrinos: The source of atmospheric neutrinos is the decay of $\pi^\pm$ (and also $K^\pm$) which are copiously produced by the collisions of high-energy cosmic rays with the atmospheric nuclei

$N + N \to N + N + \pi, \quad N + N \to N + Y + K(Y = \wedge, \sum, \equiv)$

$\pi^+ \to \mu^+ + \nu_\mu, \quad \mu^+ \to e^+ + \nu_e + \overline{\nu}_\mu$

$\pi^- \to \mu^- + \overline{\nu}_\mu, \quad \mu^- \to e^- + \overline{\nu}_e + \nu_\mu$

One expects a ratio

$$R = \frac{N(\nu_\mu) + N(\overline{\nu}_\mu)}{N(\nu_e) + N(\overline{\nu}_e)} \approx 2 \tag{14.83}$$

However, the observed ratio is significantly smaller and close to 1 in some situations. Recent data by the Kamiokande group reports zenith angle dependence of this ratio, which makes neutrino oscillations a strong candidate to understand the discrepancy between the observed ratio and the theoretically predicted ratio because the change in the neutrino flux can be related to different travel lengths for different zenith angles for the atmospheric neutrinos before they reach the ground-based detectors. The 2 main features of the Superkamiokande (SK) data are the following:

(1) It shows a deficit in the $(\nu_\mu + \overline{\nu}_\mu)$ flux, while the $(\nu_e + \overline{\nu}_e)$ flux agrees with the theoretical prediction. This favors $(\nu_\mu \leftrightarrow \nu_\tau)$ oscillation over $(\nu_\mu \leftrightarrow \nu_e)$ oscillation.

(2) The deficit is seen mostly for the upward going $\nu_\mu, \cos\theta\langle 0$ (θ = zenith angle). It may be noted that the distance travelled by the upward-going neutrino is related to earth's diameter d by

$$L = d|\cos\theta| \tag{14.84}$$

while that of the downward going neutrinos is restricted to the depth of the earth's atmosphere ($\sim$ 10 Km). Hence this result suggests the oscillation length ($L_0/2$) of the atmospheric neutrinos to be of same order as earth's diameter (10^4 Km).

(3) Reactor neutrinos: Nuclear fission reactors produce $\overline{\nu}_e$ copiously with $E_{\text{max}} = 8$ MeV. They result from the neutron β decay

$$n \rightarrow p + e^- + \overline{\nu}_e \tag{14.85}$$

This reaction was utilised by Reines and cowan in the 1950's to carry out an expt. that first experimentally confirmed the existence of the neutrino postulated by Pauli in 1930. Even if flavor oscillations take place among the neutrinos from a nuclear reactor, the available energy is not sufficient to produce charged leptons which can be observed. Hence expts. using reactor neutrinos only measure the decrease in the intensity of neutrinos of a given flavor (disappearance expts). A beam of $\overline{\nu}_e$ of known flux from a reactor is allowed to travel a certain distance and then its intensity is measured. $\overline{\nu}_e$'s from a reactor are detected through the reactions

$$\overline{\nu}_e + p \rightarrow e^+ + n, \quad e^+e^- \rightarrow \gamma + \gamma \tag{14.86}$$

An anomalous behavior in the neutrino spectrum as a function of the neutrino path length L and energy E would be indicative of neutrino oscillations. So far no reactor neutrino expt. has reported a positive evidence for neutrino oscillations.

(4) Accelerator neutrinos: The neutrinos produced in high energy accelerator expts arise in the same way as atmospheric neutrinos, mostly from the decay of charged pions produced copiously in high energy hardon-hardon collisions. Accelerator neutrinos have energies much higher than the reactor neutrinos and are thus capable of producing charged leptons which can be detected. The liquid scintillation neutrino detector (LSND) expt. uses the Los Alamos Meson Physics Facility (LAMPF) for the neutrino source. In the LSND expt. accelerated protons produce $\pi^+(p + p \rightarrow p + n + \pi^+)$ which decay via $\pi^+ \rightarrow \mu^+ + \nu_\mu$, followed by muon decay $\mu^+ \rightarrow e^+ + \nu_e + \overline{\nu}_\mu$. This supplies $\overline{\nu}_\mu$ with $E_{\text{max}} = 52.8$ MeV which enter a tank filled with a liquid scintillator located 30 m from the neutrino source. If the $\overline{\nu}_\mu$ gives rise to $\overline{\nu}_e$ due to neutrino flavor oscillations the resulting $\overline{\nu}_e$ will produce e^+ and n through the reaction, $\overline{\nu}_e + p \rightarrow e^+ + n$. The neutron is captured producing a photon of energy 2.2 MeV through the reaction

$n + p \rightarrow d + \gamma$. Thus the signature of a neutrino oscillation event is the detection of the e^+, followed by a 2.2 MeV photon correlated with the e^+ signal in both space and time. The LSND expt. is consistent with neutrino oscillations provided a proper choice of the neutrino mixing angle θ and $\Delta m^2 = |m_1^2 - m_2^2|$ is made. However, results from the LSND expt. are yet to be confirmed.

(F) Neutrino mass. The "see-saw" mechanism

The simplest and most popular way of giving mass to the neutrino is through what is known as the "see-saw" mechanism. It assumes each SM neutrino to have a right-handed singlet counterpart N_R like the other fermions (quarks and charged leptons). However, unlike the q_R and l_R, the N_R has a unique property. Its gauge charges corresponding to the SM gauge-group $SU(3)_C \times SU(2)_L \times U(1)_Y$ are all zero. Thus it carries no color, weak i-spin or weak hypercharge. It is called a Majorana fermion, since the particle and its antiparticle have the same zero gauge charges. Of course, the Majorana fermion and its antiparticle have opposite lepton number which is, however, not a gauge charge and hence need not be conserved. Consequently, one can have a diagonal mass term coupling the right-handed singlet neutrino N_R with its left-handed antiparticle N_L^C,

$$M\overline{N}_L^C N_R \tag{14.87}$$

M is called a Majorana mass. It can be large, since it does not break any known gauge symmetry. In addition, it can lead to a Dirac mass term through Yukawa interaction with the SM neutrino ν_L and SSB in the same way like other fermions in the SM, i.e. a mass term of the form

$$m(\overline{\nu}_L N_R + h \cdot c) \tag{14.88}$$

Diagonalizing the resulting mass matrix in the $\nu_L - N_R$ basis induces a tiny Dirac mass for the SM neutrino

$$m = \begin{pmatrix} o & m \\ m & M \end{pmatrix} \xrightarrow{\text{diagonalization}} \begin{pmatrix} m^2/M & 0 \\ 0 & M \end{pmatrix} \tag{14.89}$$

The larger the Majorana mass M of N_R, the smaller will be the mass m_ν of the SM neutrino where $m_\nu = m^2/M$. Hence it is called the "see-saw" mechanism. It is a clever method that solves 2 puzzles at one stroke. The large Majorana mass prevents the Majorana neutrino from being observed. It also makes the Dirac mass of the SM neutrino tiny compared to other SM fermions which are characterized by Dirac mass m. It may be noted that many grand unified theories, for example, the $SO(10)$ GUT, have a spinor representation of dimension 16, which can accomodate all the SM fermions of one generation together with a member which transforms like a $SU(5)$ singlet and has precisely the right SM quantum numbers (all zero) to play the role of the "right-handed" neutrino of the "see-saw" mechanism discussed above. Moreover, it can have a large mass $\leq M_{\text{GUT}}$. On the other hand, the Dirac mass m is $\sim yv$ where y is the magnitude of the Yukawa coupling constant and v is the vacuum expectation value (VEV) of the scalar field in the SM.

The most important test of the existence of Majorana neutrinos ($\nu = \overline{\nu}$) is the experimental detection of neutrinoless nuclear double β-decay $(Z, A) \to (Z+2, A)+2e(2n \to 2p+2e^-)$. For a Dirac neutrino, the allowed process is $2n \to 2p + 2e^- + 2\overline{\nu}_{eR}$. Neutrinoless double β-decay has been looked for but has not yet been found (for example in the double β-decay of ${}^{76}_{32}$Ge to ${}^{76}_{34}$Se, $\tau(o\nu)\rangle 5 \times 10^{24}$ yr.) Futhermore, observations of neutrinoless double β-decay will imply the existence of another scale in physics, well below the Planck scale or even the GUT scale.

Table 14.1

Neutrino Source	Neutrino Type	Neutrino Energy(E)	Propagation Distance (L) $(L \geq \frac{L_0}{2})$	Δm^2
Sun	ν_e	$\sim$ MeV	10^{11} m (= 1A.U.)	$\geq 10^{11}(\text{eV}/c^2)^2$
Atmospheric	ν_μ, ν_e	$\sim$ GeV	10^4 km	$\geq 10^{-4}\ (\text{eV}/c^2)^2$
Reactor	$\overline{\nu}_e$	$\sim$ MeV	10^2 m	$\geq 10^{-2}\ (\text{eV}/c^2)^2$
Accelerator	ν_μ	$\sim$ GeV	1 km	$\geq 1\ (\text{eV}/c^2)^2$

In Table 14.2 the neutrino flux is measured in unit of 10^6 cm^{-2} sec^{-1} and the event rate is measured in solar neutrino unit (1 SNU = 10^{-36} events per target atom/sec). The data in column 4 is flux for the radiochemical expts. (first three) and event rate for the scattering expts. (last two)

Table 14.2

Solar neutrino Expt.	Target	E_{th} (MeV)	Flux (EventRate) predicted	observed	Survival probability (observed flux/ predicted flux ratio)
Homestake Mine	Chlorine	0.8	$7.7^{+1.2}_{-1.0}$	2.65 ± 0.23	0.33 ± 0.05
GALLEX	Gallium	0.2	129^{+8}_{-6}	76 ± 8	$0.60_{\pm}0.07$
SAGE	Gallium	0.2	129^{+8}_{-6}	67 ± 8	$0.52_{\pm}0.07$
Kamio Kande(KK)	Water	7.5	$5.5^{+1.0}_{-0.7}$	2.80 ± 0.38	$0.54_{\pm}0.07$
Super Kamio Kande	Water	6.5	$5.15^{+1.0}_{-0.7}$	2.44 ± 0.10	0.47 ± 0.08

To summerize, the solar neutrino problem is best understood in terms of mixing and oscillation between the neutrino flavors ν_e and ν_μ with

$$\Delta m^2 = 8.2 \times 10^{-5}\ (\text{eV}/c^2)^2, \quad \theta = 32° \pm 4°$$

Similarly, the atmospheric neutrino problem is reasonably well understood in terms of mixing and oscillation between the neutrino flavors ν_μ and ν_τ with

$$\Delta m^2 = 2.4 \times 10^{-3} (\text{ev}/c^2), \quad \theta = 45° \pm 10°$$

Table 14.3. Neutrino-producing processes in the sun

$$p + p \to {}^2\mathrm{H} + e^+ + \nu_e \quad (E_\nu \le 0.420 \text{ MeV}) \qquad\qquad p + e^- + p \to {}^2H + \nu_e \text{ (pep)} \quad (E_\nu = 1.442 \text{ MeV})$$

$$99.4\% \qquad\qquad 0.4\%$$

$$p + {}^2\mathrm{H} \to {}^3\mathrm{He} + \gamma$$

$${}^3\mathrm{He} + {}^3\mathrm{He} \to {}^4\mathrm{He} + 2p \qquad {}^3\mathrm{He} + {}^3\mathrm{He} \to {}^7\mathrm{Be} + \gamma \qquad {}^7\mathrm{Be} + p \to {}^8\mathrm{B} + \gamma \qquad {}^3\mathrm{He} + p \to {}^4\mathrm{He} + e^+ + \nu_e(\text{hep}) \quad (E_\nu \le 18.77 \text{ MeV})$$

$$85\% \qquad\qquad 15\% \qquad\qquad 2 \times 10^{-2}\% \qquad\qquad 2 \times 10^{-5}\%$$

$${}^7\mathrm{Be} + e^- \to {}^7\mathrm{Li} + \nu_e \qquad {}^8\mathrm{B} \to {}^8\mathrm{Be} + e^+ + \nu_e$$

$$\begin{bmatrix} E_\nu = 0.861 \text{ MeV}(90\%) \\ 0.383 \text{ MeV}(10\%) \end{bmatrix} \qquad (E_\nu \le 15 \text{ MeV})$$

$${}^7\mathrm{Li} + p \to {}^4\mathrm{He} + {}^4\mathrm{He} \qquad {}^8\mathrm{Be} \to {}^4\mathrm{He} + {}^4\mathrm{He}$$

Problems

1. Assuming that the $SU(3) \times SU(2)$ decomposition of the fundamental representation 5 of $SU(5)$ is $5 = (3,1) \oplus (1,2)$ where the first number in each bracket gives the $SU(3)$ multiplet and the second number the $SU(2)$ multiplet, find the $SU(3) \times SU(2)$ contents of the $SU(5)$ representations 10 and 24. Hence show that (a) all the left-handed states in each fermion generation can be put in the reducible representation $5^* + 10$ of $SU(5)$ and (b) the adjoint representation 24 contains particles which carry both non-zero $SU(3)$ and $SU(2)$ quantum numbers.

2. Verify Eq. (14.20) regarding the value of the Weinberg angle in the $SU(5)$ symmetry limit for $SU(5)$ representation 5^* and 10 individually.

3. Show that the mixing angle θ parameterizing the orthogonal matrix

$$U = \begin{pmatrix} \cos\theta & \sin\theta \\ -\sin\theta & \cos\theta \end{pmatrix}$$

which diagonalizes the symmetric matrix

$$M = \begin{pmatrix} a & \Delta \\ \Delta & b \end{pmatrix}$$

is given by $\tan 2\theta = \dfrac{2\Delta}{b-a}$

and M has eigenvalues $M_{\pm} = \dfrac{1}{2}\left[(a+b) \pm \dfrac{2\Delta}{\sin 2\theta}\right]$

4. The general Lagrangian with both Dirac and Majorana Mass terms is

$$L = \overline{\psi}\partial\!\!\!/\psi + M_D[\overline{\psi}_L\psi_R + h\cdot c] + \frac{M_L}{2}[(\overline{\psi}_L)^c\psi_L + h\cdot c] + \frac{M_R}{2}[(\overline{\psi}_R)^c\psi_R + h\cdot c]$$

where $\partial\!\!\!/ = \gamma^\mu\partial_\mu$. M_D is the Dirac mass and M_L and M_R are the Majorana masses for the left-handed and right-handed neutrinos, respectively.

(a) Show that the above Lagrangian can be written in the form

$$L = \overline{V}\,\partial\!\!\!/V + \overline{V}MV$$

where V is a column vector

$$V = \frac{1}{2}\begin{pmatrix} \psi_L + (\psi_L)^c \\ \psi_R + (\psi_R)^c \end{pmatrix}$$

and M is the symmetric matrix

$$M = \begin{pmatrix} M_L & M_D \\ M_D & M_R \end{pmatrix}$$

M is known as the neutrino mass matrix (For $M_L = 0$, it reduces to the mass matrix in the "See Saw" model)'

(b) Let ψ_1 and ψ_2 be the eigen vector fields of the neutrino mass matrix M with eigen values λ_1 and λ_2, respectively. Then show that L can be rewritten as

$L = \overline{\psi}_1 \not\partial \psi_1 + \overline{\psi}_2 \not\partial \psi_2 + \lambda_1 \overline{\psi}_1 \psi_1 + \lambda_2 \overline{\psi}_2 \psi_2 =$ free Lagrangian for 2 Dirac fields ψ_1 and ψ_2.

Chapter 15

Particle Physics & Cosmology-Astro-Particle Physics

15.1 Introduction

Cosmology involves the study of the origin and evolution of the universe. It is one of those subjects which have interested every inquisitive human mind since the early days of civilisation. However, it is only recently that cosmology has become a reasonably precise science. This is mainly because highly improved and sophisticated instruments and techniques for astronomical observations have now become available.

Modern cosmology is based on the "Big Bang" paradigm according to which the universe originated as a highly dense and extremely hot fireball (of energy and matter) which gradually expanded and cooled over an enormously long period of time ($\sim 14 \times 10^9$ Yrs) to attain the present state and by looking outward in space we can learn about much earlier epochs. The idea was first proposed by Lemaitre, a Belgian priest in the early 1920's and revived again by Gamow, a physicist in the late 1940's. The idea appeared so weird to most people in the beginning that it prompted Fred Hoyle to sarcactically comment that the universe started with a 'big bang'. The idea and the terminology ultimately stayed on and is presently regarded as the basis of the standard model of cosmology.

The theoretical framework of the "Big Bang" cosmology is Einstein's general theory of relativity (GTR) which together with our present understanding of atomic, nuclear and particle physics as well as classical physics not only explained but also correctly predicted many observed features of the universe. However, inspite of many remarkable successes there are some features of the universe which still remain to be explained. These include dark matter and dark energy, baryon-antibaryon asymmetry, inflation and black hole thermodynamics.

In the Big Bang model, one can describe the evolution of the universe starting from a nonzero initial time $t = t_i$ ($t_i > 0$). If we go further back in time, we reach what is known as the point of "singularity" (the point $t = 0$) where everything (temperature and energy density) diverges. Classical gravity (as described by Einstein's GTR) breaks down as one approaches the point of singularity due to quantum effects. However, unlike QED, a quantum theory of gravity (quantum gravity) has not yet been successfully worked out. Many models have been suggested and they all involve higher symmetries and/or higher space-time dimensions.

Big Bang cosmology at present are strongly supported by four well-established astronomical observations. These are:

(1) Expansion of the universe and the Hubble law.

(2) Cosmic microwave background radiation (CMBR)—the "red-shifted" relic radiation of some distant past epoch when matter and radiation decoupled.

(3) Primordial nucleosynthesis and the cosmic abundance of light elements.

(4) Anisotropies in the CMBR which turn out to be of the right magnitude to seed the formation of large-scale structures in the universe (galaxies and clusters of galaxies).

15.2 Expansion of the Universe and the Hubble Law

In the beginning of the 20th century, the universe was thought to be static. Einstein realised that his GTR field equation did not allow for a static solution. The same is the case in Newtonian gravity. This can be easily understood if one imagines a static universe consisting of a number of uniformly distributed galaxies, then since gravitation is an attractive force there will always be a collapse and hence the universe cannot be static. Einstein found that his GTR allowed for another force, which is repulsive and behaves like Λr, over and above the familiar attractive force—GM_1M_2/r^2. The constant Λ is known as the "cosmological constant". For Λ sufficiently small, the repulsive force is not significant at 'small' distances relevant for the solar system but at 'large' distances of cosmological interest this force can provide enough repulsion to stabilise the situation and provide a static universe.

In the 1920's astronomers started to compute the relative velocities of distant galaxies from measurement of Doppler "redshifts" of their spectra. Using the data from astronomical observations Hubble in 1928 proposed a linear relation between velocity and distance. This is the famous Hubble law,

$$v = H_0 r \tag{15.1}$$

where v is the velocity of a galaxy and r is its distance from the earth (or some other reference point). The quantity H_0 is called the Hubble "constant" or more appropriately the Hubble "parameter" since from GR it follows that H_0 is not a constant but can change over cosmological time scales. H_0 may be taken to be the value of a quantity $H(t)$ at present time $(t = t_0)$. Clearly H_0 has the dimension $(\text{time})^{-1}$. It may be mentioned here that after Hubble's discovery, Einstein regretted the introduction of the cosmological constant term in his GTR equations. However, as we shall see later that the cosmological constant has now assumed a great significance from another point of view namely the dark energy problem.

One consequence of the Hubble law is that every galaxy recedes from an observer on the earth as also from any other galaxy with a velocity proportional to its distance from the earth or another galaxy. If one reads this picture backward in time, it follows that there must have been some initial state, where the universe collapsed to a point. One can identify this initial state with the "Big Bang". Taking Eq. (15.1) seriously,

$$t_0 \sim \frac{1}{H_0} \tag{15.2}$$

may then be taken as the lifetime of the universe. According to the GTR, the situation is more complicated (the singularity problem), but it turns out that $1/H_0$ is indeed an upper limit on the lifetime of our universe.

Eq. (15.1) can be written in a different form. Doppler law gives

$$\frac{\lambda_0}{\lambda_e} = \sqrt{\frac{1+\beta}{1-\beta}} \approx 1 + \beta = 1 + \frac{v}{c} \tag{15.3}$$

where λ_e is the wavelength of some electromagnetic radiation emitted from a galaxy which is subsequently observed on earth with wavelength λ_0. From

(15.3) we get

$$z = \frac{\lambda_0 - \lambda_e}{\lambda_e} = \frac{v}{c} \tag{15.4}$$

the quantity z is known as the Doppler "red shift" ($\lambda_0 > \lambda_e$). From (15.1) and (15.4) we get

$$cz = v = H_0 r \tag{15.5}$$

From observational data one finds

$$H_0 = \frac{100h \text{ Km sec}^{-1}}{\text{Mpc}}, \quad 0.5 < h < 1 \tag{15.6}$$

Here 1 Mpc $= 10^6$ pc $= 3.2615 \times 10^6$ LY $\approx 3.1 \times 10^{19}$ km

(1 Yr. $= 3.15 \times 10^7$ sec, 1 LY $= 9.46 \times 10^{12}$ km, 1 Mpc $= 3.08 \times 10^{19}$ km)

The quantity h is a "fuzz factor" which indicates a lack of precise knowledge of H_0. Alternatively, one can express (15.6) as

$$H_0 = \frac{h}{10^{10} \text{ Yr}} \tag{15.7}$$

(15.7) simply expresses that the (upper limit on the) lifetime of the universe given by $H_0 = \frac{h}{10^{10} \text{ Yr}}$ is of order (10–20) $\times 10^9$ Yrs.

The quantity h has been the subject of much discussion in recent years and its value seems to be settling at about $h = 0.73 \pm 0.03$.

15.3 The Cosmological Principle

From the above discussion we see that the Hubble law could be interpreted by assuming that the earth is at the center of the universe and everybody else is moving away from it. This point of view would have been acceptable in the early 16th century, but after Copernicus it appears rather unlikely that the earth is in such a special position.

Another possible interpretation of the Hubble law is that every "point" (and also every direction about it) in the universe is physically equivalent to any other "point". This means that the universe may be assumed to be homogeneous and isotropic around any "point" in it. This is known as the cosmological principle. It simply states that since the earth cannot be at the center of the universe, no one else should have this honor. It is, however, an observational question at which scale the cosmological principle actually works. In other words what is the size of a "point" in the universe. It is clear from looking at the sky that even our Milky way galaxy is not a homogeneous structure. Therefore, a "point" in the universe must be taken to be at least of the size of a galaxy or even larger, perhaps of the size of a cluster of galaxies. Therefore, the cosmological principle is relevant only for a very large-scale description of the universe.

(Historically, the cosmological principle, although logical, appeared a little bizzare at the time it was first proposed. It was taking the principles enunciated by Copernicus to their logical extreme, although there was virtually no observational evidence for the principle at that time.)

In order to see how the cosmological principle leads to receeding galaxies, we consider a 3-dimensional coordinate system with units plotted along the axes. At a certain time, two galaxies ("point") have coordinates say (1, 2, 3) and $(-1, 2, 3)$ plotted in some unit a. Now suppose at some later time, the coordinates of these 2 "points" are the same, but are now plotted in some other unit a' with $a'/a > 1$. We can say that the original unit has been enlarged by a "scalefactor" α, where $\alpha > 1$. Then the two galaxies have moved away from each other with time, not because their coordinates have changed, but because the unit has been enlarged by a scale factor. Also whether one looks at the situation from the point of view of the galaxy located at $(1, 2, 3)$ or the galaxy located at $(-1, 2, 3)$, it does not matter. This means that we can write

$$r(t) = R(t)r_0 \tag{15.8}$$

where $r(t)$ is the distance between 2 "points" (galaxies) in the universe, $R(t)$ is the "scale factor" which is, in general, a function of time and r_0 is a constant. We take

$$r(t_0) = r_0, \quad R(t_0) = 1 \tag{15.9}$$

Then the relative velocity of the two "point" is given by

$$v(t) = \frac{d}{dt}r(t) = \frac{d}{dt}R(t)r_0$$

or $v = \dot{r}(t) = \dot{R}(t)r_0 \quad \left[\text{notation} \; = \frac{d}{dt}x = \dot{x}\right]$

We have

$$H(t) = \frac{\dot{r}(t)}{r(t)} = \frac{\dot{R}(t)}{R(t)} \tag{15.10}$$

$H(t)$ may be identified with the Hubble parameter mentioned before which is, in general, a function of time (because of retarding effects of gravity on the expanding material). The value of the Hubble parameter quoted in (15.6) may be taken to be its value at the present time ($t = t_0$). From (15.4) and (15.11) we can obtain a new version of the Hubble law. We have

$$z = \frac{\lambda_0 - \lambda_e}{\lambda_e} = \frac{v}{c}$$

$$\text{or} \quad \frac{\lambda_0}{\lambda_e} = 1 + z = 1 + \frac{v}{c} = 1 + \frac{H}{c} r$$

Writing $r \approx c(t_0 - t_e)$ disregarding the curvature of the universe and writing $H = H(t_0)$, (15.22) becomes ($c = 1$)

$$\frac{\lambda_0}{\lambda_e} = 1 + H(t_0)(t_0 - t_e) = 1 + \frac{\dot{R}(t_0)}{R(t_0)} \cdot (t_0 - t_e) = 1 - \frac{\dot{R}(t_0)}{R(t_0)} \cdot (t_e - t_0)$$

$$\approx \frac{1}{1 + \frac{\dot{R}(t_0)}{R(t_0)}(t_e - t_0)} = \frac{R(t_0)}{R(t_0) + \dot{R}(t_0)(t_e - t_0)} \approx \frac{R(t_0)}{R(t_e)},$$

using Taylor expansion with $(t_e - t_0)$ small. Therefore we get

$$\frac{\lambda_0}{\lambda_e} = \frac{R(t_0)}{R(t_e)} \tag{15.11}$$

Friedmann-Robertson-Walker (FRW) metric

In order to implement the cosmological principle, we begin by writing the most general form of the metric consistent with the assumption of homogeneity and isotropy of space on a large scale. A metric of this form is known as the Friedmann-Robertson-Walker (FRW) metric. We can derive it by imagining our 3-dimensional space at any instant as a hypersurface in a 4-dimensional space. There should be no preferred direction on this hypersurface. In this way, we will ensure both homogeneity and isotropy. The hypersurface will then be one of constant curvature. For simplicity, we first consider the mathematics required to describe the surface of an ordinary familiar sphere (2-sphere) which can be regarded as a 2-dimensional surface of constant curvature in ordinary 3-dimensional space. For points on the surface of an ordinary sphere, the 3 spatial coordinates satisfy

$$x_1^2 + x_2^2 + x_3^2 = KR^2 \tag{15.12}$$

where $K = +1$ or -1, corresponding to a surface of positive or negative curvature. The line element on the surface for $K = +1$ (positive curvature) is given by

$$dx^2 = dx_1^2 + dx_2^2 + dx_3^2 = (dx_1^2 + dx_2^2) + \frac{(x_1 dx_1 + x_2 dx_2)^2}{x_3^2} \tag{15.13}$$

The equation (15.12) of this hypersurface gives

$$x_3^2 = R^2 - x_1^2 - x_2^2 \tag{15.14}$$

Changing from cartesian coordinates (x_1, x_2) to polar coordinates (r', θ) we have

$$x_1 = r' \cos\theta, \quad x_2 = r' \sin\theta, \quad r' = \sqrt{x_1^2 + x_2^2}, \quad \tan\theta = \frac{x_2}{x_1} \tag{15.15}$$

From (15.13) and (15.15) we get

$$dx^2 = (dr'^2 + r'^2 d\theta^2) + \frac{(r' dr')^2}{R^2 - r'^2} = \frac{R^2 dr'^2}{R^2 - r'^2} + r'^2 d\theta^2 \tag{15.16}$$

It is natural to rescale

$$r' \to r = \frac{r'}{R} \tag{15.17}$$

where R = scale factor. (15.16) becomes after rescaling,

$$dx^2 = R^2 \left[\frac{dr^2}{1 - r^2} + r^2 d\theta^2 \right] \tag{15.18}$$

similarly for $K = -1$ (negative curvature), we get

$$dx^2 = R^2 \left[\frac{dr^2}{1 + r^2} + r^2 d\theta^2 \right] \tag{15.19}$$

We can combine (15.18) and (15.19) into a single eqn.

$$dx^2 = R^2 \left[\frac{dr^2}{1 - Kr^2} + r^2 d\theta^2 \right] \tag{15.20}$$

In (15.20), $K = +1$ for a space of positive curvature and $K = -1$ for a space of negative curvature. We may also include $K = 0$ as a special case, corresponding to a flat space. We can now easily generalize (15.20) to the surface of a 3-sphere which can similarly be regarded as a hypersurface of a constant curvature in a 4-dimensional space. Hence for our usual (3+1) dimensional space-time we can write the FRW metric in the form

$$d\tau^2 = dt^2 - dx_1^2 - dx_2^2 - dx_3^2 = dt^2 - R^2(t)\left[\frac{dr^2}{1 - Kr^2} + r^2 d\Omega^2\right] \quad (15.21)$$

In (15.21) $x_1^2 + x_2^2 + x_3^2 = r'^2 = R^2 r^2$ and $d\Omega^2 = d\theta^2 + \sin^2\theta d\varphi^2$ (x_1, x_2, x_3) are the cartesian coordinates and (r', θ, φ) are the polar coordinates of a point on a 3-sphere. In (15.21), $d\tau$ is the proper time-interval, t is the cosmological time, and $K = +1, -1$ and 0 for a closed (positive curvature) universe, an open (negative curvature) universe and flat (zero curvature) universe, respectively, In general, the scale factor $R(t)$ is a function of time, $R = R(t)$.

The polar coordinates (θ, φ) are the fixed ones and are called comoving coordinates.

By general coordinate transformations, the FRW metric (15.21) can be written in a number of other convenient and commonly used forms encountered in various cosmological models.

For a closed universe ($K = +1$), the FRW metric has a simple intuitive interpretation. Consider the surface of an ordinary 2-sphere where the galaxies (points) are situated. At each point there is isotropy with respect to the 2 dimensions spanned by the surface (i.e. with respect to θ). Now increase the radius of the sphere. Any galaxy (point on the sphere) moves away from every other galaxy (again represented by a point on the sphere), as expected. This space is clearly closed and 2-dimensional. The realistic case is a 3-sphere (described by the 2 polar angles θ, φ). In this case (closed universe), the scale factor $R(t)$ gives the size of the universe (= radius of the 3-sphere).

For an open universe ($K = 1$), there is no similar interpretation of the scale factor. The universe is always of infinite extension except exactly at the Big Bang singularity, where $R(t) = 0$, This is thus a singular point where the notion of space-time breaks down. The same is true for a flat universe ($K = 0$).

15.4 The Basic Equations of the Big Bang Cosmology

The basic equations of the standard "Big Bang" cosmology are derived from the Einstein equation in the general theory of relativity (GTR), using the Friedmann-Robertson-Walker (FRW) metric, which in turn, is a consequence of the cosmological principle (homogeneity and isotropy of space on a large scale). To derive the equations exactly, a knowledge of the GTR is required. However, they can be derived heuristically making use of non-relativistic Newtonian mechanics.

We assume that the universe on a large scale can be described by an ideal gas, the "molecules" being the galaxies. Therefore it is characterized by an energy density ρ and a pressure p. These quantities are independent of space (because of the cosmological principle) but in general can depend on time.

Einstein equations in GTR are of the form

$$R_{\mu\nu} - \frac{1}{2}g_{\mu\nu}R = \frac{8\pi G}{c^4}T_{\mu\nu} \tag{15.22}$$

where $R_{\mu\nu}$ is the symmetric curvature (Ricci) tensor and R is the scalar (Ricci) curvature. $T_{\mu\nu}$ is the energy-momentum tensor and G is the Newton's gravitational constant. We shall use natural units ($\hbar = c = 1$).

$$R_{\mu\nu} = g^{\alpha\beta}R_{\alpha\mu\beta\nu}, \quad R = g^{\mu\nu}R_{\mu\nu} \tag{15.23}$$

where $R_{\alpha\mu\beta\nu}$ is a rank-4 tensor which is symmetric in (μ, ν) and (α, β) but antisymmetric in other pairs.

For a perfect fluid,

$$T_{\mu\nu} = pg_{\mu\nu} + (p + \rho)u_\mu u_\nu$$

$$\text{where} \quad u_\mu = (1, 0, 00) = \text{velocity 4-vector} \tag{15.24}$$

In the FRW metric (Eq. 15.21)

$$R_{00} = \frac{-3\ddot{R}}{R}$$

$$\text{and} \quad R_{ij} = g_{ij}\left[\frac{\ddot{R}}{R} + 2\left(\frac{\dot{R}}{R}\right)^2 - 2\left(\frac{K}{R^2}\right)\right] \tag{15.25}$$

where $R = R(t)$ is the scale factor

$$\dot{R} = \frac{d}{dt}R(t), \; \ddot{R} = \frac{d^2}{dt^2}R(t)$$

using the FRW metric the 0-0 component of the Einstein equation gives

$$\left(\frac{\dot{R}}{R}\right)^2 + \frac{K}{R^2} = \frac{8\pi G}{3}\rho \tag{15.26}$$

and the i-j component of the Einstein equation gives

$$2\left(\frac{\ddot{R}}{R}\right)+\left(\frac{\dot{R}}{R}\right)^2+\frac{K}{R^2}=-8\pi Gp \tag{15.27}$$

using (15.26) in (15.27) one gets

$$\frac{\ddot{R}}{R}=\frac{-4\pi G}{3}(\rho+3p) \tag{15.28}$$

Eqn. (15.26) is known as the Friedmann equation. It is the equation of energy-conservation for the universe, assuming a spherical distribution of mass.

$$\frac{1}{2}\dot{R}^2-\frac{4\pi G}{3}(\rho R^3)\frac{1}{R}=\text{const}=-\frac{K}{2} \tag{15.29}$$

The lhs is the sum of the kinetic energy and potential energy of a body of unit mass ($m = 1$) and $M = \frac{4\pi}{3}\rho R^3$ = mass of the universe. Similarly, Eq. (15.28) is just the "pseudo-Newtonian force law" for a body of unit mass

$$\ddot{R}=\frac{-4\pi G}{3}[(\rho+3p)R^3]\cdot\frac{1}{r^2} \tag{15.30}$$

with $\rho \to \rho_{eff} = \rho + 3p$.

Finally, we have the thermodynamic condition for adiabatic expansion,

$$dE + pdV = TdS = 0 \tag{15.31}$$

which translates into (for a spherical universe)

$$\frac{d}{dt}(\rho R^3)+p\frac{d}{dt}(R^3)=0 \tag{15.32}$$

Since $\frac{d}{dt} = \dot{R}\frac{d}{dR}$ we can write (15.32) in the form

$$\frac{d}{dR}(\rho R^3)+3pR^2=0 \tag{15.33}$$

Eqns. (15.26), (15.28) and (15.33) are the basic equations of the Big Bang cosmology.

In addition to the above 3 equations, we need an "equation of state", which relates p and ρ. We assume a linear relation,

$$p=\omega\rho, \quad \omega=\text{constant} \tag{15.34}$$

Eq. (15.33) then becomes

$$\frac{d}{dR}(\rho R^3)+3w\rho R^2=0 \tag{15.35}$$

Eq. (15.35) can be written in the form

$$\frac{d}{dR}\left[\rho R^{3(1+w)}\right] = 0 \tag{15.36}$$

This means $\rho R^{3(1+w)} = \text{const.}$ (15.37)

We consider 3 cases:

(1) Radiation-dominated (relativistic) universe,

$$p = \rho/3, \ (w = +1/3), \quad \rho \propto 1/R^4 \tag{15.38}$$

(2) Matter-dominated (non-relativistic) universe,

$$p = 0, \ (\omega = 0), \quad \rho \propto 1/R^3 \tag{15.39}$$

(3) Vacuum-dominated (de Sitter) universe,

$$p = -\rho, \ (w = -1), \quad \rho = \text{constant} = \Lambda \tag{15.40}$$

For a de Sitter universe, the energy-momentum tensor takes the Lorentz-invariant form

$$T_{\mu\nu} = \Lambda g_{\mu\nu}, \quad \Lambda = \text{constant} \tag{15.41}$$

Λ can be identified with the cosmological constant.

15.5 Some Consequence of the Big Bang Cosmology

From the "pseudo-Newtonian" force law (Eq. 15.28) we get

$$\ddot{R} < 0, \quad \text{for } (\rho + 3p) > 0 \tag{15.42}$$

and $\ddot{R} > 0, \quad \text{for } (\rho + 3p) < 0$ (15.43)

Since the energy density ρ is, by definition, positive, the first case is true for radiation-dominated universe (early universe), and also for matter-dominated universe (later universe, after radiation and matter decoupled). On the other hand, the second case is true for vacuum-dominated (de Sitter) universe.

If (15.42) is true, then $\dot{R}(t)$ is a decreasing function of t. This means

$$R(t_0) < R(t), \ \ 0 < t \le t_0 \tag{15.44}$$

here t_0 is the "present time" and $t = 0$ corresponds to the time at which the Big Bang occurred. Eq. (15.44) allows us to find an upper limit on the age of the universe

$$\text{AGE} \approx t_0 = \int_0^{t_0} dt \doteq \int_0^{R(t_0)} \frac{dR}{\dot{R}(t)} < \frac{1}{\dot{R}(t_0)} \int_0^{R(t_0)} dR = \frac{R(t_0)}{\dot{R}(t_0)}$$

$$= \frac{1}{H(t_0)} = \frac{1}{H_0} \tag{15.45}$$

$$\text{Therefore, “Age” of the universe} \approx t_0 < \frac{1}{H_0}, \quad H_0 = H(t_0) \tag{15.46}$$

where H_0 is the value of the Hubble constant as measured at the present time t_0 (Eq. 15.1).

As mentioned before, from the measured “red shifts” of distant galaxies it follows that the upper limit on the age of the universe is estimated to be (10–20) $\times 10^9$ yrs. with hopefully more precise results coming in the near future. It will clearly be an important check on the Big Bang model of cosmology whether this limit agrees with the knowledge from other sources on the age of the universe.

From (15.39) it follows that ρ decreases faster than $1/R^3$ with increasing R for $w \geq 0$ (true for radiation-dominated universe and matter-dominated universe). Consequently, the Friedmann equation (Eq. 15.26) implies that

$$\dot{R}^2 \to -K, \quad \text{for } R \to \infty \tag{15.47}$$

This is of course, only possible in a flat ($K = 0$) or open ($K = -1$) universe ($R^2 \geq 0$). For an open universe ($K = -1$) $R(t) \propto t$ for $t \to \infty$. Thus $R \to \infty$ as $t \to \infty$. In a closed universe ($K = +1$), on the other hand, the scale factor $R(t)$ is limited. After the Big Bang R increases with t (the universe expands), but at the point $\dot{R} = 0$, it starts to decrease again (that is, the universe contracts) and “red shifts” are replaced by “blue shifts”. Also ultimately, there will be a time when the universe again collapses to a point ($R = 0$). This is sometimes called the “Big crunch”.

15.6 The Energy Density and the Age Problem

We now discuss attempts to obtain information about the universe by using astronomical observations and the Big Bang model. First, we rewrite the Friedmann equation (Eq. 15.28) in terms of the Hubble parameter, $H = \dot{R}/R$:

$$\frac{K}{H^2 R^2} = \frac{8\pi G\rho}{3H^2} - 1 \tag{15.48}$$

Next, we define a quality ρ_c called the "critical density",

$$\rho_c = \frac{3H^2}{8\pi G} \tag{15.49}$$

with $$\rho_c^0 = \frac{3H_0^2}{8\pi G} \approx 2h^2 \times 10^{-29} \text{ gm/cm}^3 \tag{15.50}$$

Finally, we define a quality Ω, which is the ratio of the density and the critical density,

$$\Omega = \frac{\rho}{\rho_c} = \frac{8\pi G\rho}{3H^2} \tag{15.51}$$

Eq. (15.48) can then be written in the form,

$$\frac{K}{R^2} = (\Omega - 1)H^2 \tag{15.52}$$

From (15.54) we see that $K = 1$ corresponds to $\Omega > 1, K = -1$ corresponds to $\Omega < 1$ and $K = 0$ corresponds to $\Omega = 1$ $(\rho = \rho_c)$. This is easily understood. Since if Ω is large $(\Omega > 1)$, there is a lot of energy, so that the gravitational attraction can counteract the expansion and the universe is closed. Similarly, if Ω is small $(\Omega < 1)$, there is not enough energy of gravitational attraction to counteract the expansion and the universe is open. If $\Omega = 1$, there is an exact balance between expansion and gravitational attraction and one gets a flat universe.

To actually determine whether the universe is open or closed from Eq. (15.54) one needs to know the value of Ω. From observation of luminous (baryonic) matter in the galaxies by astronomers one finds

$$\Omega_{\text{lm}} \approx 0.04 \tag{15.53a}$$

However, from the study of motion of spiral galaxies one knows that there must be a large amount of invisible (non-luminous) matter in the universe which cannot be identified with any known baryonic matter. This must be some non-standard unknown kind of matter and is called "dark matter". From astronomical data, it turns out that

$$\Omega_{\text{dm}} \approx 0.28 \tag{15.53b}$$

(15.53a) and (15.53b) together imply that there must be almost (5–10) times as much dark matter as luminous matter in the universe.

Moreover, from more recent observations one infers that the energy density of the universe is not all accounted for by matter, whether luminous (baryonic) or dark (non-luminous) matter. The large additional contribution to the energy density of the universe must be due to some mysterious source and is termed "dark energy". From astronomical observations one estimates that

$$\Omega_{\rm de} \approx 0.73 \tag{15.53c}$$

Adding the above 3 contributions one gets

$$\Omega = \Omega_{\rm lm} + \Omega_{\rm dm} + \Omega_{\rm de} \approx 1.05 \tag{15.54}$$

This means that contributions to Ω from luminous (baryonic) matter, dark matter and dark energy are (3–5)%, (25–30)%, and (60–70)% respectively. One also knows that Ω cannot be very large because then the expansion of the universe on a large scale would be slowed down so much that the lifetime of the universe would not be long enough to produce old stars. Using this fact one can put a bound on Ω,

$$0.1 < \Omega < 3 \tag{15.55}$$

The present value of Ω, as given by Eq. (15.54) is quite consistent with this bound.

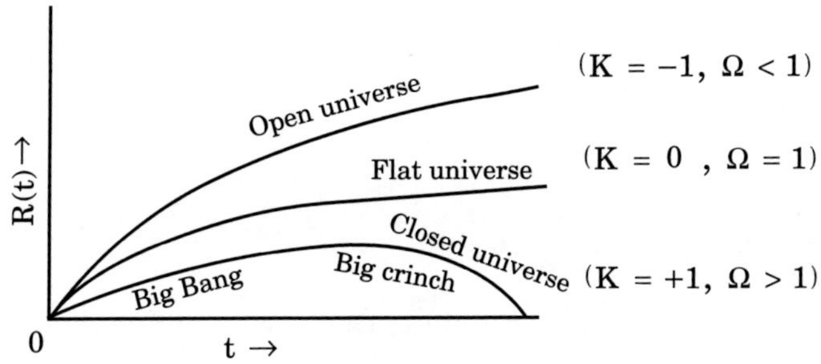

Fig. 15.1

15.7 Solutions of the Basic Equations of the Big Bang Cosmology

From the Friedmann equation (Eq. 15.52) $K = 0$ corresponds to $\Omega = 1$, which is close to the observed (present) value of Ω as given in Eq. (15.54). In our calculations we shall often take $K = 0$ as this simplifies things a lot. For $K = 0$, we get from (15.52) and (15.51).

$$K = 0, \;\; \Omega = 1 = \frac{8\pi G}{3H^2}\rho \tag{15.56}$$

(15.56) gives

$$H = \left(\frac{8\pi G}{3}\right)^{\frac{1}{2}} \rho^{\frac{1}{2}} \tag{15.57}$$

Case 1. Radiation-dominated universe: The early universe was radiation-dominated. In such a universe, one has a relativistic equation of state,

$$p = \rho/3 \quad \text{and} \quad \rho \propto \frac{1}{R^4} \tag{15.58}$$

From (15.57) and (15.38) one gets

$$H \propto \rho^{\frac{1}{2}} \propto \frac{1}{R^2} \tag{15.59}$$

Since $H = \dot{R}/R$, it follows that

$$\dot{R} \propto \frac{1}{R} \tag{15.60}$$

which in turn implies

$$R \propto t^{\frac{1}{2}} \tag{15.61}$$

$$\text{Therefore } H = \frac{\dot{R}}{R} = \frac{1}{2t} \Rightarrow t_0 = \frac{1}{2H_0}, \quad H_0 = H(t_0) \tag{15.62}$$

From (15.57) and (15.62)

$$\rho = \frac{3}{8\pi G} H^2 = \left(\frac{3}{32\pi G}\right) \frac{1}{t^2} \tag{15.63}$$

$$\text{Thus } \rho \propto \frac{1}{R^4} \propto \frac{1}{t^2}, \; H \propto \frac{1}{R^2} \propto \frac{1}{t} \tag{15.64}$$

From statistical mechanics it follows that for a relativistic weakly coupled Bose gas,

$$\rho = (kT)^4 \cdot \left(\frac{1}{\hbar^3 c^3}\right) \cdot \left(\frac{\pi^2}{30}\right) g$$

$$\text{we write } \rho = g\left(\frac{\pi^2}{30}\right) \cdot (kT)^4 \text{ (In natural units } \hbar = c = 1) \tag{15.65}$$

and for a similar Fermi gas

$$\rho = \left(\frac{7}{8}\right) g \left(\frac{\pi^2}{30}\right) \cdot (KT)^4 \tag{15.66}$$

In the above, g is a degeneracy factor, counting the number of physical helicity states of each particle type. From (15.63), (15.65) and (15.66) we get for early universe

$$KT = \left(\frac{30}{2\pi^2 g'}\right)^{1/4} \rho^{1/4} = \left(\frac{30}{2\pi^2 g'}\right)^{1/4} \cdot \left(\frac{3}{32\pi G}\right)^{\frac{1}{4}} \cdot \frac{1}{t^{\frac{1}{2}}} = \frac{A}{t^{\frac{1}{2}}}$$

where $A = \left(\frac{45}{32\pi^3 g'} \cdot \frac{1}{G}\right)^{1/4}$ (15.67)

and $2g' = g_B + \frac{7}{8} g_F$

g' is an effective degeneracy factor, taking into account both fermionic and bosonic degrees of freedom

Evaluating A one finds

$$KT \approx \frac{1 \text{ MeV}}{t^{1/2}}, \ T \sim \frac{10^{10}}{t^{1/2}} K \quad (15.68)$$

where t is in sec. It may be noted that for radiation-dominated epoch,

$$T \propto \rho^{1/4} \propto \frac{1}{R} \propto \frac{1}{t^{1/2}} \quad (15.69)$$

Thus for $t \to 0, R \to 0, H \to \infty, \rho \to \infty, T \to \infty$ (15.70)

Case 2. matter-dominated universe: In such a universe,

$$\rho \propto \frac{1}{R^3}$$

$$H \propto \rho^{1/2} \propto \frac{1}{R^{3/2}}$$

$$H = \frac{\dot{R}}{R} \propto \frac{1}{R^{3/2}}$$

or $\dot{R} \propto \frac{1}{R^{1/2}} \Rightarrow R \propto t^{2/3}$

Therefore, $H = \frac{\dot{R}}{R} = \frac{2}{3t}$ (15.71)

$$\rho = \left(\frac{3}{8\pi G}\right) H^2 = \left(\frac{1}{6\pi G}\right) \cdot \frac{1}{t^2} \quad (15.72)$$

Thus $\rho \propto \frac{1}{R^3} \propto \frac{1}{t^2}, \quad H \propto \frac{1}{R^{3/2}} \propto \frac{1}{t}$ (15.73)

Therefore the ρ and H depend on t in the same way in the 2 cases.

Case 3. Vaccum-dominated universe:

$\rho = \Lambda = \text{constant}$ (15.74)

$$H = \left(\frac{8\pi G}{3}\right)^{1/2} \rho^{1/2} = \left(\frac{8\pi G\Lambda}{3}\right)^{1/2} = \text{const} = H_D \quad (15.75)$$

$$H = \frac{\dot{R}}{R} = H_D = \text{const}$$

Therefore, $R = R_0 e^{H_D t} = R(t)$ (15.76)

The radius of the universe increases exponentially with t in this case.

15.8 The Cosmic Microwave Background Radiation (CMBR)

As discussed above, according to the Big Bang model, as the universe expands the temperature decreases, and matter and radiation cool correspondingly. At a temperature somewhat less than $T = 10^4$ K ("decoupling temperature") which corresponds to KT = 1 eV and happens at $t \approx 10^{12}$ secs. $\approx 10^5$ years after the Big Bang, matter and radiation decouple, because the free electrons combine with positively charged nuclei (ions) to form stable neutral atoms. Prior to that there was electron-photon scattering (Compton scattering) and we do not have any direct observational evidence of what exactly happened before decoupling. After decoupling, the radiation has a black-body spectrum, consisting essentially of the "red shifted" (Doppler shifted) photons left over at the decoupling time. Thus the energy distribution is given by Planck's law.

$$\rho(\nu)d\nu = \frac{h\nu}{e^{hv/KT} - 1} \frac{8\pi\nu^2 d\nu}{c^3} \tag{15.77}$$

where we know how the frequencies red-shift from Eq. (15.23). Assuming thermal equilibrium we therefore have the important relation

$$\frac{T_0}{T_1} = \frac{R(t_1)}{R(t_0)} \tag{15.78}$$

as follows from Eq. (15.69) which says that the temperature behaves like the inverse of the scale parameter. Thus expansion means lower temperature as expected. This relationship was used to predict that at present the temperature of this CMBR should be approximately 5K. In 1965 Penzias and Wilson first detected this radiation as a hiss in their microwave detector independent of the direction of their antenna. NASA'S COBE (Cosmic Background Explorer) satellite found a marvelous agreement with the Planck's law (15.77) for $T = 2.735 \pm 0.06$ K, which is therefore the temperature of the present universe. The CMBR is thus a significant evidence for the correctness of the Big Bang scenario.

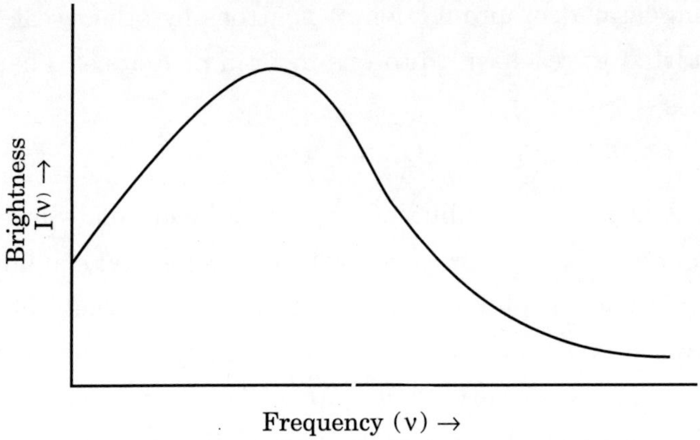

Fig. 15.2: CMBR spectrum measured by the far-infrared Absolute spectrometer aboard the COBE satellite

The angular distribution of the CMBR on earth shows a significant (10^{-3}) anisotropy. The bulk of this can be ascribed to the velocity of the earth relative to the galactic cluster. After allowing for this, a tiny anisotropy still remains. Such an anisotropy is of fundamental importance, as it reflects fluctuations at the level of $\sim 10^{-4}$ in the matter density of the universe on a large scale. Such fluctuations have been found to be exactly of the right order of magnitude to seed the large-scale structure in the universe which developed after radiation and matter decoupled in the distant past.

15.9 Primordial (Big Bang) Nucleosynthesis

From the Big Bang model of cosmology one can compute the expansion rate of the universe, how fast the universe cools, how fast it is slowed down by gravity etc. Knowledge of nuclear physics then allows a calculation of the rates of different nuclear reactions in the early universe.

In the very early universe there were no atomic nuclei, only free quarks, gluons, photons, leptons and weak gauge bosons. The quark confinement mechanism was not yet operative because of very high temperature. When the nucleons started to form after the QCD phase transition, the universe was still too hot ($KT - 1$ GeV) to be energetically favorable to form nuclei. At about $t \sim 1$ sec. after the Big Bang, $KT \sim 1$ MeV, neutrinos decoupled from the evolution process. The loss of neutrons by the decay process was not compensated by production of neutrons by other weak processes. Neutrons started to react with protons to form deuterons. That is to say, nucleosynthesis started

$$n + p \rightleftarrows 2_H + \gamma + Q \tag{15.79}$$

where $Q = 2.22$ MeV = binding energy of the deuteron. Since the cross section is of order 0.1 mb, 2 processes will stay in thermal equilibrium. The huge preponderance of photons over nucleons implied that the deuterons were not "frozen out" until the temperature dropped to about that given by $KT \approx \Delta m/25 \approx 0.05$ MeV ($\Delta m = M_n - M_p$) corresponding to $t \approx 400$ secs after the Big Bang. As soon as the photodisintegration of the deuteron stopped, competing processes leading to ^{4}He production took over.

$$\begin{aligned}
&^2\text{H} + n \to {}^3\text{H} + \gamma \\
&^2\text{H} + p \to {}^3\text{He} + \gamma \\
&^3\text{H} + p \to {}^4\text{He} + \gamma \\
&^3\text{He} + n \to {}^4\text{He} + \gamma
\end{aligned} \tag{15.80}$$

At this time (temperature) the n-p ratio is found to be

$$r = \frac{N_n}{N_p} \sim 0.14 \tag{15.81}$$

The important point is that once neutrons are bound inside deuterons or heavier nuclei they no longer decay and the ratio gets fixed. One finds for the ratio

$$\begin{aligned}
R &= \frac{\text{Amount of } ^4\text{He}}{\text{Amount of } ^1\text{H}} = \frac{\text{No of } ^4\text{He nuclei}}{\text{No of } ^1\text{H nuclei}} \cdot \frac{\text{mass of } ^4\text{He nucleus}}{\text{mass of } ^1\text{H nucleus}} \\
&= \frac{\text{No of } ^4\text{He nuclei}}{\text{No of neutrons}} \cdot \frac{\text{No of neutrons}}{\text{No of protons}} \cdot \frac{\text{mass of } ^4\text{He}}{\text{mass of } ^1\text{H}} \\
&= \frac{1}{2} \cdot r \cdot 4 = 2r
\end{aligned} \tag{15.82}$$

$$\begin{aligned}
f &= \frac{\text{Amount of } ^4\text{He}}{\text{Total mass of nuclear matter}} \approx \frac{\text{Amount of } ^4\text{He}}{\text{Amount of } ^1\text{H} + \text{ Amount of } ^4\text{He}} \\
&= \frac{R}{1+R} = \frac{2r}{1+2r} \approx 0.22
\end{aligned} \tag{15.83}$$

Most of the nuclear matter in the early universe was thus hydrogen. Only around 25% (by mass) was due to Helium.

The ^{4}He abundance has been measured in many celestial sites (the sun, the Jupiter, the solar wind, stellar atmosphere, planetary nebulae, globular clusters and blue compact galaxies). The results are of different precisions but all are consistent. The resultant representative value is $f = 0.24 \pm 0.01$ (after allowing for a small contribution of order 0.04 in the Helium mass fraction from the helium produced subsequently in thermonuclear reactions in stars). The close agreement between the observed value and the theoretically calculated value of the Helium abundance was indeed one of the early successes of the Big Bang model of cosmology. If we did not have this framework there is no reason for this particular relative abundance of hydrogen and helium in the universe.

Some light nuclei heavier than helium (Li, Be...) were also formed by nucleosynthesis in the early universe. However because of mass gaps in the sequence of stable light nuclei (no stable nuclei with A=5 and A=8) inhibited higher mass nuclei formation. Small amounts of primordial Li and Be were produced via the nuclear reactions

$$\begin{aligned} {}^4\text{He} + {}^3\text{H} &\to {}^7\text{Li} + \gamma \\ {}^4\text{He} + {}^3\text{He} &\to {}^7\text{Be} + \gamma \end{aligned} \tag{15.84}$$

The expected amount of 7_{Li} and 7_{Be} and of residual 2_{H}, 3_{H} and 3_{He} are more model-dependent than that of ^{4}He. For ^{3}He, ^{2}H and ^{7}Li a model calculation gives mass fractions $\approx 10^{-5}$, 10^{-4}, and 10^{-9} respectively. These numbers agree well with the observed abundance of these nuclei in the universe. These results of mass abundances of light nuclei are functions of baryon-to-photon ratio (n_B/n_γ) and the number (N_ν) of light neutrinos. It was predicted that $N_\nu = 3$ or 4. This was confirmed in the 1990's by LEP experiments at CERN.

15.10 Baryon-Antibaryon Asymmetry

In the very early universe following the Big Bang hadrons were being continually created and annihilated, the population of any type being determined by statistical weight factors (spin and *i*-spin). The only survivors of the hadron era were the nucleons and antinucleons, the rest disappearing by decay processes. At temperatures well below $KT = m_\pi c^2 \approx 140$ MeV, a population of N and $\overline{N}$ was in equilibrium with radiation, according to the reaction

$$N + \overline{N} \rightleftarrows \gamma + \gamma \tag{15.85}$$

Eventually, a critical temperature T_c was reached ($KT_c \sim 20$ MeV) at which the $N\overline{N}$ annihilation rate fell below the expansion rate of the universe. The antinucleons could no longer find nucleons to annihilate with (since the nucleons were now getting bound in nuclei and no longer created by pair creation) and a residue of nucleons and antinucleons (the only baryons and antibaryons still left) was "frozen out". From the experimental data on $N\overline{N}$ annihilation cross sections, the model expansion rate and knowledge of statistical mechanics one can calculate N_B/N_Y (baryon-to-photon ratio). One also expects $N_{\overline{B}} \approx N_B$. Since there was no further baryon-antibaryon pair creation and annihilation, this ratio should hold even to-day. However, one finds that the calculated value is off by a factor of 10^9 from the observed value and moreover $(N_{\overline{B}}/N_B)_{\text{obs}} < 10^{-4}$. The number is estimated from the observed absence of antinucleons in cosmic rays, in our milky way galaxy and intense gamma ray bursts (GRB) that would have resulted from the annihilation of distant galaxies in collision with antimatter.

The $(N_B/N_\gamma)_{\text{obs}}$ is obtained from the measured photon density in the 2.7 K CMBR and the observed value $\Omega_{lm}(\Omega = \rho/\rho_c)$ of luminous matter. One has

$$(N_\gamma)_{\text{obs}} = 411 \text{ cm}^{-3}$$

and $$(\Omega_B)_{\text{obs}} = (\rho_B)_{\text{obs}}/\rho_c \approx 0.04$$

with $$\rho_c = 2h^2 \times 10^{-29} \text{ gm cm}^{-3} \approx 1.06 \times 10^{-29} \text{ gm} \cdot \text{cm}^{-3}$$

one gets $$(\rho_B)_{\text{obs}} \approx 4 \times 10^{-31} \text{ gm} \cdot \text{cm}^{-3}$$

and $$(N_B)_{\text{obs}} = \text{No. of baryons} = \rho_B/m_N \approx 2.4 \times 10^{-7} \text{ cm}^{-3}$$

Therefore $$\left(\frac{N_B}{N_\gamma}\right)_{\text{obs}} \approx 0.6 \times 10^{-9}$$

A more detailed estimate gives

$$\left(\frac{N_B}{N_\gamma}\right)_{\text{obs}} \approx (4 \pm 2) \times 10^{-9} \tag{15.86}$$

The observed large value of the baryon-antibaryon asymmetry is one of the unsolved problems of modern cosmology.

The Sakharov Criteria

It was first pointed out by Sakharov in 1969 that assuming $B = N_B - N_{\overline{B}} = 0$ initially following the "Big Bang", a baryon-antibaryon asymmetry could develop if the following 3 conditions are satisfied;

(a) presence of Baryon number (B) violating interaction

(b) C and CP violation (to avoid the cancellation of the asymmetry created by B-violating interaction by the inverse process).

(c) freeze-out from thermal equilibrium state.

We have seen in the discussion of BSM theories that GUTs naturally allow baryon number non-conservation (because of the presence of "lepto-quark gauge bosons). C violation is familar to us in weak decay and CP violation has been seen in neutral kaon decays ($K_S - K_L$ system) and is also a natural feature in GUTS. In the discussion of GUTs, we also saw that due to variation of gauge couplings with energy we should expect rates for B-violating processes to vary with energy with GUT symmetry breaking scale $M_{\text{GUT}} \sim 10^{15}$ GeV providing a natural mechanism for decoupling from thermal equilibrium. Thus at least qualitatively, GUTs offer a possible solution to the baryon-antibaryon asymmetry problem. However, the quantitative development of these ideas is not simple. GUTs currently do not predict the asymmetry exactly but they can certainly accomodate the observed baryon-antibaryon asymmetry.

(It is a surprising fact that in the very early universe all matter mostly annihilated with antimatter into photons (or neutral gauge bosons) while only the existence of a tiny CP asymmetry allowed the survival of some matter from which all the structure in our present universe originated.)

15.11 Problems with the Standard Big Bang Model

Although the Big Bang model of cosmology has been highly successful in accounting for some of the most striking features of the universe-it suffers from some serious basic difficulties, the solution of which requires an important addition or modification of the standard model. Many of these are concerned with the early universe. These difficulties can be enumerated as follows.

(a) The Horizon (causality) problem

The Big Bang model is based on the cosmological principle according to which the universe is homogeneous and isotropic on a large scale. The correctness of this assumption is strikingly reflected in the COBE data on the CMBR which reveal an anisotrophy in the CMBR to about 1 part in $10^5 (\Delta T/T \leq 10^{-5})$. This appears very surprising. As we look out at distances 14×10^9 light yrs. away, points seperated by a tiny fraction of a degree were seperated at 10^5 yrs. after the Big Bang by an enormous distance compared to the "horizon" (the physical distance out to which we can see) at that time. So if these points have not been in causal contact by recombination time, there is no reason for their having almost identical temperatures. The question is more serious if we consider the early universe. This difficulty is known as the horizen (causality) problem.

If a light signal can propagate from one point to another point, it means that the physical processes at the second point can be influenced by conditions at the first point. In this way one can understand how the universe can be homogeneous and isotropic on a large scale, provided the universe is "contained" within a causal distance. However, one can show that this is not case in early universe and this gives rise to the casuality problem.

For propagation of light in flat space one has $dl = cdt$ (in natural units, $h = c = 1$). However, in cosmology one must distinguish between the physical distance and the comoving distance. If dl is the comoving distance then the physical distance is comoving distance scaled by the scale factor $R(t)$. So for propagation of light we therefore have

$$R(t)dl = dt$$

Integrating we get

$$l_{12} = \int_1^2 dl = \int_1^2 \frac{dt}{R(t)} \tag{15.87}$$

The physical distance is then given by

$$d_{12} = R(t_2)l_{12} = R(t_2)\int_{t_1}^{t_2} \frac{dt}{R(t)} \tag{15.88}$$

For radiation-dominated (relativistic) universe

$$R(t) = At^{1/2} \tag{15.86}$$

Substituting (15.61) in (15.88) we get

$$d_{12} = 2t_2 \left[1 - \left(\frac{t_1}{t_2}\right)^{1/2}\right] \approx 2t_2, \quad \text{for } t_2 \gg t_1 \tag{15.90}$$

Now in the early universe, all times are small ($t \ll 1$ sec) and therefore

$$d_{12} \sim 2t_2 \ll R(t_2)$$

$$\frac{R(t_2)}{d_{12}} \sim \frac{1}{\sqrt{t_2}} \tag{15.85}$$

Since R sets the scale of the physical distance, it follows that the early universe is "causally disconnected", since there can be a large number ($\sim 1/\sqrt{t}$) of causal domains inside a typical scale. Therefore one may ask how it is possible for two points in the early universe to know that they are supposed to have the same temperature. This difficulty is particularly transparent if we consider photons emitted from the two opposite sides of the sky. As already mentioned, it is observationally found that there is thermal equilibrium at the same temperature to a very high precision (COBE data on the CMBR), although there was no possibility for these regions of space to have causal contact before the photons were emitted from them. The Big Bang model does not explain this.

(b) The Flatness Problem

Let us consider the neighbourhood of the value $\Omega = 1$ ($\Omega = \rho/\rho_c$). For this we write the Friedman equation (Eq. 15.52) in the form

$$\frac{K}{R^2} = (\Omega - 1)H^2 \tag{15.92}$$

From the "pseudo-Newtonian" force law (Eq. 15.44) we know that $R(t)$ increases as we move backwards in time. Eq. (15.92) then implies that $|\Omega - 1|$ must decrease as we move backwards in time. Since we know from astronomical data that at the present time Ω is not far from 1, it follows that in the distant past (early universe) Ω must have been extremely close to 1.

For example, since for radiation-dominated universe, $R(t) \propto \sqrt{t}$, $R(t) \propto 1/\sqrt{t}$ and therefore

$$|\Omega - 1| \propto \frac{1}{t}$$

therefore if $|\Omega - 1| = \delta (\delta < 1)$ at the present time ($t = 10^{17}$ sec after the Big Bang), $|\Omega - 1|$ should have been $10^{-17\delta}$, $10^{-50\delta}$ and $10^{-60\delta}$ at primordial nucleosynthesis time ($t = 1$ sec), GUT time ($t \approx 10^{-33}$ sec) and Planck time ($t \approx 10^{-43}$ sec), respectively. This extreme "fine tuning" of the value of Ω (the range of the values is definitely enormous) is clearly not explained by the standard Big Bang model of cosmology. Since $\Omega = 1$ corresponds to a flat universe ($K = 0$), this difficulty is often referred to as the "Flatness problem".

(c) The Monopole Problem

Grand unified theories (GUT) predict the existence of magnetic monopoles. They have mass (M_m) of order $M_X/(g^2/4\pi)$, where M_X is the symmetry breaking scale, $SU(5) \xrightarrow{M_X} SU(3) \times SU(2) \times U(1)$, and g is the $SU(5)$ gauge coupling. So unless the monopole density (n_m) is many orders of magnitude smaller than the baryon density (n_B), their energy density ($\rho_m = n_m M_m$) will be much larger than the observed energy density of the universe. Astrophysical limits turn out to be even smaller. Passing through a galaxy, monopoles would deplete the galactic magnetic field. This sets a limit, known as the Parker bound, on the monopole flux in the galaxy.

$$F_m \leq 10^{-16} \left(\frac{M_m}{10^{17}\ \text{GeV}}\right) \text{cm}^{-2}\text{s}^{-1}\text{sr}^{-1} \qquad (15.93)$$

On the other hand, we might expect in GUTs quite extensive monopole production. We know that monopoles like other topological objects are defects which are produced during the phase transition from the symmetry phase to the broken symmetry phase in the GUTs. One expects twists in the fields during the phase transition on scales of order the horizen distance at that time (GUT time) and a monopole density (n_m) of order one per horizen volume. If the phase transition occurs at $KT = M_X$, the horizen distance (d) is order $d \sim M_P/(KT)^2$, where M_P = Planck mass and the monopole density (n_m), will be

$$n_m \sim \frac{1}{\text{horizen volume at GUT time}} \sim \frac{1}{d^3} \sim \frac{(KT)^6}{M_P^3}$$

Since the photon density, $n_\gamma \sim (KT)^3$,

$$\text{we have } \frac{n_m}{n_\gamma} \sim \frac{(KT)^3}{M_P^3} \sim \left(\frac{M_X}{M_P}\right)^3 \qquad (15.94)$$

and can be larger than n_B/n_γ depending on the value of M_X.

Taking $M_X \sim 10^{15}$ GeV, $M_P \sim 10^{19}$ GeV

$$\frac{n_m}{n_\gamma} \geq 10^{-12} \sim 10^{-3}\left(\frac{n_B}{n_\gamma}\right) \tag{15.95}$$

The monopole energy density is then

$$\rho_m = n_m M_m = \frac{M_X}{\alpha} \cdot 10^{-3} n_B \sim 10^{11} \rho_c \ (\alpha = g^2/4\pi) \tag{15.96}$$

a number totally inconsistent with the density (and age) of the universe.

(d) The Small-scale Inhomogeneity Problem

Most of our present knowledge of the universe depends on the fact that on the "small" scale the universe is not homogeneous but consists of stars, planetary nebulae, galaxies, galaxy clusters and many other interesting astronomical objects. Such structures require density perturbations $(\delta\rho/\rho)$ of magnitudes $\sim 10^{-3}$ on the scale of galaxies at decoupling time ($t \sim 10^5$ yrs. after the Big Bang) which would thereafter grow due to gravitational forces. The standard Big Bang cosmology does not account for such primordial density perturbations. Despite intense theoretical effort this is still an open question in cosmology.

(e) The cosmological constant problem

The observed energy density of the universe is not all accounted for by radiation and matter (both baryonic or luminous and dark matter). The difference is attributed to some mysterious source known as "dark energy". One usually associates dark energy with the cosmological constant (Λ) first introduced by Einstein in GTR in a different context (the stability of the universe).

We know that renormalizable gauge theories need spontaneous symmetry breaking (SSB) to generate mass terms in the Lagrangian which in turn requires scalar fields (Higgs fields) having a nonzero vacuum energy density. When gravity is included in the overall theory such a vacuum energy density generates an effective cosmological constant (Λ). A proper analysis shows that if SSB occurs at a mass scale M_X then the corresponding vacuum energy density $\rho_\nu \propto M_X^4$. For GUT phase transition $M_X \sim 10^{14} - 10^{16}$ GeV. However, we know that the present value (Λ_0) of the cosmological constant is $\leq \rho_{co}$, where ρ_{co} is the present value of the critical density of the universe

$$\rho_{co} = \frac{8\pi G}{3H_0^2} = 2h_0^2 \times 10^{-29}\ \text{gm/cm}^3 \sim 10^{-46}\ (\text{GeV})^4$$

Therefore, $\dfrac{\Lambda_0}{M_X^4} \leq 10^{-110}$, assuming $M_X = 10^{16}$ GeV (15.97)

One could imagine that in addition to an energy density generated by the Higgs field there was a pre-existing density which precisely cancelled the Higgs density. But the "fine tuning" required is again highly implausible unless there is some underlying mechanism which ensures absolute cancellation. This required fine tuning is known as the cosmological constant problem.

15.12 Inflation

The nature of some of the problems outlined above suggests that they may be overcome if there was a brief period in the early history of the universe when the rather smooth evolution postulated in the standard Big Bang model was interrupted by much more dramatic and rapid change. This idea is the basis of the "Inflation" hypothesis first proposed by Guth and Kazanas in the early 1980's and subsequently developed by them and others. Inflation requires a period in the early universe when the expansion was "exponential" instead of following a simple power law. As we have already seen this happens in the case of a vacuum-dominated universe which is described by a Lorentz-invariant energy momentum tensor

$$T_{\mu\nu} = \Lambda g_{\mu\nu} \tag{15.97}$$

with $T_{00} = \rho = \Lambda$ and $T_{ii} = p = -\Lambda$. Hence the equation of state for such a universe is

$$p = -\rho \tag{15.98}$$

Thus a vacuum-dominated universe is driven by a "negative" pressure. In inflationary theories this vacuum is known as the "false" vacuum to distinguish it from the true vacuum which is the state of the lowest possible energy.

The eqn. of state (15.98) has profound consequences upsetting the standard notion that gravity is attractive. To see this consider the pseudo-Newtonian eqn. of motion (Eq. 15.28)

$$\ddot{R} = -\frac{4\pi G}{3}(\rho + 3p)R \tag{15.28}$$

Usually (15.28) ensures that $\ddot{R} < 0$ when the pressure is positive $(p > o)$ since ρ is by definition positive. However, for the false vacuum (15.28) becomes

$$\ddot{R} = +\left(\frac{8\pi G\Lambda}{3}\right) R = H_D^2 R \tag{15.99}$$

$$\text{where} \quad H_D = \left(\frac{8\pi G\Lambda}{3}\right)^{1/2} = \text{constant} \tag{15.100}$$

We see that gravity is repulsive as long as the false vacuum is the energetically accessible state of the universe. (15.99) can be easily solved and we obtain

$$R(t) = R_0 e^{H_D(t-to)}, \quad R_0 = R(t_0) \tag{15.101}$$

where we have ignored a possible exponentially decreasing term. This "exponential expansion" is clearly driven by negative pressure.

In the above, we have assumed that Λ and hence H_D is a constant. This also follows from the thermodynamic condition for adiabatic expansion (Eq. 15.33). The condition gives for a spherical universe

$$\frac{d}{dt}(\rho R^3) + p\frac{d}{dt}(R^3) = 0 \tag{15.102}$$

using (15.98) we get from (15.102) the condition

$$\frac{d\rho}{dt} = 0 \tag{15.103}$$

i.e. the density of such a universe does not change with time. In natural units, any energy density has dimension $(\text{mass})^4$. Thus we can estimate $\Lambda \sim M_X^4$, where M_X would be a typical mass (or energy) in the early universe. It is natural to think of the GUT scale with $M_X \sim 10^{15}$ GeV. This gives

$$\Lambda \sim (10^{15}\ \text{GeV})^4 \tag{15.104}$$

This is an enormous energy density. If the sun should have this energy density, it would be compressed to the size of a proton.

From Eq. (15.101) we can compute the Hubble parameter

$$H = \frac{\dot{R}}{R} = H_D = \left(\frac{8\pi G\Lambda}{3}\right)^{1/2} = \text{constant} \tag{15.105}$$

So in the inflation scenario the Hubble parameter is a genuine constant. Inserting (15.104) in (15.105) we get

$$H = H_D \approx 10^{10}\ \text{GeV} = 10^{34}\ \text{sec}^{-1} \tag{15.106}$$

This is a correct estimate, if inflation is valid upto the GUT scale. During inflation the universe continues to supercool and expand exponentially but the energy density remains fixed at $\rho = \Lambda$. Thus the total energy ($\sim \rho R^3$) increases with time. However, there is no violation of conservation of energy principle. The extra energy comes from gravity doing work against the negative pressure.

If the duration of the exponential expansion is Δt, then during the inflation, the universe expands by a factor

$$z = \frac{R(t+\Delta t)}{R(t)} = e^{H_D(\Delta t)} \tag{15.107}$$

and the curvature is reduced by a factor $1/z^2$ (with the value of H_D given above one finds $z \sim 2 \times 10^4$ for $\Delta t = 10^{-33}$ sec)

"Inflation" solves some of the difficulties of the standard Big Bang model.

The Flatness problem is solved because it arises from the smallness of the ratio of the curvature term to the energy density term. During the inflation the energy density remains approximately constant but the curvature term decreases by a factor $1/z^2$ so that the problem is solved if $z \geq 10^{30}$. After the inflation Λ must decrease to its present value which is extremely small compared to M_X^4 and so $\Omega \approx 1$ within some small limits.

The Horizon problem is also solved. Our observable universe at the present time has a size $\sim 10^{26}$ cm. At the start of the inflation at a temperature corresponding to about 10^{14} GeV ($t \sim 10^{-34}$ sec) the size was only about 10 cm within which volume the thermal equilibrium reigned.

The Monopole problem is solved since within the region of small volume at the start of the inflation the Higgs field would have been aligned in the lower energy configuration so that there would have been $\leq$ one monopole within the universe.

Inflation models are currently grappling with the small-scale inhomogeneity problem. Quantum fluctuations may possibly seed the inhomogenei ties. However, the mechanisms required to account for the observed situation are not as yet established.

Inflation does not offer a solution to the cosmological constant problem.

15.13 Models of Inflation

A variety of models exists which generate inflation by means of evolution of a weakly coupled scalar field ϕ which was initially displaced from the minimum of the scalar potential $V(\phi)$. A mechanism of this kind was proposed originally by Guth and others who suggested that at the GUT phase transition, $SU(5) \to SU(3) \times SU(2) \times U(1)$, a suitable choice of the potential $V(\phi)$ leads to the universe being trapped in the symmetric phase for a period before passing to the broken symmetry phase. The universe is thus supercooled.

Let us consider a single scalar field ϕ moving in the Higgs potential $V(\phi)$,

$$\pounds = \frac{1}{2}(\partial^\mu \phi)(\partial_\mu \phi) - V(\phi)$$

$$V(\phi) = \frac{\mu^2}{2}\phi^2 + \frac{\lambda}{4}\phi^4 + \frac{\mu^4}{4\lambda} = \frac{\lambda}{4}\left(\phi^2 + \frac{\mu^2}{\lambda}\right)^2 \tag{15.108}$$

The eqn. of motion

$$\partial_\mu \left\{ \frac{\partial \pounds}{\partial(\partial_\mu \phi)} - \frac{\partial \pounds}{\partial \phi} \right\} = 0$$

becomes

$$\Box\phi + \frac{\partial V}{\partial \phi} = 0, \ \Box = \partial_\mu \partial^\mu \tag{15.109}$$

From (15.108) we get

$$\frac{\partial V}{\partial \phi} = \mu^2\phi + \lambda\phi^3 = \lambda\phi\left(\phi^2 + \frac{\mu^2}{\lambda}\right)$$

$$\frac{\partial^2 V}{\partial \phi^2} = \mu^2 + 3\lambda\phi^2 \tag{15.110}$$

$$\text{For } \ \mu^2 > 0, \ \frac{\partial V}{\partial \phi} = 0 \ \Rightarrow \ \phi = 0, \left(\frac{\partial^2 V}{\partial \phi^2}\right) = \mu^2 > 0 \tag{15.111}$$

Thus $V(\phi)$ has a minimum at $\phi = 0$

For $\mu^2 < 0, \ \mu^2 = -|\mu|^2$

$$\frac{\partial V}{\partial \phi} = 0 \Rightarrow \phi = 0,\ \phi = \pm v,\ v = +\frac{\sqrt{-\mu^2}}{\lambda} \tag{15.112}$$

$$\left(\frac{\partial^2 V}{\partial \phi^2}\right)_{\phi=0} = \mu^2 < 0$$

$$\left(\frac{\partial^2 V}{\partial \phi^2}\right)_{\phi=\pm v} = \mu^2 + 3\lambda v^2 = -|\mu^2| + 3\lambda\frac{|\mu|^2}{\lambda} = 2|\mu|^2 > 0 \tag{15.113}$$

Thus $V(\phi)$ has minimum at $\phi = \pm v$ and a local maximum at $\phi = 0$.

$$V(\phi)|_{\phi=0} = \frac{|\mu|^4}{4\lambda},\ V(\phi)|_{\phi=\pm v} = 0 \tag{15.114}$$

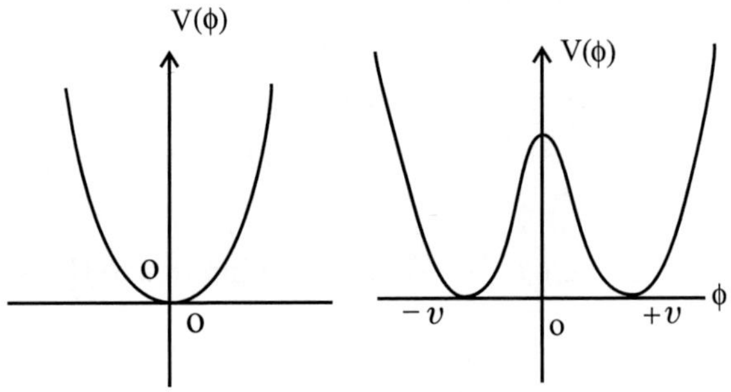

Fig. 15.3

In Inflationary scenario, $\phi = 0$ is known as the "false vacuum" and $\phi = v$ as "true vacuum". The potential energy associated with the false vacuum is identified with the cosmological constant Λ with $|\mu| \sim M_X$ (GUT scale) and $\lambda \sim 1$.

The above considerations apply to field theories at zero temperature ($T = 0$). For finite temperature field theories, $N(\phi)$ is modified to be of the form

$$V(\phi) = \frac{1}{2}\left[\left(\frac{T}{T_c}\right)^2 - 1\right]|\mu|^2\phi^2 + \frac{\lambda}{4}\phi^4 + \frac{|\mu|^4}{4\lambda} \tag{15.115}$$

where T_c is the critical temperature at which the phase transition occurs ($T_c \sim |\mu|\sqrt{\lambda}$).

Restricting to the region $\phi > 0$, the forms of $V(\phi)$ at different temperatures are shown in figure 15.4 below.

Before inflation ($T > T_c$), the vacuum energy density has a minimum with a value corresponding to $V(\phi = 0) = |\mu|^4/4\lambda$, with $|\mu|^2 \sim M_X$ and hence a large cosmological constant $\Lambda \sim M_X^4 \sim 10^{60}$ (GeV)^4 leading to inflation. The phase transition occurs at $T = T_c$ with $KT_c \sim |\mu|\sqrt{\lambda}$ and a sufficient expansion will take place if the inflation lasted for a time $\Delta t \geq 10^{-32}$ sec. The inflation gives the universe a fresh start and the nature of the original configuration becomes largely irrelevant to the subsequent evolution of the universe.

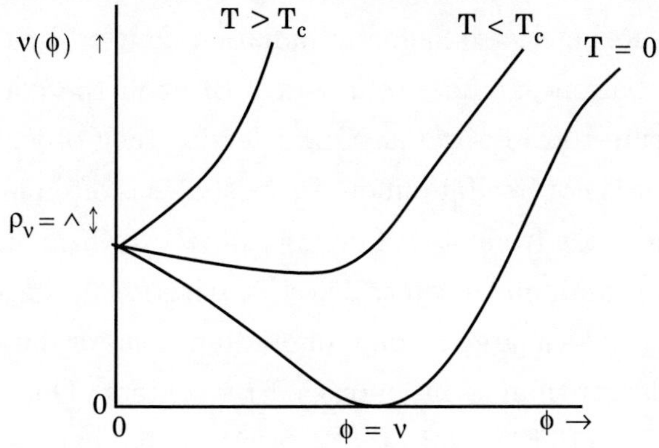

Fig. 15.4

As T falls values $\ll T_c$ the asymmetric minimum at $\phi = 0$ becomes metastable. Inflation occurs when the "false vacuum" or the supercooled state decays by penetrating the potential barrier by means of either quantum tunnelling or as a result of thermal fluctuations.

After the phase transition the latent heat released reheats the universe, Λ becomes small and the evolution of the universe takes place in accordance with the standard Big Bang model. The effective source of matter and radiation in the universe is thus the expansion of the "false vacuum". Gravitational energy is transformed into the energy of the "false vacuum" which in turn is converted into matter-antimatter (particle-antiparticle) and radiation following the phase transition.

There are several classical analogies for this picture. For example, we can envisage bubbles of the broken symmetry phase (true vacuum) being formed within the symmetric phase (false vacuum) in the same way as in a condensed matter system. However as it involves potential barrier penetration, the nucleation rate is very slow compared to the exponential expansion rate of the environment. As a result the bubbles of the broken phase do not coalesce and the transition is never completed. This problem is avoided in more recent models by Linde, Albrecht, Steinhardt and others.

15.14 Dark Matter

One of the most intriguing and important questions for modern cosmology concerns the existence of what is commonly known as "dark matter".

Direct observations of astronomical objects and associated phenomena depend largely on the electromagnetic radiation emitted by the object concerned. The radiation extends from γ-rays through the complete electromagnetic spectrum to radio waves. One refers to such objects as luminous matter. If matter is not itself luminous its existence is often inferred from its absorption of radiation from other luminous objects or from its gravitational interactions with luminous matter. These gravitational effects provide evidence that there exists a large amount of non-luminous or dark matter in the universe, much larger than its luminous matter content. One useful quantity is Ω defined in Eq. (15.51), which is the ratio of the density to the critical density. Astronomical data indicate that while $\Omega_{lm} \sim 0.04$, $\Omega_{dm} \sim 0.28$.

The most convincing evidence of the existence of dark matter comes from the observation of the velocity of the luminous matter as a function of the distance from the centre (the rotation curve) for spiral galaxies. Spiral galaxies are chosen because they happen to have their planes tilted at an angle which allows measurement of the circumferential velocity along the line of sight from the observed Doppler shifts. The luminosity falls off exponentially with radial distance from the centre of the galaxy. Beyond the central region the measured velocity is found to be nearly constant out to 30 Kpc or more. For constant v, Newtonian gravity requires the mass density $\rho(r)$ to be proportional to $1/r^2$.

The natural interpretation of the observed variation of v as a function of r is that there is a halo of dark matter around the galaxy and that this halo extends to a large radius with the ratio of luminous matter to dark matter decreasing with increasing r.

Evidence of the existence of dark matter also comes from the relative motion of the components in apparently bound galaxy clusters. These studies require masses greater by a factor of ~ 10 than the masses inferred from the internal dynamics of the luminous parts of the galaxies.

A further indicator of the existence of dark matter on the large scale is the ratio of mass to luminosity. If we normalize to the sun $M_\odot/L_\odot = 1$ then it is found that M/L ratio increases from values (1–10) for smaller-scale structures (galaxies) to values (50–300) for larger scale structures (galaxy clusters and superclusters) and the corresponding values of Ω increases from 0.03 to 0.3.

Nature of the Dark Matter

There seems little doubt that dark matter exists and indeed dominates the mass of the universe. A natural question concerns the constituents of the dark matter.

One possibility to consider is the dust cloud or high mass star remnants (while dwarfs, brown dwarfs, neutron stars, black holes etc.). Astronomers reject this possibility on the ground that the existence of such matter implies an abundance of high-Z nuclear matter inconsistent with the observed composition of visible stars.

Another hypothesis in terms of known phenomena explains the dark matter as consisting of very large numbers of low mass low luminosity stars or objects with mass much lower than the solar mass $M_{\ominus}$ such that they do not form stars. However such objects have not been found.

Primordial black holes have also been proposed as a solution to the dark matter problem. Small black holes would have evaporated by the present era and there is no evidence for the necessary number of large black holes to explain the observed amount of dark matter.

Although none of the above possibilities is a plausible solution to the dark matter problem on its own, each of them might make some small contributions. Nevertheless, the dark matter phenomenon seems to demand a more radical and unconventional explanation. Four types of Beyond-standard model (BSM) particles are presently being considered as candidates for non-baryonic dark matter. These are (a) massive neutrinos (b) axions (c) lightest supersymmetric particles (LSP) and (d) weakly interacting massive particles (WIMPS).

At one time neutrinos with mass seemed attractive candidates for the non baryonic dark matter on the grounds that at least the existence of neutrinos are well established and the neutrino mass is not required to be exactly zero by any over-riding theoretical principle. However, there are now reasons to believe that neutrinos do not constitute dark matter. From the fact that neutrinos decoupled at $KT \sim 3$ MeV and $t \sim 1$ sec. one can estimate the present neutrino density in the universe. If the three neutrino species have masses $m_{\nu_i}(i = 1, 2, 3)$ then for massive neutrinos one has a mass density.

$$\rho_\nu = n_\nu \sum_i m_{\nu_i} \approx 2 \times 10^{-31} \sum_i m_{\nu_i} \text{ (eV)gm cm}^{-3}$$

The present observed value of Ω_{dm} then requires

$$\sum_i m_{\nu_i} = (25 - 100) \text{ eV/c}^2.$$

Present limits on m_{ν_e} appear to exclude the ν_e as a possible constituent of dark matter. However ν_μ and ν_τ are not yet ruled out.

There is another argument which appears to disfavour the massive neutrinos as dark matter candidates. This argument is related to the problem of galaxy formation which is not fully understood. The neutrinos were certainly relativistic at decoupling time and are "hot dark matter" (HDM) candidates as opposed to "cold dark matter" which is made of slow massive (nonrelativistic) particles. Galaxy formation in an otherwise uniform-density universe must depend on "seeding" by density fluctuations. A sufficiently large mass ($\Delta\rho/\rho \sim 10^{-3}$) will then grow by gravitional attraction. HDM such as neutrinos "streaming" through the universe with only gravitational interactions after decoupling will however wash out any incipient fluctuations.

Axions are hypothetical particles which result from arguments concerning the resolution of strong CP problem. CP violation appears naturally in weak interaction when there are at least three generations. Non-perturbative effects can then give rise to CP violation in QCD but such CP violation is not observed. In particular one would expect a neutron electric dipole ($E1$) moment much larger than the experimental upper limit. As first suggested by Peccei and Quinn, a complex pseudopotential arising from breaking $U(1)$ global symmetry can be introduced to cancel the QCD CP-violating effects but at the cost of requiring the existence of an almost massless neutral pseudoscalar boson, called the "axion".

The best limits on the axion mass are set by cosmological constraints. If the axion mass density (ρ_A) is not to exceed the present critical density (ρ_c) of the universe, then $m_A > 10^{-5}$ eV/c^2. Otherwise, axions will be copiously produced in the Big Bang. At the other limit since axions could readily carry energy away from the centres of stars, the observed longevity of helium-burning stars set a limit $m_A < 0.1$ eV/c^2, Axions couple to fermions with a coupling constant $g_{Af\overline{f}}$ (corresponding to the above masses) in the range $10^{-16} - 10^{-12}$ which is at best some 7–12 orders of magnitude weaker than weak interaction coupling. The axions are expected to be very 'cold' with kinetic energy $\sim 10^{-6}$ of rest energy and $\beta \sim 10^{-3}$. As first proposed by Sikivie, axions could be best detected by means of their interaction with a strong magnetic field. So far terrestrial experiments have failed to detect axions.

The lightest supersymmetric particle (LSP) in SUSY model with an unbroken R-parity is expected to be stable. A good candidate for the LSP is one of the neutralinos, which are linear combination of the neutral gauginos and the neutral higgsinos, the SUSY partners of the neutral gauge bosons (W^3, B^0) and the neutral Higgs bosons $(H_1^0, \overline{H}_2^0)$ of the two complex Higgs doublets in the MSSM. One of the four neutralinos could be identified with the photino, the supersymmetric partner of the photon. Calculations indicate that the LSP density could be in the suitable range to be the constituent of dark matter. The absence of exotic isotopes demonstrate that the LSP cannot interact with ordinary matter through electromagnetic or strong interactions. They can have only weak or gravitational interactions. The SUSY particle masses are not predicted by theory and only reasonable bounds can be set. Their masses are expected to be in the range 100 GeV–1 TeV. The average velocity is expected to be close to the typical velocity of visible matter in the solar neighbourhood with $\beta \sim 10^{-3}$. The LSP is possibly a cold dark matter constituent. SUSY particles are expected to be discovered in the LHC which would be operational in near future.

There could be other unknown particles which interact with ordinary matter only through weak and gravitational interactions. They are expected to have kinetic energy of 1 KeV or even less and mass in the range 5–100 GeV/c^2, i.e. comparable with light or medium nuclei. Such weakly interacting massive particles (WIMPS) are presently only in the speculation stage. They are very hard to detect as processes involving them have slow rates and for which background minimisation is a serious problem.

15.15 Dark Energy

Results from astronomical observations of distant supernovae in the late 1990's suggested that the expansion of the universe instead of being slowed down by gravitational forces as was expected had instead accelerated. Moreover, this acceleration had started only in comparatively recent era ($\sim 7 \times 10^9$ years ago). The physics responsible for this seeming acceleration is not known and goes under the cryptic name "dark energy". Dark energy may or may not have something to do with the "cosmological constant".

15.16 The Current Wisdom

Finally, it appears that the Big Bang cosmology is not a single theory but possibly consists of five separate theories constructed on top of one another. The first tier is a theory, historically but not fundamentally based in the GTR, to explain the observed Doppler Redshifts of distant galaxies. This is the Hubble expansion of the universe, which also explains the cosmic microwave background radiation (CMBR) and the cosmic abundance of light elements attributed to the primordial nucleosynthesis at some stage in the Big Bang evolution. The second tier is the Inflation which solves the horizon, flatness and the monopole problems of the standard Big Bang cosmology. The third tier is the Dark Matter hypothesis required to explain the observed small scale structures of the universe, such as galaxies, galaxy clusters and super clusters. The fourth tier is some kind of description for the "seeds" from which such structures grew. The fifth and the topmost tier concerns the mysterious "Dark Energy", needed to allow for the recent acceleration of the cosmic expansion indicated by the supernova observations. Of these, the lowermost tier, namely, the "expansion" tier is the most well-supported hypothesis.

Therefore, finding additional and more direct evidence (over and above the CMBR, the helium abundance and age of the universe inferred from the oldest stars and star clusters in our neighbourhood) is of paramount importance.

Chapter 16

Experimental Particle Physics: Accelerators and Detectors

Particle physics involves the study of fundamental particles and their interactions. It turns out that for this one needs to probe the subnuclear region ($r \leq 1\,fm$) of matter. This is accomplished by performing high-energy collision experiments. In a conventional fixed-target collision experiment, a beam of particles moving with high velocities (projectile) is made to collide with a stationary target (usually made of atoms with isoscalar nuclei, $2_{\rm H}, 12_{\rm C}, 40_{\rm Ca}, 56_{\rm Fe}$ etc.) and the resulting final states ($a + b \rightarrow f$) are studied using various methods. From the Heisenberg uncertainty principle in quantum mechanics one knows that

$$\text{momentum transfer in the collision} \geq \frac{1}{\text{size of the system being probed}} \tag{16.1}$$

Thus to probe the subnuclear region (for example the inside of a proton) with say electrons, the electron beam energy must be ≥ 1 GeV. Most of the particles produced in high-energy collisions are unstable and decay into lighter more stable particles. It is necessary to detect and identify the particles produced in a collision and use conservation laws to reconstruct the various production and decay processes. Physical quantities that one measures in high-energy physics experiments are total cross sections as a function of energy, differential cross sections (angular distributions and energy distributions of one or more final state particles), multiplicities of the produced particles,decay rates and branching fractions in particle decays and many such quantities. It thus turns out that Accelerators and Detectors are essential ingredients of particle physics experiments.

16.1 Accelerators

Only stable charged particles (e and p) can be accelerated to high energies in a conventional accelerator. All other high-energy particle beams ($\pi^{\pm}$,

$K^{\pm}, \mu^{\pm}, \nu_{\mu}, \overline{\nu}_{\mu}, e^{+}, \overline{p}, \gamma$) are produced as secondary beams using e and p as primary beams in a collision process. At present, two kinds of accelerators are in use—linear accelerators (linacs) and cyclic accelerators (synchrotrons). Both of them accelerate stable charged particles in multiple steps by means of an oscillating electric field operating at radio frequencies (rf). However, in a linear accelerator the charged particles move in a straight line, whereas in a cyclic accelerator the charged particles are constrained to move in a curved path (circular or spiral) by applying a continuous bending force due to a magnetic field produced by a conventional or superconducting electromagnet.

16.1.1 Linear Accelerators (Linacs)

We first consider the principle of a linear accelerator used to accelerate protons (H-ions) or heavier ions. Such a machine consists of a vacuum pipe containing a series of coaxial metal cylinders (drift tubes) along the axis of which the charged particles travel. One set of alternate drift tubes are connected to one terminal of the rf power supply, while the other set of the alternate drift tubes are connected to the other terminal of the oscillator power supply. In a linac for ions (nonrelativistic particles) the successive drift tubes are of gradually increasing lengths to compensate for the increasing velocity of the charge particles so that the travel time through a drift tube remains the same, being equal to half the period of the oscillating electric field. The ion source is continuous but only those ions inside a certain time bunch will be accelerated. Such ions traverse the gap between the successive drift tubes when the electric field direction is from left to right and are inside a drift tube (field-free region) when the oscillating voltage changes sign. If the increase in length of each drift tube along the accelerator system is correctly chosen, then as the ion velocity increases as a result of the acceleration, the ions in a bunch receive a continuous acceleration. Typical electric fields employed are a few MeV/metre.

The basic resonance condition is that the constant travel time through each of the drift tubes must be equal to half the period of the oscillating electric field,

$$\frac{L_n}{v_n} = \frac{T}{2} = \frac{\lambda}{2c} \tag{16.1}$$

Here L_n is the effective length of the nth drift tube and v_n the velocity of the charged particle while moving through this tube.

If n is the number of gaps between the drift tubes that have been traversed then we can write, considering only those ions whose initial phase is such that they are optimally accelerated,

$$n(ze)V_0 = \frac{1}{2}Mv_n^2 \tag{16.2}$$

Here ze = charge and M = mass of the ion ($z = 1$ for the proton) and V_0 = accelerating potential.

Eliminating v_n between (16.2) and (16.3) we get

$$L_n = \lambda\left[\frac{n(ze)V_0}{2Mc^2}\right]^{1/2} \tag{16.3}$$

The total accelerator length is given by

$$L = \sum_{n=1}^{N} L_n \approx \lambda\left[\frac{(ze)V_0}{2Mc^2}\right]^{1/2}\int_0^N n^{1/2}dn \approx \lambda\left[\frac{(ze)V_0}{2Mc^2}\right]^{1/2}\frac{2}{3}N^{3/2} \tag{16.5}$$

The final energy of the accelerated ions is given by

$$E = N(ze)V_0 \approx \left[\frac{3L(ze)V_0}{2\lambda}\right]^{2/3}(2Mc^2)^{1/3} \tag{16.6}$$

This approximate formula shows that in order to get high-energy particle beams, we should choose heavy ions of high atomic number, use high frequency oscillating electric field, large accelerating potential and make the machine very long.

At present, proton linacs producing proton beams of energies (10–50) MeV are mostly used as injectors for higher energy ($\geq$ 1 GeV) proton synchrotrons.

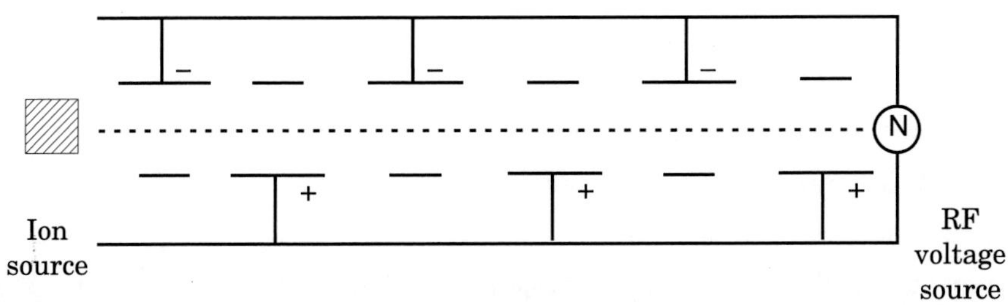

Fig. 16.1

Electron Linacs

Electrons with energies above a few MeV travel with almost the velocity of light so that after travelling the initial 1 metre or so an electron linac has drift tubes of uniform length. (For electrons of $E = 2$ MeV, $\beta \approx 0.97$). In practice, microwave frequencies (10^9 Hz) are employed, the tubes being resonant cavities of dimensions of order a few cms. fed by a series of klystron oscillators, synchronised in time to provide a continuous acceleration.

The famous electron linear accelerator at stanford, USA (SLAC) is 3 km long and was built to accelerate electrons to 25 GeV energy using 240 klystrons to give short ($2\mu s$) bursts of intense power with frequency 60 Hz, with a similar bunch structure for the beam.

The advantage of linear accelerators include a relatively high beam current and the simplicity of injection and extraction of the particle beam. Another great advantage is that there is practically no radiation loss, even for the highest energy electron beams available from a linac since the electrons travel in a straight line. This is in sharp contrast to the cyclic accelerators (synchrotrons) in which the radiation loss becomes prohibitively large above several hundred MeV energy for electrons.

16.1.2 Cyclic Accelerators (Synchrotrons)

At present all "high-energy proton accelerators and a majority of electron accelerators are cyclic accelerators, in particular" synchrotrons. In a cyclic accelerator, the charged particles are constrained to move in a vacuum pipe bent in the form of a torus that threads a series of electromagnets, providing a magnetic field perpendicular to the plane of the orbit.

The first successfully operated cyclic accelerator was the Berkeley cyclotron built by E.O. Lawrence and M.S. Livingston in the early 1930's at the Berkeley campus of the university of California, USA. It accelerated protons to 10 MeV energy.

A typical cyclotron consists of a vacuum chamber containing two hollow metal tubes shaped in the form of 'D' ('dee'). The "dees" form part of a rf oscillator power supply tuned sharply at an angular frequency ω. The ion source is located centrally between the "dees" which in turn are placed between the poles of a large powerful electromagnet.

The principle adopted for accelerating charged particles in a cyclotron (cyclotron principle) is that the time taken by a charged particle in describing a semicircle of radius r within a vacuum cavity must be equal to the period of the rf voltage source. This is known as the resonance condition. Therefore

$$t = \frac{\pi r}{v} = \frac{T}{2} = \frac{\pi}{\omega}, \; \omega = 2\pi/T$$

$$\text{or} \quad \omega = \frac{v}{r} \tag{16.7}$$

Also for motion of a charged particle of momentum p in a magnetic field B, the path is a circle of radius r in a plane perpendicular to the direction of the magnetic field such that

$$p = \gamma m v = \frac{\text{eBr}}{c} \tag{16.8}$$

where $\gamma = 1/\sqrt{1-\beta^2}$, $\beta = v/c$

From (16.7) and (16.8) we get

$$\omega = \frac{eB}{\gamma mc} = \frac{eBc}{E} \tag{16.9}$$

where $E = \gamma mc^2$ = energy of the particle. ω given by Eqn. (16.9) is known as the cyclotron frequency.

For p in GeV/c, B in Tesla ($1T = 10^4$ gauss) and r in metres, Eqn. (16.8) becomes

$$p = 0.3 \text{ Br} \tag{16.10}$$

For low-energy (non-reletivistic) particles, $\gamma \approx 1 =$ constant and hence the resonance condition can be maintained with constant ω and B. The radius of the charged particle orbit keeps increasing as v increases each time the charged particle crosses the gap between the "dees" due to acceleration by the electric field. Thus the path of a charged particle in a cyclotron is a spiral with increasing r. The particles are accelerated to high energies by the time they reach the orbit near the periphery of the "dees".

To accelerate charged particles to very high energies by the cyclotron principle it is necessary to maintain the resonance condition as the velocity of the particle increases and it becomes relativistic ($\gamma > 1$). This can be achieved in two ways:

(1) B is increased with time as v increases so that B/γ remains constant (frequency ω of the oscillating electric field remaining constant)

(2) ω is decreased with time as v increases so that $\gamma\omega$ remains constant (magnetic field B remaining constant)

The first one forms the basis of the working principle of synchrotrons while the second one is used in synchrocyclotrons (frequency modulated cyclotrons). Both kinds of machines use the principle of phase stability.

In proton synchrotrons both the magnetic field B and the frequency of the oscillating electric field are increased with time in order to maintain the resonance condition ($r = R =$ constant). Protons are injected from a linac source at low energy and low magnetic field, which increases to its maximum value over the accelerating cycle lasting for a few seconds. Then the cycle begins again.Thus the final accelerated beam arrives in discrete pulses.

Since for electrons $v \approx c =$ constant even at energies of order 2 MeV, $\omega \approx c/R$ ($R =$ constant) is constant. Thus in electron synchrotrons the resonance condition is achieved by only increasing the magnetic field with time.

In a linear accelerator the final beam energy depends on the voltage V_0 across successive drift tubes and the total length of the accelerator (Eq. 16.6), while in a synchrotron it is determined by the ring radius R and the maximum value B_m of the magnetic field.

$$E = \sqrt{p^2c^2 + m^2c^4} = \sqrt{e^2B_m^2R^2 + m^2c^4} \tag{16.11}$$

For conventional electromagnets using cu coils, B_m is of order 1.4T while if superconducting coils are used B_m can be as high as 9T. For Example, the FNAL synchrotron (Tevatron) has 1 Km radius and achieves 400 GeV proton energy with conventional electromagnets and 1 TeV energy with superconducting magnets. In recent years, superconducting rf cavities which can achieve electric field gradients upto 7 MeV/metre, have been increasingly employed.

Focussing and Beam Stability

In a cyclic accelerator the stability and focussing of the charged particle beam is very important. Otherwise the particles will quickly deviate from their prescribed paths and eventually get lost. In most synchrotrons the principle of strong focussing is utilised. For this, two types of magnets are employed-bending magnets and focussing magnets. The bending magnets produce a uniform vertical dipole ($M1$) field over the width of the beam pipe and constrain the charged particles in a circular path. On the other hand, the focussing magnets produce a quandrupole ($M2$) field. Alternate quadrupoles have their poles reversed, so that both horizontally and vertically one obtains alternate focussing and defocussing effects. As in the case of a light beam passing through a succession of diverging and converging lenses of equal focal length along the path, the net effect is focussing in both horizontal and vertical planes.

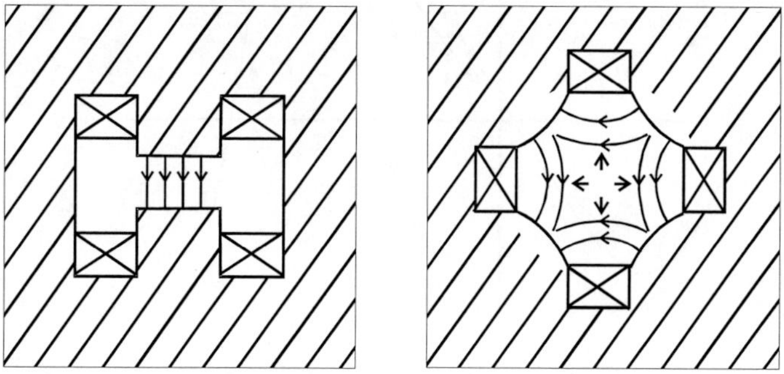

(a) (b)

Fig. 16.2: cross section of (a) conventional bending (dipole) magnet and (b) focussing (quadrupole) magnet. The light arrows indicate field directions and the heavy arrows indicate the direction of the force acting on a positively charged particle travelling into the paper.

The charged particles circulating in a sychrotron do not travel in ideal circular paths but wonder in and out of the circular path in both horizontal and vertical planes in what are betatron oscillations which arise from the natural divergences of the injected beam and from small asymmetries in the fields and alignments of the magnets. The wave lengths of the oscillations are related to the focal lengths of the magnets producing the quadrupole fields and is short compared to the radius of the particle orbits. In addition to these transverse oscillations, longitudinal oscillations called synchrotron oscillations occur as individual particles being accelerated get out of step with the ideal synchronous phase for which increase in momentum per turn (Δp), from the rf kick exactly balances the increase (ΔB) in the magnetic field intensity. Thus a particle that lags behind an exactly synchronous particle will receive a smaller rf kick, will swing into an orbit of smaller radius and next time will arrive earlier. Conversely, a particle that arrives earlier than an exactly sychronous particle will receive a larger rf kick, will swing into an orbit of larger radius and next time will arrive later. So particles in a bunch execute synchrotron oscillations but the bunch as a whole remain stable.

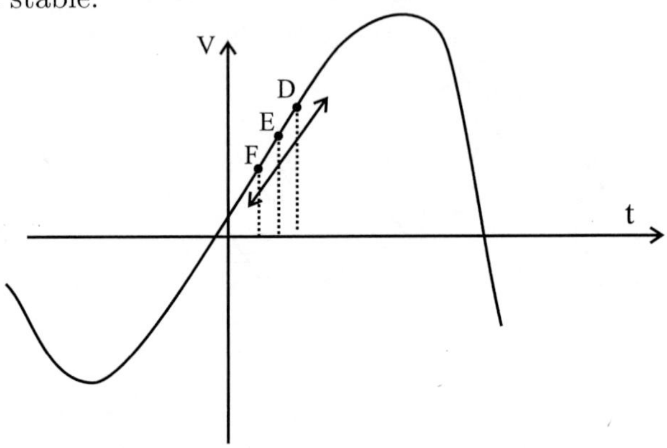

Fig. 16.3: synchrotron oscillations. *E* is a synchronous particle. *D* is a particle arriving early and *F* is a particle arriving late. The arrow indicates the amplitude of the oscillation.

Synchrotron Radiation

Electron synchrotrons suffer from an important limitation compared to proton synchrotrons as regards the attainable energy due to a phenomena known as the synchrotron radiation. An accelerated charged particle emits electromagnetic radiation. For a charged particle travelling in a circle, the energy loss per revolution due to radiation is given by

$$\Delta E = \frac{4\pi}{3} e^2 \beta^2 \frac{1}{R} \left(\frac{E}{mc^2} \right)^4 \tag{16.12}$$

In the above E is the total energy and m is the mass of the particle, R is the orbit radius, $\beta = v/c$.

If ΔE is in MeV, E in GeV and R in metres then (16.12) becomes for electrons,

$$\Delta E = 8.856 \times 10^{-2} \left(\frac{E^4}{R} \right) \tag{16.13}$$

The $1/m^4$ factor in (16.12) means that the effect is much more important (10^{13} times) for electrons than for protons at the same energy and, in practice upto energies currently attained it is only for electrons that energy loss due to synchrotron radiation is an important consideration in the accelerator design. Due to the E^4 factor the effect very rapidly becomes serious at higher energies although it can be reduced partly by increasing the orbit radius R. For energies higher than 200 GeV the energy loss due to synchrotron radiation in a cyclic electron accelerator will be so high as to require unacceptable RF power compensation and therefore linear accelerators will be the only option.

16.1.3 Colliding-beam Accelerators

Much of our present experimental knowledge about particle physics in the high energy regime has been obtained with electron and proton accelerators in which the beam has been extracted and directed onto an external target-the so called fixed-target experiments.

During the last few decades, however, colliding beam accelerators have become important. In these machines, two counter rotating beams of particles (pp, ep) or particle-antiparticles ($p\overline{p}, e^-e^+$) are made to collide at several intersection regions around the acceleration ring. The great advantage of these colliding-beam experiments is the large available CM energy for creation of new particles.

For 2 particles a and b in a fixed-target collision experiment, we have $p_b = (m_b, \vec{o})$, $p_a = (E_a^L \vec{p}_a^{\,L})$ and therefore,

$$s = W^2 = (p_a + p_b)^2 = m_a^2 + m_b^2 + 2m_b E_a^L \approx 2m_b E_a^L, \; E_a^L \gg m_a, m_b \tag{16.14}$$

whereas, in a colliding-beam experiment, we have $p_a = (E_a, \vec{p}_a)$ and $p_b = (E_b, \vec{p}_b)$ with $\vec{p}_a = |\vec{p}_a|\hat{z}$ and $\vec{p}_b = -|\vec{p}_b|\hat{z}$ and therefore

$$W^2 = s = (p_a + p_b)^2 = m_a^2 + m_b^2 + 2(E_a + E_b + |\vec{p}_a||\vec{p}_b|) \approx 4E_a E_b,$$

$$E_a \gg m_a, \; E_b \gg m_b \tag{16.15}$$

Thus while in a fixed-target expt. $W \propto \sqrt{E_a^L}$ in a colliding beam expt. $W \propto E_a$ assuming $E_a = E_b$ (equal mass case). Indeed for the value of W to be the same in the 2 types of expts. we must have

$$E_a^L = 2E_a E_b / m_b \tag{16.16}$$

Colliding-beam accelerators, however, have one big disadvantage, namely, the event rate is much smaller compared to a fixed-target machine. The event rate R is the product of cross section σ and luminosity L,

$$R = \sigma \cdot L \tag{16.17}$$

Small event rate is thus attributed to small luminosity For a Fixed-target machine

$$L = N_p \cdot N_T \cdot n \tag{16.18}$$

where N_p = number of incident (projectile) particles/pulse, N_T = no. of target particles/area and n = no. of pulse/sec.

Furthermore, $N_T = \dfrac{N_A}{A} \rho \cdot t \cdot N$ (16.19)

where N_A = Avogadro number, A = atomic weight, ρ = target density t = target thickness and N = no of target particles/ atom. A typical fixed-target expt. using a 1 metre long liquid hydrogen target bombarded with 10^{13} particles per pulse every 10 secs will yield $L \sim 10^{38}$ cm^{-2}sec^{-1}.

For a colliding beam machine

$$L = \frac{n_1 n_2}{a} b \cdot f \tag{16.20}$$

where n_1 and n_2 are no of particles/bunch in the 2 beams, a = cross sectional area of the beam (assuming them to overlap completely), b = no. of bunches in each beam and f is the revolution frequency. Obviously, L is large if the cross sectional area is small. The luminosity is, however, limited by the beam-beam interactions. Typical Luminosities are $\sim 10^{31}$ cm^{-2}sec^{-1} for e^-e^+ colliders, $\sim 10^{30}$ cm^{-2}sec^{-1} for $p\bar{p}$ colliders and $\sim 10^{33}$ cm^{-2}sec^{-1} for pp colliders.

In a pp or ep colllider, two separate beam pipes and two sets of magnets are required while $p\bar{p}$ and e^-e^+ colliders require only a single vacuum pipe and a single magnet ring (because of charge conjugation invariance the particles and the antiparticles circulate in opposite directions under the action of the same magnetic field. The two beams are placed one above the other and collide at the intersection regions). In the LHC (*pp* collider) a single super conducting ring is used but there are 2 sets of magnets with the directions of the magnetic fields in opposite directions.

To avoid synchrotron radiation losses, linear colliders are the only practical possibility at high energies for the e^-e^+ system. The first linear e^-e^+ collider at stanford (SLC) accelerated e^- and e^+ to 50 GeV in a linear accelerator and the 2 beams then diverged and were brought into head-on collisions after magnetic deflection in circular arcs. A higher energy linear collider is planned with international collaboration (ILC).

Cooling in $p\overline{p}$ colliders (stochastic cooling)

The generation of an intense beam of $\overline{p}$ to be used in a $p\overline{p}$ collider is extremely difficult. The antiprotons are produced in pairs along with protons in high-energy *p*-*p* or *p*-nucleus collitions ($p + p \to p + p + p + \overline{p}$) with a low yield and a large spread in momentum and angle of emission from the target. In other words, viewed from the CM frame of the system of produced antiprotons, individual particles behave like molecules in a very hot gas with random motions described by a "temperature". To store enough antiprotons, the antiproton beam must be "cooled" to reduce its angular divergence and longitudinal momentum spread. This is a serious technical problem, the solution of which was crucial in the discovery of the weak gauge bosons in $p\overline{p}$ collisions at CERN.

One approach, employed at the CERN $p\overline{p}$ collider is to use a statistical method, known as "stochastic cooling", first suggested by the Dutch physicist Van der Meyer. A bunch containing about 10^7 antiprotons of momentum of order 3.5 GeV/c emerges from a cu target (bombarded by a pulse of 10^{13} protons of energy 26 GeV from the CERN PS) and injected in the outer half of a wide-aperture toroidal vacuum chamber divided by a mechanical shutter into 2 regions and placed inside a special accumulator-magnet ring. A pick-up coil in one section of the ring senses the average deviation of the particles from the ideal orbit and a correction signal is sent at nearly the velocity of light across a kicker, in time to deflect them, as they come round.After about 2 secs. of circulation, longitudinal spreads are reduced by an order of magnitude. The shutter is then opened and the "cooled" bunch of antiprotons is magnetically deflected and stacked in the lower half of the vacuum chamber when they are further cooled. The process is repeated until, after a day or so 10^{12} antiprotons have been collected. They are then extracted for acceleration in the main collider ring.

16.1.4 Accelerator Complexes

All high-energy accelerators involve several stages of acceleration. The upcoming LHC at CERN is a good example. Protons emerge at 750 KeV energy from a radio-frequency quadrupole accelerator and are then boosted to 50 MeV energy in a linac. From the linac they are transferred to the PS booster ring where they achieve an energy of 1.4 GeV. They are then injected into the main PS ring and are accelerated to 26 GeV. From the SPS the protons are finally injected into the LHC superconducting magnet ring, 27 km in circumstance to be accelerated to 7 TeV energy per beam.

16.1.5 Secondary Particle Beams. Particle Separators

Fixed target machines are often utilised to produce secondary beams of various stable or unstable particles, directly or via the decay of other particles, for example, π, K, μ, e^+, ν and γ.

In a collision of accelerated protons of energies greater than a few GeV with a solid target either inside or outside the accelerator vacuum chamber many kinds of particles will be produced. With positive charge we expect p, π^+, K^+ and due to their decay μ^+ and e^+. With negative charge we will get $\overline{p}$, π^-, K^- and due to their decay μ^- and e^-. Neutrals will consist of n, π^0, K^0, ν and γ. Some of the produced particles have very short lifetimes at rest ($\sim 10^{-10}$ sec) and at low energies will decay close to the primary target. At high energies, however the lifetime of unstable particles is relativistically dilated and beams of even short-lived hyperons have been used as projectiles. Nucleons and pions are by far the most prolifically directly-produced particles. Experiments requiring pure or enriched beams of secondary particles need an efficient way of rejecting the unwanted particles.

Dipole bending magnets combined with collimators are used to select particle charge and momentum. Quadrupole magnets are used to focus the particle beam. While bending magnets are equivalent to optical prisms the quadrapole magnets are equivalent to optical lenses.

Separation of charged particles of different mass by using combinations of electric and magnetic field has been practised since the early days of atomic physics. For particle beams of energies upto about 7 GeV, electrostatic separation is commonly used, while in the range between 10 GeV and 40 GeV radiofrequency separation is effective. In both cases the beam entering the separator must be of well-defined momentum so that particles of different masses will have different velocities.

The electrostatic separator consists of a pair of parallel metal plates maintained at a high potential difference. Particles passing into the gap between the plates experience a transverse force $(ze)E$ where E is the electrostatic field and ze is the electric charge. The angular deflection is

$$\theta = \frac{p_T}{p} = \frac{(ze)EL}{p\beta} \tag{16.21}$$

where p is the beam momentum, p_T is the transverse momentum, L is the separator length, $\beta = v/c$ ($c = \hbar = 1$). Thus the difference $\Delta\theta$ in angular deflection of two relativistic particles of masses m_1 and m_2 but same momentum p and same electric charge ze after traversing an electric field of strength E and length L is given by

$$\Delta\theta = \frac{(ze)EL}{p}\left(\frac{1}{\beta_1} - \frac{1}{\beta_2}\right) \approx \frac{(ze)EL}{2p^3}\Delta m^2 \tag{16.22}$$

We have used

$$\beta = p/(m^2+p^2)^{1/2} = \left(1+\frac{m^2}{p^2}\right)^{-1/2} \approx 1 - \frac{m^2}{2p^2}.$$

The deflection of the wanted particles is generally compensated by vertical bending magnets at the end of the separators. The angular separation is then converted to a spatial separation at the mass slit by a converging quadrupole lens focussing at the slit through which the wanted particles are allowed to pass.

Because the angular seperation varies as the inverse cube of the momentum (Eq. 16.22) the electrostatic separation method is not suitable at high momentum (momentum $\geq$ few GeV/c). For beams of higher momentum, radiofrequency separation method is used, which depends on the time of flight of charged particles between 2 radiofrequency cavities. The travel time depends on the velocity and hence mass of the particle for fixed momentum. The oscillation frequency of the cavities can be adjusted so that sideways deflection of the wanted particles is doubled when they reach the second cavity, while the deflection of the unwanted particles in the first cavity is reduced by the second cavity. The required condition is

$$\frac{L}{\beta c} = \frac{1}{2f} \tag{16.23}$$

For separation of muons and neutrinos produced in the decay of charged kaons and pions which are generated by collisions of high energy proton beams with an external target, the charged pions and kaons are made to traverse a decay channel several hundred metres in length. A muon beam of fixed momentum is produced by a subsequent system of bending and focussing magnets. Pure neutrino beams are produced by using a thick concrete or steel shielding to filter out hadrons (by nuclear interactions) and muons (by ionisation loss). For example, neutrinos from a 200 GeV pion beam require a filter of about 200 metres of steel.

Table 16.1 Some present day Accelerators

(a) Proton Accelerators (Synchrotrons)

Accelerator	**Location**	**Beam Energy** (GeV)
CERN PS	Geneva, Europe	28
BNL AGS	Brookhaven, USA	32
KEK	Tokyo, Japan	12
Serpukhov	Russia	76
SPS	Geneva, Europe	450
Tevatron II	FNAL, USA	1000

(b) Electron Accelerators

SLAC (Linac)	Stanford, USA	25–50
DESY (synchrotron)	Hamburg, Germany	7

(c) Colliding Beam Accelerators

Accelerator	Beams	Location	Energy/beam (GeV)
PETRA	e^-e^+	DESY, Hamburg	22
PEP	e^-e^+	Stanford, USA	18
CESR	e^-e^+	Comell, USA	8
TRISTAN	e^-e^+	Tokyo, Japan	30
SLC	e^-e^+	Stanford, USA	50
LEP I	e^-e^+	CERN, Geneva	50
LEP II	e^-e^+	CERN, Geneva	100
ISR	pp	CERN, Geneva	30
Sp $\overline{p}$S	$p\overline{p}$	CERN, Geneva	310
Tevatron I	$p\overline{p}$	FNAL, USA	1000
HERA	ep	Hamburg, Germany	30 GeV (e) and 820 GeV (p)
LHC	pp	CERN, Geneva	7000
ILC	e^-e^+		250

16.2 Detectors

The particles produced in high-energy collisions between a projectile and a target or between two fast moving particles can be either charged or neutral. The **detection of charged particles** depends on the fact that, directly or indirectly, they transfer energy to the medium they traverse by electromagnetic interactions, mostly by **ionisation** and excitation of the consituent atoms of the matter of the medium and partly by coulomb scattering with the electrons and positive ions resulting from the ionisation process. For relativistic electrons, there is another process, namely **bremsstrahlung**. A relativistic electron travelling in the electric field of atomic nuclei gets decelerated and emits radiation. At very high energies, radiation loss dominates over ionisation loss in the case of electrons. Electrically **neutral particles** are detected either through their decay ($K^0 \to \pi^+\pi^-, \Lambda \to \pi^-p$) or through interaction with matter ($\pi^0 \to 2\gamma, \gamma \to e^-e^+$) leading to production of secondary charged particles.

High energy **electromagnetic radiation** (γ rays) traversing a material medium lose energy mostly through 3 types of processes-**photoelectric effect** ($\gamma + A \to e^- + A$), **Compton scattering** ($\gamma + e^- \to \gamma' + e^-$) and pair production ($\gamma \rightarrow e^-e^+$). The threshold for pair production process is $E_{\text{th}} = 2m_ec^2 = 1.02$ MeV. The process can take place only in the electric field of an atomic nucleus to conserve momentum and is closely related to electron bremsstrahlung. The photoelectric crosssection varies as $1/E^2$ and the Compton scattering crossection varies as $1/E$, where E is the photon energy but the pair production cross section is essentially independent of photon energy. Therefore for **photon energy $E >$ 10 MeV**, the pair production process dominates.

The objectives of detectors in particle physics experiments are:

(a) to localize the particle trajectory in space: If this can be done in a region of known magnetic field then for a charged particle the momentum can be obtained from the relation $p = eBR/c$, where p is the particle momentum, B the magnetic field intensity, R the radius of curvature of the particle trajectory, e the particle charge and c is the velocity of light. The higher the field and the smaller the distortions, the more precisely the sagitta and hence the particle momentum can be measured.

(b) to measure the particle energy: This may be done by means of calorimeters where a particle loses all its energy or for charged particles in some circumstances by measurement of the rate of energy loss by ionisation of the atoms of the medium they traverse. The rate of ionisation loss of charged particles dE/dx is independent of the mass M of the particle, varies as $1/v^2$ at non relativistic energies and after passing through a minimum for $E \approx 3Mc^2$ increases logarithmically with energy E. In some circumstances it is possible to measure the particle velocity directly by time-of-flight method or by emission of Cerenkov radiation.

(c) to localize the particle in time: This is not only necessary in the case of the time-of-flight method for measurement of particle velocities but more frequently because it is necessary to associate signals due to the same particle from the same event in a number of different detectors against a big background of signals from unassociated particles. Signals within an appropriate time gate (with proper allowance for time-of-flight) can be used to eliminate unassociated particles.

(d) to identify the particle: This may be done by determining the particle mass from measurements of its momentum and energy or from measurements of its momentum and velocity. In certain cases this is also done by the nature of its interactions with matter or by its decays. For example, muons do not produce nuclear interactions and are thus distinguished from hadrons. On the other hand, muons are so much heavier (200 times) than electrons, that they produce very little bremsstrahlung and do not give rise to electromagnetic showers and are thus distinguished from electrons.

No single detector is in general able to meet all these requirements and a combination of detectors of different types is often required.

Furthermore, the direct loss of energy by ionisation for a charged particle is always very small. It is thus apparent that particle detectors must depend on some means of amplification of the primary signal to render it measurable.

In one class of detectors the amplification process depends on the use of a detector material in a semi-stable state. The energy released by ionisation then triggers the instability along the track of the ions left by the parent particle and so renders the track observable. In the **bubble chamber** the unstable state is that of a **superheated liquid** and we get boiling of the liquid along the track which thus appears as a string of bubbles. In the **cloud chamber** the unstable state is that of a **supersaturated vapour** so that condensation of droplets is triggered off along the track. In the **photographic emulsion** the instability is chemical instesd of physical and the ions give rise to the **development of grains** of silver bromide (AgBr) to provide a latent image of grains along the track.

A **second approach** which has been in use since the early clays of nuclear and particle physics, is to detect the ionisation in gases produced by a passing high-energy charged particle by collection of ions **onto electrodes** using an electric field and subsequently amplifying the signal electronically. Sometimes the ionisation signal is amplified inside the detector by avalanche formation. This approach is used in detectors like proportional counters, spark chambers and drift chambers.

A **third class of detectors** exploits the **energy loss by** charged particles either by **Cerenkov radiation** when the particle velocity exceeds the phase velocity of light in the medium through which the particle is passing through as in **Cerenkov detectors** or by excitation of the constituent atoms of the material of the medium which then deexcite to emit light as in **scintillation counters**. In both cases the emitted primary light energy is small but a photomultiplier tube amplifies the signal to a detectable level. Over the last few decades, very few radically new concepts in basic detection methods have been developed. However, there has been a great advancement in the electronics and the computer technology for handling and analysing huge amounts of data at very high speed, often in real time. Furthermore, hybrid complex systems involving many different types of detector have been developed at par with great progress in high-energy accelerator technology.

16.2.1 Bubble Chambers

The bubble chamber was an indispensable tool in high-energy physics experiments for over 30 years. Conceived by Glaser in 1952, it relied for its operation on the fact that in a superheated liquid boiling will start with the formation of gas bubbles at nucleation centers in the liquid, and particularly along tracks of ions. The liquid filling was maintained at a pressure higher than the atmospheric pressure (5–20 atmospheres), and the superheating was achieved by the sudden release of a piston, bellows or diaphragm placed at the rear of the bubble chamber. After expansion, bubbles along the tracks of the particles were allowed to grow over a period of about 10 ms and were photographed by an array of stereo cameras using flash illumination. The bubbles then collapsed under a recompression stroke. Since the cycle time was of order 1 sec., the bubble chamber was well suited for use with pulsed, cyclic accelerators and secondary beams from fixed targets, with repetition rates of the same order.

The liquids commonly used in bubble chambers were liquid hydrogen, neon-hydrogen mixture, propane (C_3H_8) or Freon (CF_3 Br).. The entire chamber was placed in a strong magnetic field (2-3.5T) provided by an electromagnet with conventional or superconducting coils to permit momentum measurement from curvature of tracks. Subsequent measurement of bubble images on photographic films were digitised, and a geometry program was used to reconstruct the tracks and even vertices in three dimensions.

The main disadvantages of a bubble chamber are—(1) it has a low repetition rate, and the analysis of films is a very lengthy process (involving more than 10^5 useful events). (2) many present-day accelerators are colliding-beam type, with an effectively d.c. interaction rate. Thus the very low duty cycle of a bubble chamber, that is the ratio of the sensitive time to the total time ($\sim 10^{-2}$), as well as the interaction geometry of the beams, exclude their use in colliding-beam experiments. Nevertheless, bubble chambers although now almost redundant, have played a notable role in experimental particle physics-in cataloguing baryon and meson resonances in the 1960s, and in the early experiments in neutrino physics in the 1970s including experiments (BEBC experiments) leading to the discovery of the weak neutral currents, which gave a great impetus to the standard model of particle physics.

16.2.2 Proportional Counters

The proportional counter is one of the oldest devices that have been used to record ionisation. Single counters consist of a glass-filled cylindrical metal or glass tube of radius maintained at a negative potential with a fine central anode wire of radius r_1 maintained at a positive potential. The electric field in the gas for a potential difference V_0 between the cathode and the anode is then

$$E(r) = \frac{V_0}{r \ln(r_2/r_1)} \tag{16.22}$$

An electron liberated by ionisation at radius r_a will drift towards the anode gaining an energy

$$W = e\int_{r_a}^{r_b} E(r)dr \tag{16.22a}$$

when it reaches radius r_b. If W exceeds the ionisation potential of the gas atoms, then fresh ions are liberated and a chain of such secondary ionisation processes leads to an avalanche of electrons and positive ions. The gas amplification factor, equal to the total number of electrons reaching the anode per initial electron-ion pair is typically of order 10^5 and is independent of the number of primary charged particles. Hence the name "proportional counters".

The most significant recent advance in this field was the introduction of the **multiwire proportional counter (MWPC)** by Charpak in the late 1960's. This device consists of many parallel anode wires stretched in a plane between two cathode planes. The different anode wires act as independent detectors. A typical structure has anode wires of 20 μm diameter with 2 mm spacing, between cathode planes 12 mm apart, operating at a potential difference of 5 KV between cathode and anode (producing electric field of order $10^4 - 10^5$ volt per cm) and containing an argon-isobutane gas mixture. In general, several electrons from the primary ionisation will drift toward an anode wire and create separate avalanches yielding negative pulses with very fast rise times ($\sim$ 0.1 ns). The decay time of pulses which is dependent on ion mobility is of order 20 ns. The positive ions have much lower mobility and induce pulses on both the cathode and the neighbouring anode wires of duration of order 30 ns. The effective spatial resolution from the anode pulses is of order 0.7 mm. If the cathode is in the form of strips,the centre of gravity of the cathode pulses may be used to obtain an accurate spatial position of the avalanche.

Successive layers of MWPCS with wires oriented in different directions can be used to measure the trajectories of a charged particle. For a single particle the signals in two orthogonally oriented layers give a unique space point. However, when more than one particle is present ambiguities arise in associating the signals in different layers and it is customary to use modules with planes oriented along x,y and tilted u and v axes. Such arrays of MWPCS are frequently used either within magnetic fields or on either side of a magnet to measure particle momentum. The usefulness of systems of such MWPCS which may contain many thousands of wires depends on the availability of relatively cheap transistor electronic technology.

16.2.3 Spark and Streamer Chambers

Proportional counters are operated with anode-cathode potential difference of order 5 KV, where the efficiency for recording charged particles reaches a plateau of almost 100%. Further increase in voltage eventually leads to electric field breakdown of the gas. This takes place when the space charge inside the avalanche is strong enough to shield the external electric field.

Recombination of ions then occurs, resulting in photon emission and the generation of secondary ionisation and new avalanches over and above the initial one. This process propagates until an ion column links the anode and the cathode and a spark discharge occurs. The counter is then said to operate in the Geiger region and is known as the **spark chamber**.

If, however, a short (10 ns) high-voltage (10–50 KV cm^{-1}) pulse is applied between transparent parallel-plate electrodes, then only short (2–3 mm) streamer discharges develop from the ion trail of a passing charged particle, and a direct track image, similar to that in a bubble chamber, can be obtained and photographed through the electrodes. Such ionisation detectors are known as **streamer chambers**. They possess good multi-track efficiency and space resolution. It was a widely used technique in the 1960s but was later replaced by other methods because of its long recovery time and because the analysis of optical images (as for a bubble chamber) is required.

16.2.4 Drift Chambers

The MWPC has a typical spatial resolution of 1 mm or less and a time resolution of about 30 ns. However, to achieve this resolution over large areas an enormous number of wires is required. A great reduction in cost for building an ionisation detector can be achieved by drifting the electron from the point of primary ionision over a distance of order 10 cm in a low field ($\sim$ 1 KV/cm) region before going to the high-field amplification region near the anode wire. The collection time of the avalanche then gives a measure of the position of the ionisation column. A planar arrangement of anode and cathode wires with field wires at graded potentials, as required to give a uniform drift field is known as a **drift chamber**. A drift chamber thus operates with fewer wires than a MWPC by measuring the time for the ions to drift to the wire in a uniform field. If this drift velocity is known, this time can then yield the coordinate to high precision. A start signal from some other detector in the system is needed for the timing. The multi-wire electronics is then replaced by a limited number of timing channels. The uniform drift field in a drift chamber is of order 1 KV/cm and drift velocities in argon-hydrocarbon gas mixtures is typically of order 50 μm/ns almost independent of the field intensity. For a timing precision of a few ns the equivalent space precision is thus of order 100 μm. Drift distances of 10 cm or more are commonly used.

An elaborate development of the drift chamber is the **Time-Projection Chamber** (TPC) which consists of a gas volume subject to a uniform electric field along one dimension. On the plane orthogonal to the drift direction is a 2-dimensional detector such as a MWPC with 2-dimensional capability afforded for instance by anode wires and cathode pads. Along the drift direction the coordinate is measured by timing so that for a particle track in the gas a string of pulses is picked up at different positions on the end plate detector from which the trajectory can be uniquely reconstructed in 3-dimensions. Such detectors are becoming increasingly popular for studies of complex multi-particle interactions, particularly at colliding-beam machines. They can be placed in magnetic fields to give a measure of the particle momentum while the ionisation can be sampled many times along the track to give information on particle identification over a range of velocities , including the relativistic-rise region.

16.2.5 Scintillation Counters

The scintillation counter has been a universal detector in a wide range of particle physics experiments since the early days. The excitation of atoms of some materials by ionising particles results in luminescence (scintillation) and the light emitted can be recorded by a photomultiplier. The scintillators commonly used are inorganic crystals (NAI), organic crystals (anthracene), organic liquids (toluene) or plastics (solid solutions). The decay times of the fastest scintillators are of order 1 ns.

Inorganic crystal scintillators (NAI) are often doped with activaton centres (Thallium). A charged particle passing through the crystal produces free electron-hole pairs which drift around until captured by activation centres. These are excited to higher energy states which subsequently decay to ground states with the emission of light over a broad spectrum in the visible region with a decay time of order 200 ns. In organic materials (solid/liquid), however, the scintilation mechanism is the excitation of molecular energy levels which then decay to ground levels with the emission of light in the UV region. This light is then converted to visible (blue/green) wave lengths by fluorescent excitation of certain dye molecules known as wavelength-shifters which are incorporated into the primary scintillator medium.

The light emitted from the scintillator medium is detected and recorded by a photomultiplier tube. This consists of a photocathode coated with alkali metals (Cesium) where electrons are liberated by the photoelectric effect. The electrons travel to a chain of secondary-emission electrodes (dynodes) at successively higher potentials. Since about 4 secondary electrons are emitted for each incident electron, amplification factors of about 10^8 are achieved with 14 dynodes ($4^{14} = 2^{28} = 2.7 \times 10^{28}$). The transit time from the photocathode to the output dynode (anode) is typically 50 ns, with a jitter of about 1 ns. The jitter is determined mostly by the variations in the transit times to the first dynode from different points on the cathode. Photocathode quantum efficiencies are typically of order 25% peaking at $\lambda = 400$ nm.

Although inorganic crystals are most useful for γ-ray detection, such crystals are expensive and most of the large scintillation counters used in high energy physics experiments are usually made of organic plastic scintillators. The plastics may be used either in the form of large-slabs or in other shapes or sometimes in the form of a matrix of thin finger counters, according to the application mode. It is often inconvenient or even impossible to place the scintillator in direct contact with the photomultiplier tube, in which case lucite light-pipes are employed to link them. By this means photomultiplier tubes may be kept away from the strong magnetic fields which affect their operation. An alternative method is to use bars of wavelength-shifter material which shift the primary UV light into the green region. The light output is much less than using light guides but there is a large saving in space, number of photomultipliers and construction of complex light guide structures.

The output pulse from the photomultiplier is finally fed into an array of suitable amplifiers, discriminators and scalars, the essential function of which is to store the number of photomultiplier pulses of a particular amplitude.

A particularly important use of scintillation counters is in time-of-flight measurement coupled with a measurement of particle momentum. Such a measurement can distinguish between particles of different masses. The momentum p and the time-of-flight t are related by

$$p = m\beta\gamma = m\frac{v}{c}\left(1 - \frac{v^2}{c^2}\right)^{-1/2} = \frac{mL}{tc}\left(1 - \frac{L^2}{t^2c^2}\right)^{-1/2} \tag{16.23}$$

where L is the flight path, m is the mass and c is the velocity of light. In a small fast scintillator, measurement errors as low as 0.1 ns have been achieved using the rising edge of the pulse. If a 3σ separation is required then πs and Ks can be distinguished upto momentum of about 1 GeV/c.

16.2.6 Cerenkov Counters

When high-energy charged particles traverse a dielectric medium with velocities greater than the velocity of light in the medium ($v > c/n$), part of the radiation emitted by the excited atoms of the medium appears in the form of a coherent wave front at a fixed angle with respect to the trajectory of the passing charged particle. This phenomena is known as **Cerenkov effect**, after the name of the discoverer. Cerenkov counters use photomultiplier tubes to detect the Cerenkov radiation and thus to record the passage of a charged particle. Compared with scintillation counters Cerenkov counters are less sensitive and more clumsy, but they possess the property of discriminating between particles of different velocity and thus, if the momentum is known by other methods, of different mass.

From the Huygens construction one knows that the wave front of the emitted light forms the surface of a cone about the particle trajectory as axis, such that the angle of the cone is given by

$$\cos\theta = \frac{(c/n)t}{vt} = \frac{1}{\beta n}, \quad \beta = \frac{v}{c}$$

$$\cos\theta < 1 \Rightarrow \beta > \frac{1}{n} \quad \text{or} \quad v > \frac{c}{n} \tag{16.24}$$

The emitted light is plane-polarized with the electric vector in the plane containing the emitted photon and the incident charged particle.

Basically, Cerenkov counters consist of a container filled with a gas, liquid or solid material transparent to the Cerenkov light and for which the index of refraction is suitably chosen (Eq. 16.24). This simple condition underlies the operation of the threshold Cerenkov detector. Suppose we wish to distinguish K^+ mesons from protons and π^+s at the same momentum, say 1 GeV/c. The values of β for p, K^+ and π^+ at this momentum are 0.73, 0.89 and 0.99 respectively. The corresponding values for n are 1.37, 1.12 and 1.01 ($n = 1/\beta$). If we arrange 2 Cerenkov counters, one filled with water ($n = 1.33$) and the other with at certain pressure and temperature such that $n = 1.05$, then the protons will be recorded in neither of the counters, kaons only in the water Cerenkov counter and the pions in both the counters.

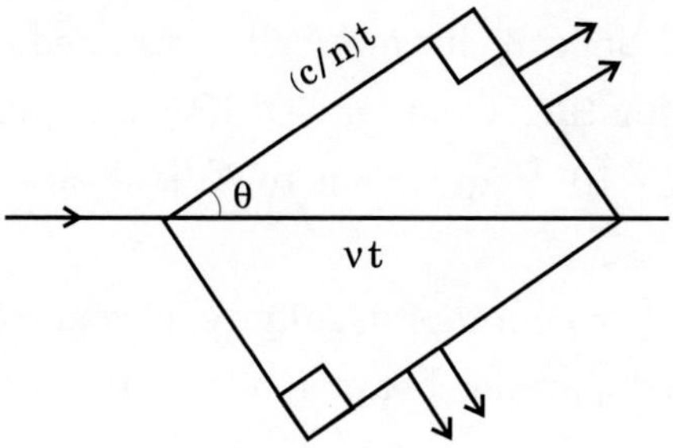

Fig. 16.4

Cerenkov radiation appears as a continuous spectum. In a dispersive medium, both the refractive index n and the angle of emission of the photon will be a function of the light frequency ν. The number of photons emitted with energy between E and $E + dE$ ($E = h\nu$) by a particle of charge (ze) in a track length dx between x and $x + dx$ is given by

$$\frac{d^2N}{dxdE} = \left(\frac{ze}{\hbar c}\right)^2 \left(1 - \frac{1}{\beta^2 n^2})\right), \quad n = n(v) \tag{16.25}$$

The number of photons emitted with wave length between λ and $\lambda + d\lambda$ is proportional to $d\lambda/\lambda^2$ so that blue light dominates. The total rate of particle energy loss by Cerenkov radiation is given by

$$\frac{dW}{dx} = \left(\frac{ze}{\hbar c}\right)^2 \int_{\nu_1}^{\nu_2} \left(1 - \frac{1}{\beta^2 n^2})\right) (h\nu)(hd\nu)$$

$$= \frac{1}{2}\left(\frac{ze}{\hbar c}\right)^2 [(h\nu_2)^2 - (h\nu_1)^2] \cdot \left(1 - \frac{1}{\beta^2 n^2})\right)_{aV} \tag{16.26}$$

neglecting the variation in refractive index n over the frequency range ($\nu_2 - \nu_1$). Here $\alpha = e^2/\hbar c = 1/137$

For a particle with $z = 1$ and $\beta \approx 1$ in water ($n = 1.33$) Eq. (16.26) gives $dW/dx \approx 525$ eV/cm for visible light ($\lambda = 400 - 700$ nm) which is carried by a total of 215 photons/cm. Thus the energy loss by Cerenkov radiation is extremely small ($\sim 2 \times 10^{-4}$) compared to ionisation loss (~ 2 MeV/cm) and is of order 1/10 of the rate of photon production in a typical scintillation counter.

The angular condition given by Eq. 16.24 is exploited in differential Cerenkov counters which accept light over a narrow angular interval and hence over a narrow range of velocity. The resolution is limited by finite divergence in the particle directions, multiple scattering in the medium and chromatic dispersion. Special chromatically-corrected counters (Directional Isochronous Self-collimating Counters DISC) have been developed which achieve $\Delta v/v = 10^{-6} - 10^{-7}$ equivalent to K/π separation upto momentum of order 500 GeV/c.

Differential Cerenkov counters usually require all the particles to travel in a closely-collimated parallel beam if they are to achieve the ultimate resolution. An important development designed to allow the use of such counters in the identification of particles emerging from an interaction over a wide range of angles is the Ring Image Cerenkov Counters (RICH). In such a device an optical system is required to focus the light emitted along the Cerenkov cone on to a detector in the form of a ring the radius of which is a measure of the cone angle and hence the velocity of the particle. It may be noted that in the minimum ionisation range Cerenkov counters offer the only method of particle identification.

16.2.7 Solid-state Counters

Semiconductor solid-state detectors have been widely used in nuclear physics experiments for several decades, on account of their very good energy resolution. With the advent of high-energy colliding beam machines and the production of heavy (mass > 1.86 GeV) short-lived ($\tau < 10^{-12}$ s) charm and bottom mesons, and with the need to record the details of particle tracks and interaction vertices from the production and decay of these hadrons, solid-state microdetectors with spatial resolution of order 5 μm have now become indispensable in particle physics experiments.

One form of the solid-state counters is the silicon-strip detector. This consists of a wafer of n-type silicon typically 250 μm thick, one side of which is conducting and the other side of which is etched with many thin parallel strips of p-type materials, separated by gaps of order 20 μm. Upon application of a (reverse bias) voltage the system operates like a solid-state version of a gas counter, with electron-hole pairs instead of electron-ion pairs. Since the average energy needed to form an electron-hole pair is only 3.6 eV, compared with 20–40 eV for producing an electron-ion pair in a gas, adequate signals ($\sim 2 \times 10^4$ pairs) can be collected within the fast rise-time (10 ns) from such thin wafers. The spatial resolution can be of order 5 μm in a field-free region, and of order 10 μm in a $2T$ magnetic field region. High resolution silicon microvertex detectors (MVD) have been widely used in LEP collider experiments at CERN in the 1990s.

16.2.8 Calorimeters

The energy and position coordinates of the secondary particles produced in high energy collisions can also under suitable conditions, be measured by total absorption methods. For this one employs devices known as "calorimeters". In a calorimeter the incident particle loses all or almost all its energy in a large detector mass by processes which include a stage of ionisation and ultimately result in heat (hence the name calorimeter). The measurement of energy of the incident particle is achieved by measuring the energy deposited by the ionisation and excitation stage of its energy loss. Since the particle is no longer available for further study because of total energy loss in contrast to situation in ionisation chambers or even bubble chambers, calorimeters are some times called "destructive" detectors.

The characteristics of a calorimeter depend on the nature of the dominant processes responsible for the particle energy loss. For electrons and γ-rays the energy loss is dominated by electromagnetic processes-bremsstrahlung ($e^- \rightarrow e^- + \gamma$), pair production ($\gamma \rightarrow e^- + e^+$) and Compton scattering ($\gamma + e^- \rightarrow \gamma + e^-$), the first 2 processes taking place in the electric field of the atomic nucleus for conservation of momentum. For hadrons on the other hand, strong interactions are primarily responsible for the loss of energy of the incident particle. Neutrinos, which experience weak interactions only, pass through a calorimeter without any signal. Muons do not produce electromagnetic showers due to their high mass compared to the electron and also do not have strong interactions. Thus muons also pass through calorimeters without any signal (Muon counters mostly employ ionisation or Cerenkov radiation and energy carried away by neutrinos (missing energy) is computed by using energy conservation principle to the reconstructed reaction process at a vertex). Calorimeters thus fall into 2 categories—(1) Electromagnetic Calorimeters designed to measure electron and γ-ray energies and (2) Hadron calorimeters designed to measure the energies of hadrons by strong interaction processes.

Electromagnetic Calorimeters

A high-energy electron traversing the detector material will radiate photons due to the bremsstrahlung process. The produced photons will create further electrons by the pair production process and will also transfer energy to electrons (produced by ionisation of the detector material) via Compton scattering. The second-generation electrons will in turn generate further photons which in turn will produce further electron-positron pairs and so on to produce what is known as an electromagnetic shower consisting of photons and electron-positron pairs. The number of shower particles increases exponentially with depth in the detector medium and soon the maximum number is reached when the average particle energy is no longer high enough to continue the multiplication process. Beyond this point the shower decays as the electrons lose energy by ionisation and the photons by Compton scattering.

The energy loss by radiation of a high energy electron in traversing a thickness of the medium is given by

$$\Delta E = -E(\Delta x)/X_0 \tag{16.27}$$

where X_0 is a constant for the medium and is known as the **radiation length**. Integrating Eq. (16.27) we get

$$E(x) = E_0 e^{-x/X_0} \tag{16.28}$$

where E_0 is the initial energy of the electron and $E(x)$ is its average energy after traversing a thickness x of the material. Thus the energy of an electron will drop by a factor e in a thickness X_0. The attenuation of a beam of high energy photons by pair production in traversing a thickness x of the detector medium is similarly approximate by

$$I = I_0 e^{-7x/9X_0} \tag{16.29}$$

where I_0 is the initial photon energy. Therefore the photon intensity is reduced by a factor e in a distance $(9/7)X_0$, sometimes called the **conversion length**.

X_0 is in general a function of electron (photon) energy and of the nature of the detector material. However, for energies > 1 GeV, X_0 is almost energy independent. The dependence of X_0 on the properties of the medium is given, to a good approximation, by the expression

$$\frac{1}{X_0} = 4z^2 \left(\frac{N_0}{A}\right) \alpha^3 \left(\frac{\hbar c}{mc^2}\right)^2 \ln \left(\frac{183}{z^{1/3}}\right) \tag{16.30}$$

where N_0 is the Avogadro, number z and A are the atomic number and mass number of the atoms of the medium, α is the fine structure constant and m is the electron mass. X_0 is usually quoted in gmcm^{-2}. Eq. (16.30) shows that the radiation length X_0 decrease as z^{-2} so that electromagnetic calorimeters generally employ high z material such as Pb in order to minimize the overall size.

In the design of a calorimeter one would like to obtain maximum resolution and locate the position of the incident electron or photon as precisely as possible. Therefore one needs to understand the properties of an electromagnetic shower in terms of its longitudinal extent, transverse spread and rate of energy deposition along the shower path.

The number of particles in a shower increases to reach a maximum and then decreases when the energy of the shower particle drops below a critical value. We define the critical energy as the electron energy E_c for which the radiation loss equals the ionisation loss. Typical values of E_c are in the range 7 MeV (Pb) to 30 MeV (Ar). To a rough approximation the number of shower particles (electrons plus photons) doubles for each radiation length X_0 upto the shower maximum at thickness $t_{\max}$ where the energy per particle equals E_c. Thus if $t_{\max}$ corresponds to $n_{\max}$ radiation lengths then the number of shower particles present at $t_{\max}$ is $2^{n_{\max}}$ and

$$E_c = E(t_{\max}) = \frac{E_0}{2^{n_{\max}}}, \quad E_0 = \text{energy of the incident electron}$$

$$\text{and} \quad N_{\max} = N(t_{\max}) \approx 2^{n_{\max}} \tag{16.31}$$

Therefore

$$t_{\text{max}} \text{ (in unit of } X_0) \approx n_{\text{max}} \approx \frac{\ln(E_0/E_c)}{\ln 2}$$

and $$N_{\text{max}} \approx 2^{n_{\text{max}}} \approx \frac{E_0}{E_c} \approx e^{t_{\text{max}} \ln 2} \tag{16.32}$$

we can write

$$N(t) \approx e^{t \ln 2} \tag{16.33}$$

Thus t_{max} increases logarithmically with the incident electron energy A more exact calculation gives

$$t_{\text{max}} = \ln(E_0/E_c) - 1.0 \tag{16.34}$$

The total longitudinal depth L beyond which there is no measurable increase in signal is given by

$$L = t_{\text{max}} + x \tag{16.35}$$

where x measures the attenuation phase of the shower which is approximately exponential. For containment of 95% of the incident energy it is found that

$$L_0 = L(0.95E_0) \approx [\ln(E_0/E) - 1.0] + 12X_0 \tag{16.36}$$

The transverse spread of the shower which determines the transverse dimension of the calorimeter is primarily determined by multiple coulomb scattering of the electrons which spread the shower. The radial spread of the shower is determined by the radiation length X_0 together with the angular deflection per radiation length at the critical energy E_c. In all materials, the spread is of order one Moliere unit $R_m = (21/E_c)X_0$, where E_c is in MeV.

The energy resolution depends primarily on the fluctuations of the rate of energy deposit in the active material. Most electromagnetic calorimeters are built of layers of high-z absorbing materials of small radiation length interlaced with layers of detectors such as scintillation or ionisation detectors and it is the statistical variations in the length of the electron path in the active material which determines the energy resolution. The energy resolution is typically

$$\frac{\Delta E}{E} \approx \frac{0.05}{\sqrt{E}} \tag{16.37}$$

where E is in GeV.

Hadron Calorimeters

The process of degradation and measurement of energies of hadrons produced in high-energy collisions is more complex than electrons and photons which interact only electromagnetically. In a hadron calorimeter a hadron shower results when a high-energy incident hadron undergoes inelastic nuclear collisions with the production of secondary hadrons which also interact inelastically to generate further hadrons and so on. Some of the secondary hadrons also undergo decay to produce new charged or neutral particles. The scale for longitudinal development of a hadron shower is set by the nuclear absorption length which is analogous to the radiation length λ_0 in electromagnetic calorimeter. It is given by

$$\lambda_0 = [(\text{number of nuclei/volume})\ \sigma_I]^{-1} = \frac{A}{N\rho\sigma_I} \tag{16.38}$$

where A is the atomic number of the atom of the detector material, ρ is the density, N is the Avogadro number and σ_I is the inelastic interaction cross section. σ_I and hence λ_0 varies with incident hadron energy in the region of nucleon resonances but above 2 GeV, λ_0 is almost energy-independent. The nuclear absorption length X_0 is large compared to the radiation length X_0 in heavy elements such that compared to electromagnetic calorimeters, hadron calorimeters have a large size. For example, the longitudinal and transverse dimensions of an iron-scintillator sandwich are of order 2 m and 0.5 m respectively.

In an electromagnetic cascade, the bulk of the incident electron or photon energy appears eventually in the form of ionisation and excitation of atoms. However, in a hadron cascade 30% of the incident hadron energy is lost by the breakup of nuclei, nuclear excitation and production of neutral particles such as neutrinos and thermal or slow neutrons which escape from the calorimeter without leaving any signal. One successful method of compensation involves the use of ^{238}U as the calorimeter material, the extra energy released by fast neutron fission and photofission of ^{238}U making up for this "invisible" energy losses from nuclear break up. The neutron energy is thus converted into charged particle ionisation and measured. Such "fission compensation" can bring the electron/hadron ratio close to unity.

The energy resolution of a hadron calorimeter is in general, much poorer than an electromagnetic calorimeter due to much greater fluctuation in the development of the hadron shower. Depending on the number of π^0 produced in the early stages of the cascade, the shower may develop into either a predominantly electromagnetic or a hadronic mode and thus contributing substantially to the variation in energy deposit. For a hadron calorimeter not employing fission compensation the energy resolution is approximately given by

$$\frac{\Delta E}{E} \approx \frac{0.5}{\sqrt{E}} \tag{16.38}$$

where E is in GeV. The transverse spread of the shower is such that 95% of the incident hadron energy is deposited within a cylinder of radius λ_0

Like electromagnetic calorimeters hadron calorimeters are also constructed usually with a stack of alternate layers of absorbing material such as Fe or Pb and detection devices such as scintillator counters or multiwire proportional counters. Large-scale calorimeters, as required for full angular coverage at colliders often use liquid argon as the detection medium. In detectors employing a magnetic field, the iron return yoke of the magnet may be instrumented in this way to provide a hadron calorimeter.

Large Hybrid Detectors

Typical HEP experiments involve the simultaneous detection, measurement and identification of many particles both charged and neutral from each interaction that occurs in a high-energy collision. They therefore involve incorporation of a large number of different types of detectors in a single array. An example of a hybrid detector is the detector array used by the CDMS collaboration at the CERN SPS. It is essentially a large magnetised calorimeter and consists of a succession of iron plates, separated by scintillation counters and drift chambers. The latter record the charged secondaries resulting from the neutrino interactions, while the integrated pulse height provided by the scintillation counters indicates the hadron shower energy. High energy secondary muons are identified by their penetration depth and their momentum is determined from their deflection in the magnetised iron.

Other examples of large hybrid detectors are the spectrometer detector used by the JADE collaboration at the e^-e^+ collider PETRA at DESY,

Hamburg, Germany, the CDF detector at the $p\bar{p}$ collider at the FNAL, USA in the discovery of the t-quark and the DELPHI detector at the e^-e^+ collider LEP at CERN for the study of the Z^0 weak neutral boson production and decay.

16.3 Electronics and computers

Every particle physics experiment at present is highly dependent on sophisticated electronics and computers. Detectors produce small electric signals which require fast amplication, discrimination and often analog-to-digital conversion (ADC) and/or time-to-digital conversion (TDC). Signals will often be combined or composed in simple coincidence units or more complicated matrices which can be programmed to accept only the desired combinations.

In most particle physics experiments conditions are imposed on the produced signals to select a particular class of events from a much more large background. Selection at levels of 1 in 10^5 events are quite common. An array of scintillation counters and Cerenkov detectors with appropriate trigger electronics demanding signals in the desired combinations of counters would be necessary. On-line computers to control the data collection, analysis and display of events and data and even to make complicated trigger decisions are often used in more elaborate experiments. Since it is not possible to make the trigger decisions sufficiently tight without losing genuine events it is often necessary to carry out further off-line filtering using more sophisticated algorithims. Subsequent stages in off-line analysis will generally include a pattern recognition stage to identify particle trajectories, geometrical reconstruction to obtain accurate positions, angles and curvatures and sometimes also kinematical fitting to calculate missing energies and momenta.

The technologies of electronics, on-line and off-line computing and data analysis are in themselves big specialised subjects beyond the scope of this book.

Appendix: The Dirac Equation

A.1 The Klein-Gordon (KG) Equation

The Dirac equation evolved out of attempts to obtain a relativistic wave equation for the electron. The natural first attempt would be to consider the simplest physical system, that is, an isolated free particle and use the relativistic energy-momentum relation with the same prescriptions that are used to obtain the non-relativistic Schrodinger equation. Thus the required wave equation will be

$$(E^2 - \vec{p}^{\,2} c^2 - m^2c^4)\phi = 0 \tag{A.1}$$

$$\text{with} \quad E \to i\hbar\frac{\partial}{\partial t} \quad \text{and} \quad \vec{p} \to \frac{\hbar}{i}\vec{\nabla}_x \tag{A.2}$$

using (A.2), (A.1) becomes

$$\left(-\hbar^2\frac{\partial}{\partial t^2} + \hbar^2c^2\nabla_x^2 - m^2c^4\right)\phi(x) = 0$$

which can be written in the form

$$\left[\Box + \left(\frac{mc}{\hbar}\right)^2\right]\phi(x) = 0 \tag{A.3}$$

where $\Box = \dfrac{1}{c^2}\dfrac{\partial^2}{\partial t^2} - \vec{\nabla}^2$

Eq. (A.3) is known as the Klein-Gordon equation.

The general plane wave solution of (A.3) will be

$$f(x) = Ae^{-i(Ept - \vec{p}\cdot\vec{x})} + Be^{+i(Ept - \vec{p}\cdot\vec{x})} \tag{A.4}$$

which is a linear superposition of positive-energy and negative-energy solutions, that is, solutions corresponding, respectively, to $E = +E_p$ and $E = -E_p$ where $E_p = +\sqrt{\vec{p}^{\,2}c^2 + m^2c^4}$.

In analogy to the Schrodinger equation, Eq. (A.3) can be used to derive an equation of continuity. Multiplying Eq. (A.3) with ϕ^* and the equation obtained by complex conjugation by ϕ and then subtracting one from the other, one gets

$$\phi^*\left[\Box + \left(\frac{mc}{\hbar}\right)^2\right]\phi - \phi\left[\Box + \left(\frac{mc}{\hbar}\right)^2\right]\phi^* = 0$$

$$\text{or,}\quad \frac{1}{c^2}\frac{\partial}{\partial t}\left(\phi^*\frac{\partial\phi}{\partial t} - \phi\frac{\partial\phi^*}{\partial t}\right) - \vec{\nabla}\cdot[\phi^*(\vec{\nabla}\phi) - \phi(\vec{\nabla}\phi^*)] = 0 \tag{A.5}$$

we can write (A.5) in the form

$$\frac{\partial\rho}{\partial t} + \vec{\nabla}\cdot\vec{J} = 0 \tag{A.6}$$

$$\text{where}\ \rho = \frac{i\hbar}{2mc^2}\left(\phi^*\frac{\partial\phi}{\partial t} - \phi\frac{\partial\phi^*}{\partial t}\right)$$

$$\text{and}\quad \vec{J} = \frac{\hbar}{2im}[\phi^*(\vec{\nabla}\phi) - \phi^*(\vec{\nabla}\phi^*)] \tag{A.7}$$

$\vec{J}$ has the same form as in the Schrodinger case. We can interpret ρ as the probability density and $\vec{j}$ as the probability current density. However, this is highly unsatisfactory since ρ is not positive-definite.

Thus the K-G equation has 2 difficulties: (1) the problem of negative energy solutions which is characteristic of any relativistic wave equation and are found to be associated with antiparticles in a field theoretic (many body) description and (2) absence of a positive-definite probability density so that one cannot have a probabilistic interpretation of the relativistic wave function. Hence the KG equation was abandoned soon after it was first obtained. However, later on, the KG equation turned out to be the Euler-Lagrange equation for a free scalar field in a field theoretic framework.

A.2 The Dirac Equation

Dirac found a relativistically covariant wave equation that leads to a positive-definite probability density. He argued that as in the case of the Schrodinger equation such an equation must be linear in the time derivative and for relativistic covariance must also be linear in the space derivatives. Hence he wrote down a Schrodinger-like equation

$$i\hbar\frac{\partial}{\partial t}\psi(\vec{x},t) = H_D\psi(\vec{x},t) \tag{A.8}$$

where H_D is the Hamiltonian operator for a relativistic particle. From the above arguments H_D must be linear in the 3-momentum. Hence for a relativistic free particle, the most general form of H_D will be

$$H_D = c\vec{\alpha} \cdot \vec{p} + \beta mc^2 \tag{A.9}$$

where $\vec{p}$ is the 3-momentum, m is the particle mass and c is the velocity of light. The coefficients $\vec{\alpha} = (\alpha_1, \alpha_2, \alpha_3)$ and β must be space-time independent and also independent of the energy-momentum of the particle. Now, in order to satisfy the relativistic energy-momentum relation, any solution of the Dirac equation must also be a solution of the K-G eqn. From (A.8) one gets

$$i\hbar\frac{\partial}{\partial t}\left(i\hbar\frac{d\psi}{\partial t}\right) = \left(i\hbar\frac{d}{\partial t}\right)(H_D\psi) = H_D\left(i\hbar\frac{d}{\partial t}\right)\psi = H_D^2\psi$$

$$\text{or} \quad -\hbar^2\frac{\partial^2\psi}{\partial t^2} = H_D^2\psi \tag{A.10}$$

In order that (A.10) is the same form as (A.3) one must have

$$H_D^2 = -\hbar^2c^2\nabla^2 + m^2c^4 \tag{A.11}$$

In terms of momentum operators, (A.11) becomes

$$(c\vec{\alpha} \cdot \vec{p} + \beta mc^2)^2 = c^2\vec{p}^{\,2} + m^2c^4 \tag{A.12}$$

$$\text{or} \quad c^2\sum_{j,k=1}^{3}\alpha_j\alpha_k p_j p_k + \sum_{j=1}^{3}(\alpha_j\beta + \beta\alpha_j)mc^2p_j + \beta^2m^2c^4$$

$$= c^2\sum_{j=1}^{3}p_j^2 + m^2c^4 \tag{A.13}$$

since p_j, p_k commute we can write

$$\sum_{j,k=1}^{3}\alpha_j\alpha_k p_j p_k = \frac{1}{2}\sum_{j,k}(\alpha_j\alpha_k + \alpha_k\alpha_j)p_j p_k \tag{A.14}$$

Substituting (A.14) in (A.13) we see that α_j's and β must satisfy the relations

$$\begin{aligned} \alpha_j\alpha_k + \alpha_k\alpha_j &= 2\delta_{jk} \\ \alpha_j\beta + \beta\alpha_j &= 0 \\ \beta^2 &= 1 \end{aligned} \tag{A.15}$$

$(j, k = 1, 2, 3)$

Since α_j's and β anticommute they may be regarded as matrices and hence ψ is a column-vector.

From (A.15) one can derive additional properties of α_j and β. We have

$\mathrm{Tr}\alpha_j = \mathrm{Tr}(\alpha_j\beta^2) = \mathrm{Tr}(\beta\alpha_j\beta) = -\mathrm{Tr}(\beta, \beta\alpha_j) = -\mathrm{Tr}(\beta^2\alpha_j) = -\mathrm{Tr}\alpha_j$

which means $\mathrm{Tr}\alpha_j = 0$

similarly, $\mathrm{Tr}\beta = 0$ (A.16)

Thus α_1, α_2, α_3, β are traceless hermitian matrices.

Since $\alpha_j^2 = 1$, $\beta^2 = 1$ they have eigenvalues ± 1. Futhermore, the trace of a matrix is the sum of its eigenvalues. Thus for $\mathrm{Tr}\alpha_j$ and $\mathrm{Tr}\beta$ to be zero the number of positive eigenvalues $(+1)$ must be equal to the number of negative eigenvalues (-1). Therefore, α_js and β must be traceless hermitian $N \times N$ matrices where N is an even integer; $N = 2, 4, 6, \ldots$ However, $N = 2$ is ruled out, since it can accomodate only 3 anticommuting linearly independent hermitian matrices, namely the Pauli matrices. Thus the smallest value of N that can accomodate 4 such matrices must be 4. Thus α_js and β are 4×4 traceless hermitian matrices and hence ψ must be a 4 component column vector, known as the Dirac spinor.

Any 4×4 traceless hermitian matrix can be written as the direct product of 2 sets of 2×2 linearly independent matrices, namely, the Pauli matrices $(\sigma_1, \sigma_2, \sigma_3)$, (ρ_1, ρ_2, ρ_3) and the 2×2 unit matrix (1). There are many such choices and one such choice is the following

$$\beta = 1 \otimes \rho_3 \quad \text{and} \quad \alpha_j = \sigma_j \otimes \rho_1 \tag{A.17}$$

and one gets from (A.17)

$$\beta = \left(\begin{array}{c|c} 1 & 0 \\ \hline 0 & -1 \end{array}\right), \quad \alpha_j = \left(\begin{array}{c|c} 0 & \sigma_j \\ \hline \sigma_j & 0 \end{array}\right) \quad (j = 1, 2, 3) \tag{A.18}$$

where each of the submatrices is a 2×2 matrix. The matrices (A.18) constitute the Dirac representation.

From the Dirac equation we can derive an equation of continuity. Multiplying Eq. (A.8) from the left by we get

$$i\hbar\psi^+ \frac{\partial\psi}{\partial t} = \psi^+(H_D\psi) \tag{A.19}$$

Taking the hermitian conjugate of (A.19) we get

$$i\hbar\frac{\partial\psi^+}{\partial t}\psi = (H_D\psi)^+\psi \tag{A.20}$$

Subtracting (A.20) from (A.19) we get

$$i\hbar\frac{\partial}{\partial t}(\psi^+\psi) = \psi^+(H_D\psi) - (H_D\psi)^+\psi = \frac{\hbar c}{i}\sum_j \frac{\partial}{\partial x^j}(\psi^+\alpha_j\psi) \tag{A.21}$$

we can write (A.21) in the form

$$\frac{\partial\rho}{\partial t} + \vec{\nabla}\cdot\vec{J} = 0 \tag{A.22}$$

where $\rho = \psi^+\psi = \sum_{j=1}^{4}|\psi_j|^2$

and $\quad \vec{J} = c\psi^+\vec{\alpha}\psi \qquad \text{(A.23)}$

Since ρ is positive-definite, it can be interpreted as a probability density.

It can be easily checked that $[\vec{L}, H_D] \neq 0$ but

$$\left[\vec{L} + \frac{\hbar}{2}\vec{\Sigma}, H_D\right] = 0$$

where $\vec{\Sigma} = \left(\begin{array}{c|c} \vec{\sigma} & 0 \\ \hline 0 & \vec{\sigma} \end{array}\right) \qquad \text{(A.24)}$

Thus for the Dirac Hamiltonian (and hence the Dirac equation) to be rotationally invariant one must take the components of $\vec{J} = \vec{L} + \frac{\hbar}{2}\vec{\Sigma}$ as the generators of spatial rotation. Since the generators of spatial rotation are identified with the angular momentum operators, the angular momentum of a Dirac particle consists of 2 parts, orbital angular momentum and an intrinsic angular momentum ("spin") which exists even in the rest frame of the particle. Thus it is said that the Dirac equation describes a spin $-\frac{1}{2}$ particle, the electron.

It turned out that in order to avoid the difficulties associated with the negative-energy solutions one must give up the one-particle interpretation of a relativistic wave equation and therefore the necessity of having a positive-definite probability density no longer arises. Thus a relativistic wave equation necessarily describes a system with infinitely many degrees of freedom, that is, a field. In particular, the Dirac equation is now regarded as the

Euler-Lagrange equation (Field equation) of a (spinor) field that describes a spin $-\frac{1}{2}$ particle.

A.3 The Dirac Equation in a Covariant Form

We have the Dirac equation,

$$i\hbar\frac{\partial}{\partial t}\psi(\vec{x},t) = H_D\psi = (c\vec{\alpha}\cdot\vec{p} + \beta mc^2)\psi(\vec{x},t) \quad \text{(A.25)}$$

with $\vec{p} \to \frac{\hbar}{i}\vec{\nabla}$

Multiplying both sides of (A.25) from the left by β, we get

$$\text{or} \quad i\hbar\beta\frac{d\psi}{\partial t} = \beta\left(\frac{\hbar c}{i}\vec{\alpha}\cdot\vec{\nabla} + \beta mc^2\right)\psi(\vec{x},t) = 0$$

$$\text{or} \quad \left[i\hbar\left(\beta\frac{1}{c}\frac{\partial}{\partial t} + \sum_{j=1}^{3}\beta\alpha_j\frac{\partial}{\partial x^j}\right) - mc\right]\psi(\vec{x},t) = 0 \quad \text{(A.26)}$$

Writing $\gamma^0 = \beta$, $\gamma^j = \beta\alpha_j$, $\frac{\partial}{\partial x^0} = \frac{1}{c}\frac{\partial}{\partial t}$ (A.26) becomes

$$\left[i\hbar\left(\gamma^0\frac{\partial}{\partial x^0} + \sum_{j=1}^{3}\gamma^j\frac{\partial}{\partial x^j}\right) - mc\right]\psi(\vec{x},t) = 0 \quad \text{(A.27)}$$

Using the 4-vector notation and the summation convention we can write (A.27) in the form

$$(i\hbar\gamma^\mu\partial_\mu - mc)\psi(x) = 0 \quad \text{(A.28)}$$

where $\gamma^\mu = (\gamma^0, \gamma^1, \gamma^2, \gamma^3) = (\gamma^0, \vec{\gamma})$

$$x^\mu = (x^0, x^1, x^2, x^3) = (x^0 = ct, \vec{x})$$

$$\partial_\mu = \left(\frac{\partial}{\partial x^0}, \frac{\partial}{\partial x^1}, \frac{\partial}{\partial x^2}, \frac{\partial}{\partial x^3}\right) = \left(\frac{1}{c}\frac{\partial}{\partial t}, \vec{\nabla}\right)$$

Eq. (A.28) is the Dirac equation written in a covariant form γ^μs are the Dirac matrices in a 4-vector notation. We shall often use the notation

$$\not{a} = \gamma^\mu a_\mu \quad \text{(A.29)}$$

From their definition and the properties of the original Dirac matrices β and α_js one finds that the γ-matrices have the following properties,

$$\gamma^\mu\gamma^\nu + \gamma^\nu\gamma^\mu = 2g^{\mu\nu}\cdot 1 \quad \text{(A.30)}$$

$$\text{and} \quad (\gamma^\mu)^+ = \gamma^0\gamma^\mu\gamma^0 \quad \text{(A.31)}$$

In the above $g^{oo} = 1$, $g^{jj} = -1$ $(j = 1, 2, 3)$, $g^{\mu\nu} = o$, $\mu \neq \nu$.

In the Dirac representation (A.18) of the β and α_j matrices, one has

$$\gamma^o = \beta = \begin{pmatrix} 1 & 0 \\ 0 & -1 \end{pmatrix}, \quad \gamma^j = \beta\alpha_j = \begin{pmatrix} 0 & \sigma_j \\ -\sigma_j & 0 \end{pmatrix} \tag{A.32}$$

A.4 Lorentz Covariance of the Dirac Equation

Under Lorentz transformation, $x \to x' = \wedge x$, such that

$$x' \cdot x' = x \cdot x \tag{A.33}$$

In component form,

$$x'^{\mu} = \wedge^{\mu}_{\nu} x^{\nu}$$

Now $x' \cdot x' = g_{\mu\nu} x'^{\mu} x'^{\nu} = g_{\mu\nu} \wedge^{\mu}_{\alpha} \wedge^{\nu}_{\beta} x^{\alpha} x^{\beta}$

$$x' \cdot x' = g_{\mu\nu} x'^{\mu} x'^{\nu} g_{\mu\nu} \wedge^{\mu}_{\alpha} \wedge^{\nu}_{\beta} x^{\alpha} x^{\beta} \quad \text{and} \quad x \cdot x = g_{\alpha\beta} x^{\alpha} x^{\beta}$$

Therefore, $x' \cdot x' = x \cdot x$ requires

$$g_{\mu\nu} \wedge^{\mu}_{\alpha} \wedge^{\nu}_{\beta}$$

or $\quad \widetilde{\wedge} g \wedge = g, \; \widetilde{\wedge}^{\alpha}_{\mu} = \wedge^{\mu}_{\alpha} \qquad$ (A.34)

In natural units ($\hbar = c = 1$) (A.28) becomes

$$(i\gamma^{\mu}\partial_{\mu} - m)\psi(x) = 0 \tag{A.35}$$

under Lorentz transformation,

$$\psi(x) \xrightarrow{\wedge} \psi'(x') = S(\wedge)\psi(x) \tag{A.36}$$

Lorentz covariance requires that $\psi'(x')$ satisfies the same eqn. as $\psi(x)$. Thus

$$(i\gamma^{\nu}\partial'_{\nu} - m)\psi'(x') = o \tag{A.37}$$

Multiplying (A.35) from the left by $S(\wedge)$, we get

$$(iS\gamma^{\mu}S^{-1}\partial_{\mu} - m)S\psi(x) = o \tag{A.38}$$

Now $\partial_{\mu} = \dfrac{\partial}{\partial x^{\mu}} = \dfrac{\partial x'^{\nu}}{\partial x^{\mu}} \dfrac{\partial}{\partial x'^{\nu}} = \wedge^{\nu}_{\mu} \partial'_{\nu}$

Therefore, we can write (A.38) as

$$(iS\gamma^{\mu}S^{-1} \wedge^{\nu}_{\mu} \partial'_{\nu} - m)\psi'(x') = o \tag{A.39}$$

In deriving (A.39) we have used (A.36). Comparing (A.39) with (A.37) we see that Lorentz covariance requires

$$S\gamma^\mu S^{-1}\wedge^\nu_\mu = \gamma^\nu \tag{A.40}$$

or equivalently,

$$S^{-1}\gamma^\mu S = \wedge^\mu_\nu \gamma^\nu \tag{A.41}$$

Finding S amounts to finding a solution of (A.41).

We first construct S for an infinitesimal Lorentz transformation,

$$\wedge^\mu_\nu = g^\mu_\nu + w^\mu_\nu \tag{A.42}$$

where $g^\mu_\nu = 1$, for $\mu = \nu$ and $g^\mu_\nu = o$, for $\mu \neq \nu$. w^μ_ν is an infinitesimal quantity so that one can neglect terms of order w^2 and higher. From (A.34) and (A.42) one gets

$$w_{\alpha\beta} = -w_{\beta\alpha},$$

where $w_{\alpha\beta} = g_{\alpha\lambda} w^\lambda_\beta$.

We write

$$S = 1 - \frac{i}{4} w_{\alpha\beta}\Sigma^{\alpha\beta} \tag{A.43}$$

where $\sum^{\alpha\beta}$s are 4×4 antisymmetric matrices, $\sum^{\alpha\beta} = -\sum^{\beta\alpha}$. There can be 6 such matrices.

Substituting (A.43) in (A.41) one gets

$$\left(1 + \frac{i}{4} w_{\alpha\beta}\Sigma^{\alpha\beta}\right)\gamma^\mu\left(1 - \frac{i}{4} w_{\alpha\beta}\Sigma^{\alpha\beta}\right) = (g^\mu_\nu + w^\mu_\nu)\gamma^\nu \tag{A.44}$$

Assuming that $w_{\alpha\beta}$s are infinitesimal, this gives

$$[\Sigma^{\alpha\beta}, \gamma^\mu] = -2i(g^{\mu\alpha}\gamma^\beta - g^{\mu\beta}\gamma^\alpha) \tag{A.45}$$

One easily finds that the solution for (A.45) is

$$\Sigma^{\alpha\beta} = \frac{i}{2}(\gamma^\alpha\gamma^\beta - \gamma^\beta\gamma^\alpha) \tag{A.46}$$

That is, $\sum^{\alpha\beta}$s are antisymmetric products of 2 γ-matrices.

One can choose the 6 linearly independent $\sum^{\alpha\beta}$s as follows

$$(\Sigma^{23}, \Sigma^{31}, \Sigma^{12}) = (\Sigma_1, \Sigma_2, \Sigma_3) = \vec{\Sigma}$$

where $\sum_j = \begin{pmatrix} \sigma_j & o \\ o & \sigma_j \end{pmatrix}$

and $\quad (\sum^{o1}, \sum^{o2}, \sum^{o3}) = (i\alpha_1, i\alpha_2, i\alpha_3) = i\vec{\alpha}$

where $\alpha_j = \begin{pmatrix} o & \sigma_j \\ \sigma_j & o \end{pmatrix}$

Similarly, one can choose the 6 parameters of proper Lorentz transformation as follows

$(w^{23}, w^{31}, w^{12}) = (\theta_1, \theta_2, \theta_3) = \vec{\theta}$ and $(w^{o1}, w^{o2}, w^{o3}) = (\phi_1, \phi_2, \phi_3) = \vec{\phi}$

clearly,

$\sum_{ij} = \sum^{ij}, \sum_{oj} = -\sum^{oj}, w_{ij} = w^{ij}, w_{oj} = -w^{oj}$

Then one can write

$$S = S_R(\vec{\theta})S_L(\vec{\theta}) \tag{A.47}$$

$$\text{where } S_R(\vec{\theta}) = 1 - \frac{i}{2}\vec{\theta}\cdot\vec{\Sigma},\ S_L(\vec{\theta}) = 1 - \frac{1}{2}\vec{\phi}\cdot\vec{\alpha} \tag{A.48}$$

for infinitesimal $\vec{\theta}$ and $\vec{\phi}$

$S_R(\vec{\theta})$ corresponds to spatial rotation and $S_L(\vec{\theta})$ to pure Lorentz (boost) transformation. For finite $\vec{\theta}$ and $\vec{\phi}$ one has

$$S_R(\vec{\theta}) = e^{-\frac{1}{2}\vec{\theta}\cdot\vec{\Sigma}}, \quad S_L(\vec{\theta}) = e^{-\frac{1}{2}\vec{\theta}\cdot\vec{\alpha}}$$

Further, one finds

$$S^{-1} = \gamma^o S^+ \gamma^o \tag{A.49}$$

One can also write the equation of continuity (Eq. A.22) corresponding to the Dirac equation in a covariant form:

$$\partial_\mu J^\mu = o \tag{A.50}$$

$$\text{where } J^\mu = (J^0, J^1, J^3, J^4) = \overline{\psi}\gamma^\mu\psi \tag{A.51}$$

is a 4-vector. In (A.48)

$$\overline{\psi} = \psi^+\gamma^o \tag{A.52}$$

which is known as the adjoint spinor. Under Lorentz transformation

$$\begin{aligned}\overline{\psi}(x) &= \psi^+(x)\gamma^o \to \psi^{+'}(x')\gamma^o \\ &= \psi^+(x)S^+\gamma^o = \overline{\psi}(x)\gamma^o S^+\gamma^o = \overline{\psi}(x)S^{-1}\end{aligned}$$

Thus under Lorentz transformation,

$$\overline{\psi}(x) \to \vec{\psi}'(x') = \vec{\psi}(x)S^{-1} \tag{A.53}$$

under Lorentz transformation,

(1) $\overline{\psi}(x)\psi(x) \to \overline{\psi}(x)S^{-1}S\psi(x) = \overline{\psi}(x)\psi(x)$

(2) $J^\mu(x) = \overline{\psi}(x)\gamma^\mu\psi(x)$

$$\rightarrow \overline{\psi}(x)S^{-1}\gamma^\mu S\psi(x) = \Lambda^\mu_\nu\overline{\psi}(x)\gamma^\nu\psi(x) = \Lambda^\mu_\nu J^\nu(x) \quad \text{(A.54)}$$

Thus $\overline{\psi}(x)\psi(x)$ is invariant under Lorentz transformation and $\overline{\psi}(x)\gamma^\mu\psi(x)$ transforms like a 4-vector under Lorentz transformation.

The γ_5 matrix

One useful matrix in connection with the Dirac equation is the γ_5 matrix

$$\gamma_5 = \gamma^5 = i\gamma^o\gamma^1\gamma^2\gamma^3 = \frac{i}{4}\in_{\mu\nu\alpha\beta}\gamma^\mu\gamma^\nu\gamma^\alpha\gamma^\beta \quad \text{(A.55)}$$

In the Dirac representation,

$$\gamma_5 = \begin{pmatrix} 0 & 1 \\ 1 & 0 \end{pmatrix} \quad \text{(A.56)}$$

one easily finds that

$$\gamma_5^2 = 1, \quad \gamma_5^+ = \gamma_5 \quad \text{and} \quad \gamma^\mu\gamma_5 + \gamma_5\gamma^\mu \quad \text{(A.57)}$$

A.5 Bilinear Covariants

Bilinear covariants are expressions of the form $\overline{\psi}(x)\Gamma\psi(x)$ which transforms in a definite way under Lorentz transformations. Here Γ is a 4×4 matrix which can be written as an antisymmetric product of n γ-matrices with $n \leq 4$. There are 16 such linearly independent matrices and accordingly there are 16 bilinear covariants which are written in the following standard forms.

Bilinear covariant	**Lorentz Transformation property**	**No. of matrices**
1) $\overline{\psi}\psi$	Scalar (S)	1
2) $\overline{\psi}\gamma^\mu\psi$	Vector (V)	4
3) $\overline{\psi}\Sigma^{\mu\nu}\psi$	Tensor (T)	6
4) $\overline{\psi}\gamma^\mu\gamma_5\psi$	Axial vector (A)	4
5) $\overline{\psi}\gamma_5\psi$	Pseudoscalar (P)	1

Axial vectors (A) and the pseudoscalar (P) differ from the vectors (V) and the scalar (S), respectively, only in their behaviour under parity transformation.

A.6 Parity Transformation

Under parity transformation,

$$x = \{x^\mu\} \xrightarrow{P} x' = \{x'^\mu\}, \quad x'^\mu = x_\mu, \ [(t, \vec{x}) \xrightarrow{P} (t, -\vec{x})] \tag{A.58}$$

$$\psi(x) \xrightarrow{P} \psi'(x') = S\psi(x) \tag{A.59}$$

we have the Dirac eqn.

$$(i\gamma^\mu \partial_\mu - m)\psi(x) = o \tag{A.35}$$

Multiplying from the left by S, we get

$$(iS\gamma^\mu S^{-1}\partial_\mu - m)S\psi(x) = o \tag{A.60}$$

For parity invariance, $\psi'(x')$ must satisfy the eqn.

$$(i\gamma^\mu \partial'_\mu - m)\psi'(x') = o \tag{A.61}$$

where $\partial'_\mu = \dfrac{\partial}{\partial x'^\mu} = \dfrac{\partial}{\partial x_\mu} = \partial^\mu$

In order that eqn. (A.61) and (A.60) be the same,

$$S\gamma^\mu S^{-1} = \gamma_\mu \tag{A.62}$$

Eq. (A.62) is satisfied if

$$S = \gamma^o \tag{A.63}$$

Thus under parity transformation,

$$\psi(x) \xrightarrow{P} \psi'(x') = \gamma^o\psi(x) \tag{A.64}$$

where $x'^\mu = x_\mu$

$$\text{That is } \psi(t, \vec{x}) \xrightarrow{P} \psi'(t, -\vec{x}) = \gamma^o\psi(t, x) \tag{A.65}$$

A.7 Plane Wave Solution of the Dirac Equation

The Plane wave solution of the Dirac equation

$$(i\gamma^\mu \partial_\mu - m)\psi(x) = o \tag{A.35}$$

will be of the form

$$\psi(x) = w(p)e^{-ip\cdot x} \tag{A.66}$$

where $w(p)$ is a 4-component column vector, known as the Dirac spinor.

Substituting (A.66) in (A.35) one gets

$$(\not{p} - m)w(p) = o, \quad \not{p} = \gamma^\mu p_\mu = \gamma^o p^o - \vec{\gamma} \cdot \vec{p} \tag{A.67}$$

multiplying from the left by $(\not{p} + m)$ we get

$$(\not{p} + m)(\not{p} - m)w(p) = o \quad \text{or} \quad (p^2 - m^2)w(p) = o$$

For nontrivial solution, one must have

$$p^2 = m^2 \quad \text{or} \quad (p^o)^2 - \vec{p}^{\,2} = m^2 \quad \text{or} \quad p^o = \pm\sqrt{\vec{p}^{\,2} + m^2} = \pm E_p$$

$$\text{where } E_p = +\sqrt{\vec{p}^{\,2} + m^2} \tag{A.68}$$

Hence the plane wave solution of the Dirac equation is

$$\psi(x) = w(E_p, \vec{p})e^{-i(E_pt - \vec{p}\cdot\vec{x})} + w(-E_p, \vec{p})e^{-i(-E_pt - \vec{p}\cdot\vec{x})}$$

without any loss of generality, one can change $\vec{p}$ to $-\vec{p}$ in the second (negative-energy) term. Thus the plane-wave solution becomes

$$\psi(x) = w(E_p, \vec{p})e^{-i(E_{pt} - \vec{p}\cdot\vec{x})} + w(-E_p, -\vec{p})e^{-i(+E_{pt} + \vec{p}\cdot\vec{x})}$$

$$\text{or} \qquad \psi(x) = u(p)e^{-ip\cdot x} + v(p)e^{+ip\cdot x} \tag{A.69}$$

$$\text{where } p = (E_p, \vec{p}),\ u(p) = w(p),\ v(p) = w(-p) \tag{A.70}$$

From (A.35) or (A.67) and (A.69) it is easily seen that $u(p)$ and $v(p)$ are solutions of the equations

$$(\not{p} - m)u(p) = o$$

$$\text{and} \quad (-\not{p} - m)v(p) = o \tag{A.71}$$

$u(p)$ and $v(p)$ are known as the positive energy Dirac spinor and the negative energy Dirac spinor, respectively.

In the rest frame of the particle, $p = (m, \vec{o})$. Calling the solutions in the particle rest frame $u(o)$ and $v(o)$ one sees that they satisfy the equations

$$(\gamma^o - 1)u(o) = o$$

$$\text{and} \quad (-\gamma^o - 1)v(o) = o \tag{A.72}$$

That is, $u(o)$ and $v(o)$ are eigenvectors of the parity operators with eigenvalues $+1$ and -1, respectively. Since the positive energy solutions correspond to the particle and the negative energy solutions to the antiparticle, it is often said that a Dirac (spin $\frac{1}{2}$) particle and its antiparticle have opposite intrinsic parities. The 4 independent solutions of the Dirac equation in the rest frame of the particle are, therefore, (in the Dirac representation of the γ matrices).

$$\underbrace{\begin{pmatrix}1\\0\\0\\0\end{pmatrix} \begin{pmatrix}0\\1\\0\\0\end{pmatrix}}_{u(o)}, \quad \underbrace{\begin{pmatrix}0\\0\\1\\0\end{pmatrix} \begin{pmatrix}0\\0\\0\\1\end{pmatrix}}_{v(o)} \tag{A.73}$$

The first 2 are positive-energy solutions and the last 2 are the negative energy solutions, respectively. In each case one has 2 independent solutions corresponding to 2 possible values of spin orientations $\left(S_z = \pm\frac{1}{2}\right)$.

In a moving frame ($\vec{p} \neq 0$), the solutions are of the form,

$$u(p) \sim (\not{p} \ \ +m)u(o) = N \begin{pmatrix} \chi \\ \dfrac{\vec{\sigma}\cdot\vec{p}}{E_p + m}\chi \end{pmatrix} \tag{A.74}$$

and $$v(p) \sim (-\not{p} \ \ +m)v(o) = N' \begin{pmatrix} \dfrac{\vec{\sigma}\cdot\vec{p}}{E_p + m}\chi \\ \chi \end{pmatrix}$$

where χ is a 2-component Pauli spinor.

The normalization of the Dirac spinors is usually chosen to be

$$\overline{u}(p,s)u(p,s) = 1 = -\overline{v}(p,s)v(p,s) \tag{A.75}$$

The gives

$$N = N' = \left(\frac{E_p + m}{2m}\right)^{1/2} \tag{A.76}$$

Accordingly, one has the

Orthogonality relations

$$\begin{aligned} \overline{u}(p,s)u(p,s') &= \delta_{ss'} = -\overline{v}(p,s)v(p,s') \\ \overline{v}(p,s)u(p,s') &= o = v^{+}(p,s)u(-p,s') \end{aligned} \tag{A.77}$$

Completeness relation

$$\sum_s [u(p,s)\overline{u}(p,s) - v(p,s)\overline{v}(p,s)] = 1 \tag{A.78}$$

Furthermore, one can define the energy projection operators

$$\begin{aligned}\wedge_+(p) &= \sum_s u(p,s)\overline{u}(p,s) = \frac{\not{p}+m}{2m}\\ \wedge_-(p) &= -\sum_s v(p,s)\overline{v}(p,s) = \frac{-\not{p}+m}{2m}\end{aligned} \tag{A.79}$$

$\wedge_+(p)$ and $\wedge_-(p)$ project out the positive-energy spinor and the negative energy spinor, respectively.

A.8 Chirality Operators. Weyl Spinors

Let us define

$$\begin{aligned}\psi_R &= \frac{1}{2}(1+\gamma_5)\psi = P_R\psi, \quad P_R = \frac{1}{2}(1+\gamma_5)\\ \psi_L &= \frac{1}{2}(1-\gamma_5)\psi = P_L\psi, \quad P_L = \frac{1}{2}(1-\gamma_5)\end{aligned} \tag{A.80}$$

one finds

$$\gamma_5\psi_R = +\psi_R, \quad \gamma_5\psi_L = -\psi_L \tag{A.81}$$

Therefore, ψ_R and ψ_L are eigenstates of γ_5 with eigenvalues $+1$ and -1, respectively. For a massless Dirac particle, the action of γ_5 is the same as the helicity operator. γ_5 is known as the chirality (handedness) operator. P_R and P_L are the chirality projection operators.

In the Weyl representation of the Dirac matrices.

$$\gamma^o = \begin{pmatrix} 0 & 1 \\ 1 & 0 \end{pmatrix}, \quad \gamma^j = \begin{pmatrix} 0 & \sigma_j \\ -\sigma_j & 0 \end{pmatrix}, \quad \gamma^5 = \begin{pmatrix} 1 & 0 \\ 0 & -1 \end{pmatrix} \tag{A.82}$$

$$P_R = \frac{1}{2}(1+\gamma_5) = \begin{pmatrix} 0 & 1 \\ 1 & 0 \end{pmatrix}, \quad P_L = \frac{1}{2}(1-\gamma_5) = \begin{pmatrix} 0 & 0 \\ 0 & 1 \end{pmatrix} \tag{A.83}$$

since general solution of the Dirac equation is of the form

$$\psi(x) = u(p)e^{-ip\cdot x} + v(p)e^{ip\cdot x}$$

$$\begin{aligned}&p = (E_p, \vec{p}), \quad E_p = +\sqrt{\vec{p}^{\,2} + m^2}\\ &u(p) = \begin{pmatrix} \chi \\ \xi \end{pmatrix}, \quad v(p) = \begin{pmatrix} \xi \\ \chi \end{pmatrix}, \quad \xi = \frac{\vec{\sigma}\cdot\vec{p}}{E_p + m}\chi\end{aligned} \tag{A.84}$$

in the Weyl representation,

$$\begin{aligned} u_R = P_R u = \begin{pmatrix} \chi \\ o \end{pmatrix}, \qquad u_L = P_L u = \begin{pmatrix} o \\ \xi \end{pmatrix} \\ v_R = P_R v = \begin{pmatrix} \xi \\ o \end{pmatrix}, \qquad v_L = P_L v = \begin{pmatrix} o \\ \chi \end{pmatrix} \end{aligned} \tag{A.85}$$

u_R, u_L, v_R and v_L are known as Weyl spinors. A Dirac spinor is a direct sum of 2 Weyl spinors.

A.9 Charge Conjugation (C) Invariance

The problem of interpreting the negative-energy solutions of the Dirac equation and of understanding their physical implications remained a stumbling block for some time after the equation was first introduced. Dirac formulated his "hole theory" in order to resolve the dilemma posed by the negative-energy solutions. He introduced the notion of a filled negative-energy sea of electrons. A "hole" or vacancy created by interaction with an electromagnetic field was interpreted as a positive-energy particle with positive electric charge and was named "positron" by Dirac. But the success of the hole theory depended on the applicability of the Pauli exclusion principle which is valid only for half-integral spin particles like the electron. Hence the problem of interpreting the negative energy solutions of the relativistic wave equation corresponding to integral-spin particles (bosons) like the pion remained. However, there emerges from the hole theory a fundamental new symmetry in nature, known as the charge conjugation symmetry or C-invariance according to which for each particle there corresponds an antiparticle with the same mass and spin as the particle but opposite charges. Negative-energy solutions of the wave equation for a particle then correspond to the positive-energy solutions of its antiparticle and the correspondence is one-to-one.

Consider the Dirac equation for a particle in an electromagnetic field.

$$(i\gamma^\mu D_\mu - m)\psi(x) = o \tag{A.86}$$

where $D_\mu = \partial_\mu + ieA_\mu(x)$ (A.87)

By definition, the antiparticle wave function $\psi_c(x)$ will be a positive-energy solution of the equation obtained by changing the sign of the electric charge in (A.54), that is, the equation.

$$(i\gamma^\mu D^c_\mu - m)\psi_c(x) = o \tag{A.88}$$

where $D^c_\mu = \partial_\mu - ieA_\mu(x)$ (A.89)

Taking the complex conjugate of (A.86) we get

$$(-i\gamma^{\mu^*} D^*_\mu - m)\psi^*(x) = o$$

or $$(-i\gamma^{\mu^*} D^c_\mu - m)\psi^*(x) = o \tag{A.90}$$

Multiplying (A.90) from the left by a 4×4 matrix C, we get

$$(-iC\gamma^{\mu^*} C^{-1} D^c_\mu - m)C\psi^*(x) = 0 \tag{A.91}$$

(A.91) and (A.88) will be the same, provided

$$C\gamma^{\mu^*} C^{-1} = -\gamma^\mu \quad \text{and} \quad \psi_c(x) = C\psi^*(x) \tag{A.92}$$

That there exists such a matrix C may be verified by explicit construction. In the Dirac representation.

$$\begin{aligned} \gamma^{\mu^*} &= \gamma^\mu, \quad \text{for } \mu = 0, 1, 3 \\ \text{and} \quad \gamma^{\mu^*} &= -\gamma^\mu, \quad \text{for } \mu = 2 \end{aligned} \tag{A.93}$$

Thus C commutes with γ^2 and anticommutes with γ^o, γ^1 and γ^3. This means

$$C = a\gamma^2, \quad a = \text{constant} \tag{A.94}$$

The transformation matrix C must be unitary, $CC^+ = 1$. This means $|a| = 1$. We choose $a = i$

Therefore the required matrix is

$$C = i\gamma^2 \tag{A.95}$$

A negative-energy electron at rest with spin down is described by the wave function

$$\psi^{(4)} = \begin{pmatrix} o \\ \chi_- \end{pmatrix} e^{\text{imt}}, \quad \chi_- = \begin{pmatrix} 0 \\ 1 \end{pmatrix} \tag{A.96}$$

The corresponding positron wave function is

$$\psi_c = i\gamma^2\psi^{(4)*} = i\begin{pmatrix} o & \sigma_2 \\ -\sigma_2 & 0 \end{pmatrix}\begin{pmatrix} 0 \\ \chi_- \end{pmatrix} e^{-\text{imt}} = \begin{pmatrix} i\sigma_2\chi_- \\ o \end{pmatrix} e^{-\text{imt}}$$

Now $i\sigma_2\chi_- = i\begin{pmatrix}0 & -i\\ i & 0\end{pmatrix}\begin{pmatrix}0\\ i\end{pmatrix} = \begin{pmatrix}1\\ 0\end{pmatrix} = \chi_+$

Therefore $\psi_c = \begin{pmatrix}\chi_+\\ o\end{pmatrix} e^{-\text{imt}} = \psi^{(1)}$ (A.97)

That is, the absence of a spin-down negative-energy electron at rest is equivalent to the presence of a spin-up positive-energy positron at rest, as interpreted by the hole theory. In the field-free case there is no difference between an electron and a positron and the transformation (A.92) has formed just another solution. In general,

$$Cu^*(p,\pm) = v(p,\mp), \quad Cv^*(p,\pm) = u(p,\mp) \tag{A.98}$$

A.10 Trace Relations

In computing crosssections and decay rates one often encounters traces of products of γ-matrices, resulting from spin summation. There are a number of properties of traces of products of γ-matrices and also some useful relations involving γ-matrices, independent of specific choice of representations of γ-matrices. We enumerate some of them.

(1) $\text{Tr}\gamma^\mu = o$, $\text{Tr}(\gamma^\mu\gamma^\nu) = 4g^{\mu\nu}$, $\text{Tr}\gamma_5 = o$ (A.99)

(2) **Trace of the product of an odd number of γ-matrices is zero.**

$$\begin{aligned}\text{Tr}(\gamma^{\mu_1}\cdots\gamma^{\mu_n}) &= \text{Tr}(\gamma^{\mu_1}\cdots\gamma^{\mu_n}\gamma_5^2) \quad [\gamma_5^2 = 1]\\ &= \text{Tr}(\gamma^{\mu_1}\cdots\gamma^{\mu_n}\gamma_5\gamma_5)\\ &= (-)^n\text{Tr}(\gamma_5\gamma^{\mu_1}\cdots\gamma^{\mu_n}\gamma_5) \quad [\gamma^\mu\gamma_5 + \gamma_5\gamma^\mu = o]\\ &= (-)^n\text{Tr}(\gamma^{\mu_1}\cdots\gamma^{\mu_n}\gamma_5^2) \quad [\text{Tr}(AB) = \text{Tr}(BA)]\\ &= (-)^n\text{Tr}(\gamma^{\mu_1}\cdots\gamma^{\mu_n})\end{aligned}$$

or $[1-(-)^n]\text{Tr}(\gamma^{\mu_1}\cdots\gamma^{\mu_n}) = o$ (A.100)

Therefore we get that for $n =$ odd integer

$$\text{Tr}(\gamma^{\mu_1}\cdots\gamma^{\mu_n}) = o \tag{A.101}$$

(3) $\text{Tr}(\not{a}\not{b}) = \text{Tr}(a_\mu\gamma^\mu b_\nu\gamma^\nu) = a_\mu b_\nu\text{Tr}(\gamma^\mu\gamma^\nu) = 4g^{\mu\nu}a_\mu v_\nu$

$= 4(a\cdot b)$

(4) $\mathrm{Tr}(\not{a}_1\cdot\not{a}_2\cdot\not{a}_3\cdot\not{a}_4) = \mathrm{Tr}[a_{1\mu}a_{2\nu}\gamma^\mu\gamma^\nu\not{a}_3\cdot\not{a}_4]$

$$= a_{1\mu}a_{2\nu}\mathrm{Tr}[(2g^{\mu\nu} - \gamma^\nu\gamma^\mu)\not{a}_3\cdot\not{a}_4]$$

$$= 2(a_1\cdot a_2)\mathrm{Tr}(\not{a}_3\cdot\not{a}_4) - \mathrm{Tr}(\not{a}_2\cdot\not{a}_1\cdot\not{a}_3\cdot\not{a}_4)$$

$$= 2(a_1\cdot a_2)\mathrm{Tr}(\not{a}_3\cdot\not{a}_4) - 2(a_1\cdot a_3)\mathrm{Tr}(\not{a}_2\cdot\not{a}_4) + \mathrm{Tr}(\not{a}_2\cdot\not{a}_3\cdot\not{a}_1\cdot\not{a}_4)$$

$$= 2(a_1\cdot a_2)\mathrm{Tr}(\not{a}_3\cdot\not{a}_4) - 2(a_1\cdot a_3)\mathrm{Tr}(\not{a}_2\cdot\not{a}_4) + 2(a_1\cdot a_4)\mathrm{Tr}(\not{a}_2\cdot\not{a}_3) - \mathrm{Tr}(\not{a}_2\cdot\not{a}_3\cdot\not{a}_4\cdot\not{a}_1)$$

$$= 2(a_1\cdot a_2)\mathrm{Tr}(\not{a}_3\cdot\not{a}_4) - 2(a_1\cdot a_3)\mathrm{Tr}(\not{a}_2\cdot\not{a}_4) + 2(a_1\cdot a_4)\mathrm{Tr}(\not{a}_2\cdot\not{a}_3) - \mathrm{Tr}(\not{a}_1\cdot\not{a}_2\cdot\not{a}_3\cdot\not{a}_4)$$

Therefore we get using (A.102)

$$\mathrm{Tr}(\not{a}_1\cdot\not{a}_2\cdot\not{a}_3\cdot\not{a}_4) = (a_1\cdot a_2)\mathrm{Tr}(\not{a}_3\cdot\not{a}_4) - (a_1\cdot a_3)\mathrm{Tr}(\not{a}_2\cdot\not{a}_3) + (a_1\cdot a_4)\mathrm{Tr}(\not{a}_2\cdot\not{a}_3)$$

$$= 4[(a_1\cdot a_2)(a_3\cdot a_4) - (a_1\cdot a_3)(a_2\cdot a_4) + (a_1\cdot a_4)(a_2\cdot a_3)] \quad \text{(A.103)}$$

The result (A..103) can be easily generalized to products of any higher even number of matrices ($n = 6, 8, \ldots$)

(5) $\mathrm{Tr}(\gamma_5\,\not{a}\,\not{b}\,\not{c}\,\not{d}) = 4i \in_{\mu\nu\alpha\beta} a^\mu b^\nu c^\alpha d^\beta$ (A.104)

Since $\gamma_5 = i\gamma^o\gamma^1\gamma^2\gamma^3$, for a non-vanishing contribution all components of the a, b, c, d must be different and the total contribution is the sum of various combinations of the components multiplied by the sign of the combination. To fix the overall sign we take

$$\mathrm{Tr}(\gamma_5\gamma_0\gamma_1\gamma_2\gamma_3)a^0b^1c^2d^3 = i \in_{0123} \mathrm{Tr}(\gamma_5^2)a^0b^1c^2d^3 = 4i\epsilon_{0123}a^0b^1c^2d^3$$

(6) $\gamma_\mu\gamma^\mu = 4$

$$\gamma_\mu\,\not{a}\,\gamma^\mu = -2\,\not{a}$$

$$\gamma_\mu\,\not{a}\,\not{b}\,\gamma^\mu = 4(a\cdot b) \quad \text{(A.105)}$$

$$\gamma_\mu\,\not{a}\,\not{b}\,\not{c}\,\gamma^\mu = -2\not{c}\,\not{b}\,\not{a}$$

$$\gamma_\mu\,\not{a}\,\not{b}\,\not{c}\,\not{d}\,\gamma^\mu = 2(\not{d}\,\not{a}\,\not{b}\,\not{c} + \not{c}\,\not{b}\,\not{a}\,\not{d})$$

Although (A.105) are not trace relations, they are often useful in reducing products of γ-matrices. These relations can be proved very easily by utilising the properties of γ-matrices.

(7) $\mathrm{Tr}(\not{a}_1 \cdot \not{a}_2 \cdots \not{a}_{2n}) = \mathrm{Tr}(\not{a}_{2n} \cdots \not{a}_1)$ (A.106)

Proof. we have

$$C\gamma^\mu C^{-1} = -(\gamma^\mu)^T$$

where C is the charge conjugation matrix, and T denotes the transpose of a matrix. Since

$$\mathrm{Tr}(SAS^{-1}) = \mathrm{Tr}\ A$$

we can write

$$\begin{aligned}
\mathrm{Tr}(\not{a} \cdots \not{a}_{2n}) &= \mathrm{Tr}[C(\not{a} \cdots \not{a}_{2n})C^{-1}] \\
&= \mathrm{Tr}[(C\not{a}_1 c^{-1})(C\not{a}_2 c^{-1}) \cdots (C\,\not{a}_{2n}\,C^{-1})] \\
&= (-)^{2n}\mathrm{Tr}[\not{a}_1^T \not{a}_2^T \cdots \not{a}_{2n}^T] \\
&= +\mathrm{Tr}[\not{a}_{2n} \cdots \not{a}_1)^T] \\
&= \mathrm{Tr}(\not{a}_{2n} \cdots \not{a}_1)
\end{aligned}$$

Bibliography

General Texts

1. D. J. Griffiths, Introduction to Elementary Particles (John Wiley, 1987).
2. I. S. Hughes, Elementary Particles (3rd ed.) Cambridge Univ Press, 1991).
3. F. Halzen and A. D. Martin, Quarks and Leptons (John Wiley, 1984).
4. D. H. Perkins, Introduction to High Energy Physics (4th ed.) (Cambridge Univ Press, 2000).

Invariance and Symmetry principles in particle physics

1. J. J. Sakurai, Invariance principles and elementary particles (Princeton Univ Press, 1964).
2. J. McL Emmerson, Symmetry principles in particle physics (Clarendon Press, Oxford, 1972).

Lie groups and Lie algebras in particle physics

1. P. A. Carruthers, Introduction to unitary symmetry (Interscience, John Wiley, 1966).
2. D. Lichtenberg, Unitary symmetry and elementary particles (Academic Press, 1978).
3. H. Georgi, Lie algebras in particles physics (Benjamin, Cummings, 1984).

The Quark Model

1. J. J. Kokkedee, The Quark Model (Benjamin, 1969).
2. F. E. close, An introduction to quarks and partons (Academic Press, 1979).
3. J. L. Rosner Quark Models in "Techniques and concepts of High Energy Physics", NATO Advanced Summer Institute Series edited by T. Ferbel (Plenum, 1981) pp. 1–181.

4. D. Flamm and E. Schorbel, Introduction to the Quark Model of elementary particles (Gordon and Breach, 1982).

Quantum Field Theories

1. J. D. Bjorken and S. D. Drell, Relativistic Quantum Mechanics vol. I and Relativistic Quantum Fields, Vol. II (Mc-Graw Hill, 1964).
2. M. Peskin and D. Schroder, Introduction to Quantum Field Theory (Addison and Wesley, 1995).

Weak Interactions

1. E. D. Commins, Weak Interactions (McGraw-Hill, 1973).
2. D. Bailin, Weak Interactions (2nd ed) (Adam and Hilger 1982).
3. L. B. Okun, Leptons and Quarks (North Holland, 1982).
4. R. E. Marshak, Riazuddin and C.P. Ryan, Theory of Weak Interactions in particle physics (Wiley Interscience, 1969).
5. H. Georgi, Weak Interactions and Modern Particle Theory (Benjamin, Cummings, 1984).

Deep Inelastic Scattering

1. R. P. Feynman, Photon-Hadron Interactions (Addison Wesley, Advanced Book Program, 1972).
2. Ta-Pei Cheng and Ling-Fong Li, Gauge Theory of Elementary Particles (Clarendon Press, Oxford, 1984).

Gauge Theories in Particles Physics

1. I. J. R. Aitchison and A. J. Hey, Gauge Theories in particle physics (Adam Hilger, 1981).
2. C. Quigg, Gauge Theories of strong weak and electromagnetic interactions (Benjamin Cummings 1983).
3. E. S. Abers and B. W. Lee, Gauge Theories Physics, Reports 9C (1973) 1–141.
4. W. Marciano and H. Pagels, Quantum Chromodynamics, Physics Reports 36C (1978) 137–236.
5. E. Leader and E. Predazzi, Introduction to Gauge theories and Modern Particle Physics (Cambridge Univ. Press, 1996).

Beyond the Standard Model

1. Michael Dine, Supersymmetry and string theory—Beyond the Standard model (Cambridge Univ. Press 2001).
2. H. E. Haber and G. L. Kane, The search for supersymmetry probing physics beyond the standard model, Physics Report 117C (1985) 75–263.

Cosmology and particle physics

1. S. Weinberg, The first three minutes (Fontana, London, 198).
2. Andrew Liddie, An introduction to modern cosmology (2nd ed.) (John Wiley, 2003).
3. Andrei Linde, Particle Physics and Inflationary Cosmology (Harwood Academic Publishers 1990).
4. E. W. Kolb and M. S. Turner, The Early Universe (Addison-Wesley, 1987).

Experimental particle physics

1. Richard C. Fernow, Introduction to Experimental Particle Physics (Cambridge Univ. Press, 1986).
2. T. J. Ferbel (ed), Experimental techniques in High Energy Physics (Addison-Wesley, 1987).
3. V. Berger and R. J. N. Philips, Collider Physics (Addison-Wesley, 1987).

Particle properties

1. Particle Data Group, Review of particle properties, Journal of physics, G 33 (2006) 1–1232.

Index